Introduction to Electromagnetic Theory

A Modern Perspective

The Quantum Challenge:
Modern Research on the Foundations of Quantum Mechanics, Second Edition

George Greenstein, Amherst College and
Arthur Zajonc, Amherst College
ISBN: 0-7637-2470-X
Hardcover • 300 Pages • ©2006

The Quantum Challenge, Second Edition is an engaging and thorough treatment of the extraordinary phenomena of quantum mechanics, and of the enormous challenge they present to our conception of the physical world.

> "This is a lovely book...pointing out the amazing and unintuitive features that make quantum theory the endlessly fascinating subject it is."
>
> **Daniel Greenberger, Professor of Physics City College of CUNY**
> Regarding First Edition

Table of Contents

New to this edition:
- Thoroughly updated to include dramatic new experiments that illustrate new challenges of quantum theory.
- A chapter on the exciting new field of Quantum Computation and Information.
- Includes an annotated bibliography of experiments accessible to the undergraduate laboratory.
- Pedagogy allows text to be accessible to a wide range of readers.

Greenstein and Zajonc present the puzzles of quantum mechanics using vivid reference to contemporary experiments and focus on the most striking and conceptually significant quantum phenomena. Ideally suited to those in physical science, mathematics and engineering fields, **The Quantum Challenge** will engage and mystify its readers from beginning to end.

JONES AND BARTLETT PUBLISHERS
BOSTON   TORONTO   LONDON   SINGAPORE

Introduction to Electromagnetic Theory

A Modern Perspective

Tai L. Chow
California State University, Stanislaus

**JONES AND BARTLETT
PUBLISHERS**

BOSTON   TORONTO   LONDON   SINGAPORE

World Headquarters
Jones and Bartlett Publishers
40 Tall Pine Drive
Sudbury, MA 01776
978-443-5000
info@jbpub.com
www.jbpub.com

Jones and Bartlett Publishers
Canada
6339 Ormindale Way
Mississauga, Ontario L5V 1J2
CANADA

Jones and Bartlett Publishers
International
Barb House, Barb Mews
London W6 7PA
UK

Jones and Bartlett's books and products are available through most bookstores and online booksellers. To contact Jones and Bartlett Publishers directly, call 800-832-0034, fax 978-443-8000, or visit our website www.jbpub.com.

Substantial discounts on bulk quantities of Jones and Bartlett's publications are available to corporations, professional associations, and other qualified organizations. For details and specific discount information, contact the special sales department at Jones and Bartlett via the above contact information or send an email to specialsales@jbpub.com.

PRODUCTION CREDITS
Chief Executive Officer: Clayton Jones
Chief Operating Officer: Don W. Jones, Jr.
President, Higher Education and Professional Publishing: Robert W. Holland, Jr.
V.P., Design and Production: Anne Spencer
V.P., Sales and Marketing: William Kane
V.P., Manufacturing and Inventory Control: Therese Connell
Executive Editor, Science: Stephen L. Weaver
Acquisitions Editor, Science: Cathleen Sether
Managing Editor, Science: Dean W. DeChambeau
Editorial Assistant, Science: Molly Steinbach
Senior Production Editor: Louis C. Bruno, Jr.
Marketing Manager: Andrea DeFronzo
Text and Cover Design: Anne Spencer
Illustrations: George Nichols
Composition: SPI Publisher Services
Printing and Binding: Malloy, Inc.
Cover Printing: Malloy, Inc.
Cover Photo: © Able Stock

Library of Congress Cataloging-in-Publication Data
Chow, Tai L.
    Introduction to electromagnetic theory: a modern perspective/
Tai L. Chow
    p. cm.
    Includes index.
    ISBN 0-7637-3827-1 (alk. paper)
    1. Electromagnetic theory—Textbooks. I. Title.
    QC670.C425 2006
    530.14′1—dc22   2005052100

Printed in the United States of America
09 08 07 06 05   10 9 8 7 6 5 4 3 2 1

To Jalen, Tailo, Daniel, and Eric

Brief Contents

Contents

Chapter 11.   Electromagnetic Waves in Matter 409

Chapter 12.   Electromagnetic Waves in Bounded Media 427

Preface

This book presents an account of classical electromagnetism with a modern perspective for advanced undergraduate physics students. It has evolved from a set of lecture notes for a course on the subject, which I have taught at California State University, Stanislaus, for many years. Prerequisites for this subject are a calculus-based general physics course (using a textbook such as that of Halliday and Resnick) and a course in calculus (including differentiation of field functions). No prior knowledge of differential equations is necessary. Differential equations and new mathematical methods are developed in the text as the occasion demands. Vectors are used from the start. Chapter 1 provides a summary of vector analysis.

The choice of topics and their treatment throughout the book are intended to emphasize the modern point of view. Applications to other branches of physics are made wherever possible. Special note is made of concepts that are important to the development of modern physics.

The student will find that a generous amount of detail is given in mathematical manipulations and that "it-may-be-shown-that's" are kept to a minimum. However, to ensure that the student does not lose sight of the development underway, some of the more lengthy and tedious algebraic manipulations were omitted when possible.

Each chapter contains a set of homework problems of varying degrees of difficulty. They are intended to supplement or amplify the material in the text and are arranged in the order in which the material is covered in the chapter.

I have restricted any discussion of the historical development of the subject to the introduction due to the length of the book and not to a lack of interest on my part. References to the original literature were omitted except for recent works to which the student may be expected to have access.

Tai L. Chow
Turlock, California
October 2005

Introduction

Classical electromagnetism embraces electric and magnetic fields and their interactions with matter on a macroscopic scale. As the world around us is made up of atoms that consist of positive and negative charges, electromagnetic phenomena play an exceedingly important role in nature. Today, four basic types of forces are recognized: gravitational, electromagnetic, strong, and weak. All other forces may be reduced to these four. For example, viscous and frictional forces are essentially electromagnetic in origin.

The gravitational and electromagnetic forces are long range and directly observable. Strong and weak forces are confined to distances of some 10^{-15} m: the strong nuclear force binds nucleons together to form nuclei of the atoms, and the weak force describes β decay in which fast electrons are emitted in the course of nuclear transformation.

As with most branches of physics, electromagnetic theory has developed in an inductive-deductive manner. A wide variety of experimental facts have been observed over a long period of time. The Greeks had some knowledge of electricity; for instance, they knew about the electrification of amber by friction. However, the Greeks did not conduct experimental tests to find out *how* nature behaves. Instead they speculated about *why* things happen. This led them more into the realm of philosophy than science and caused them to make guesses, most of which were wrong.

The Romans gave little time to science, and in the Middle Ages there was little interest in learning. Science, as we think of it today, dates from about the time of Nicolas Copernicus and Galileo Galilei (the middle of the 16th century). Scientific attention was first directed toward simple mechanical problems. A study of the more complicated phenomena of electricity and magnetism did not begin until the year 1600, when Sir William Gilbert, court physician to Queen Elizabeth, published his book *De Magnete,* in which he described his experiments in electrostatics and magnetism.

Charles Augustin Coulomb conducted the first quantitative work in electricity and magnetism, and he, just before 1790, discovered a simple relationship that governs the force between static charges. This general principle, which we call *Coulomb's law*, coordinated a large number of observed facts and opened the way for a theory of electrostatics. Coulomb discovered a similar relationship for the forces between magnetic poles and, from this, developed, in a parallel manner, the theory of magnetostatics.

With the 1800 discovery of the voltaic cell by Alessandro Volta, a source of electric current became available, enabling the properties of such currents to be investigated. An important result was the observation by Hans Christian Oersted in 1819 that current in a wire affects a compass needle placed over the wire. This was the first conclusive proof of an interrelationship between electricity and magnetism. Oersted's discovery was soon followed by quantitative experiments on the effects of magnetic force on parallel current-carrying conductors by Ampere and others. André Marie Ampere determined the general principle that explained the observations that the force is attractive if currents in the two conductors flow in the same direction and repulsive if they flow in opposite direction This fundamental law, *Ampere's law*, was first published in 1822; it expresses the force of interaction between short sections of two current carrying conductors. From this expression the force between complete current circuits can be computed.

Michael Faraday, in 1831, discovered a second interrelationship between electricity and magnetism. He reasoned that if electric currents produced magnetic effects, then magnetism should produce electric effects. After considerable experimentation, he discovered electromagnetic induction. Like Ampere, he summarized his results in the form of a general principle that we call *Faraday's law*. From this law we can compute the electromotive force induced in a circuit when the time rate of change of the magnetic flux through the circuit is known. Faraday regarded electric and magnetic forces as fields, picturing these fields by drawing lines of force to represent the direction and intensity of a field.

In 1864, James Clerk Maxwell brought the theory of electromagnetic fields to its culmination. He summarized, in the form of differential equations, the laws governing electric and magnetic fields. These electromagnetic equations embody the laws of Coulomb, Ampere, and Faraday, and they are compatible with the *continuity principle* that electric charge can neither be created nor destroyed. In his attempt to make his equations consistent with that expressing the continuity principle, Maxwell introduced the term *displacement current* into one of his equations. Furthermore, Maxwell showed that electromagnetic waves should exist. He computed the speed of such waves to be that of light and so concluded that light and electromagnetic waves must be the same thing. The experiments of Heinrich Hertz on the properties of electromagnetic waves confirmed Maxwell's theory.

Since Maxwell's time, physicists' views on electromagnetism have changed. First, because isolated magnetic poles have not been observed, all magnetic effects are attributed to charges in motion, in line with a suggestion made by Ampere. The magnetism of a piece of iron, for example, is regarded as the result of small "Amperian currents" (which are spinning electrons) in the iron. This viewpoint eliminates the need for such a concept as the magnetic pole, thus making Coulomb's law for the force between poles unnecessary. We are left with the following basic principles: (1) the law of force between static charges, (2) the law of force between moving charges (or currents), (3) electromagnetic induction plus the displacement current introduced by Maxwell (explained in greater detail in the coming chapters).

The nineteenth century viewpoint was also modified by the theory of relativity. The basic principle of relativity is that the laws of physics that apply to, for example, a rocket moving at high but *constant velocity*, are the same laws that apply to objects on (or relative to) the earth. Thus from the laws of Coulomb, Ampere, Faraday, and Maxwell alone, we cannot detect such a thing as absolute motion. In the aforementioned case of the earth and rocket, one is no more "at rest" than the other. If we change from the usual laboratory reference system (x, y, z, t) to a coordinate-time reference system (x', y', z', t') moving with the rocket, Maxwell's equations should have the same form as before, that is, they should be *invariant,* according to the relativity theory.

What links electromagnetism to Newtonian mechanics? It is the Lorentz force, $\vec{F} = q(\vec{E} + \frac{1}{c}\vec{v} \times \vec{B})$, which determines the force on a charged particle moving in an electromagnetic field.

Because all of electromagnetism is contained in the Maxwell equations, we could start the study of electromagnetism with the Maxwell equations and deduce everything from them. This approach is more suitable when the concepts of electromagnetism are already clear. We cannot fully understand the Maxwell equations until we can appreciate the meaning of electric and magnetic fields and the sources of the fields (the electric charges and currents). Also, we cannot follow a strictly historical route of development. To include every detail of the work of centuries would be very wearisome and is not the point of the text. As we are concerned with the subject as it stands today, we adopt an approach that builds a gradual understanding of the subject that finally leads to the Maxwell equations.

In the following chapters we shall divide our treatment of electromagnetism into two parts, considering (1) the laws for charged bodies in a vacuum (or in air at atmospheric pressure) and (2) the modifications in these laws when a medium possessing particular electric and magnetic properties is introduced.

A general point on the role of quantum theory: There are various electromagnetic phenomena of which proper explanation cannot be given without a quantum theory treatment. The most obvious example is the electromagnetic properties of media. Thus, in this text we give rudimentary explanations of the magnetic and dielectric properties of media without introducing quantum ideas. Many recent technical devices such as laser, the operation of which depends on quantum ideas, are not covered in this text.

1 Mathematical Preliminaries

Many mathematical techniques will be introduced as the physical ideas are developed, but vector analysis is the basic techniques set forth as a preliminary to the main discussion. Vector methods provide us a most concise and elegant way for the discussion of physical ideas and phenomena. At the end of the chapter, a brief review of mechanics is given, reviewing only those ideas or concepts that are useful for our study of electromagnetic theory.

1.1 Dimensions, Units, and Dimensional Analysis

A few words on dimensional analysis are very helpful for our readers. A dimension defines some physical characteristic, such as length, mass, or force. The dimensions of length (L), mass (M), time (T), electric current (I), temperature (T), and luminous intensity (I) are considered as the fundamental dimensions because other dimensions can be defined in terms of these six. For example, the dimensions of velocity are L/T^2 and of force are ML/T^2.

A unit is a standard of reference by which a dimension can be expressed numerically. In this book SI units (the international system of units) are used. The material in this book deals almost exclusively with the four fundamental dimensions: length, mass, time, and electric current. The four fundamental units for these dimensions are *meter*, *kilogram*, *second*, and *ampere*, respectively.

It is our experience that most students do not pay much attention to dimensional analysis. It is a necessary condition for correctness that each term in the equation has the same dimension. This dimension homogeneity is of some practical use. For example, in studying mechanics we often encounter Newton's second law:

Force = mass × acceleration

The dimension of force is ML/T^2, and the dimension of acceleration is L/T^2. So both sides of the equation have the same dimensions.

We can use this to check the correctness of equations: If different terms have different dimensions, then the equation is incorrect. Of course, the fact that the terms have the same dimension is no guarantee that the equation is correct. Thus, the dimensional homogeneity is not a sufficient condition for correctness.

Very often dimensional analysis can be used to indicate a physical relationship. For example, one might suppose that the period τ of a pendulum is dependent on the length l of the pendulum and the acceleration of gravity g. We may write this as

$$\tau = \text{const. } l^a\, g^b$$

where a and b can be determined by dimensional analysis. The dimensions of this equation are

$$T = L^a (L/T^2)^b = L^{(a\,+\,b)} T^{-2b}$$

Both sides of the equation must have the same dimensions; therefore

$$a + b = 0 \quad \text{and} \quad -2b = 1$$

This would suggest that

$$\tau = \text{const. } \sqrt{l/g}$$

Here we cannot determine the constant, which is equal to 2π.

1.2 Vector Algebra

We assume the reader is already familiar with the basic mathematics of scalar and vector fields in three dimensions, the properties of the del (∇) operator, and the integral theorems that hold for these fields. In this summary, we review some basic definitions and outline without proof those theorems we use extensively in the chapters that follow.

It is customary to represent vectors by letters in boldface type. In writing it is more convenient to use an arrow over the symbol (as $\vec{A}$). We use either representation, depending on convenience. The magnitude of a vector is represented by the symbol itself in italic (as A). A unit vector in the direction of the vector $\vec{A}$ is denoted by the corresponding letter A with a circumflex over it, as $\hat{A} = \vec{A}/A$. A vector can be specified by its components and

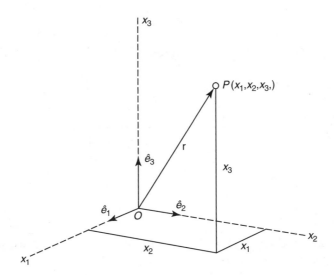

I FIGURE 1.1

the unit vectors along the coordinate axes. For example, in a Cartesian coordinate system **(Figure 1.1)** an arbitrary vector *r* can be expressed as

$$\vec{r} = x_1\hat{e}_1 + x_2\hat{e}_2 + x_3\hat{e}_3 = \sum_{i=1}^{3} x_i\hat{e}_i$$

where $\hat{e}_i$ (i = 1, 2, 3) are unit vectors along the rectangular axes x_i ($x_1 = x$, $x_2 = y$, $x_3 = z$). The component triplet (x_1, x_2, x_3) is also used as an alternative designation for vector *r*.

a. The Scalar Product of Two Vectors

The scalar or dot product of two vectors **A** and **B** is a scalar quantity of value equal to the magnitudes of **A** and **B** and the cosine of the angle between **A** and **B**. It is represented by a dot between **A** and **B**; thus

$$\vec{A} \cdot \vec{B} = AB\cos\theta \tag{1.1}$$

where θ is the angle between **A** and **B**. The dot-product operation is commutative:

$$\vec{A} \cdot \vec{B} = AB\cos\theta = \vec{B} \cdot \vec{A} \tag{1.2}$$

Figure 1.2 shows that $\vec{A} \cdot \vec{B}$ defined by equation (1.1) is the product of the magnitude of **A** and the rectangular projection of $B\cos\theta$ (or vice versa).

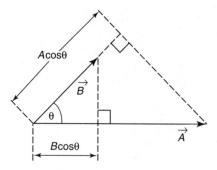

I FIGURE 1.2

The scalar product of **A** and **B** can be written in terms of the Cartesian components of these vectors:

$$\vec{A} \cdot \vec{B} = (A_1 \hat{e}_1 + A_2 \hat{e}_2 + A_3 \hat{e}_3) \cdot (B_1 \hat{e}_1 + B_2 \hat{e}_2 + B_3 \hat{e}_3)$$

Now the unit vectors $\hat{e}_1, \hat{e}_2, \hat{e}_3$ are mutually orthogonal (i.e., $\hat{e}_i \cdot \hat{e}_j = 0$ if $i \neq j$), it follows that

$$\vec{A} \cdot \vec{B} = A_1 B_1 + A_2 B_2 + A_3 B_3 \tag{1.3}$$

The magnitude of the vector **A** is written as

$$A = \left| \vec{A} \right| = \sqrt{\vec{A} \cdot \vec{A}} = \sqrt{A_1^2 + A_2^2 + A_3^2} \tag{1.4}$$

b. The Vector Product of Two Vectors

The vector product of two vectors, **A** and **B**, is another vector whose magnitude is the product of the magnitudes of the two vectors and the sine of the angle θ between the two vectors and whose direction is the direction of advance of a right-hand screw as it is turned from **A** toward **B** through the angle θ:

$$\vec{A} \times \vec{B} = AB \sin \theta \hat{i}_N \tag{1.5}$$

where $\hat{i}_N$ is a unit vector in the direction of advance of a right-hand screw as it is turned from **A** toward **B** through the angle θ, and it is perpendicular to both **A** and **B**. **Figure 1.3** shows the geometrical relationship between the vector product **A** × **B** and the vectors **A** and **B**.

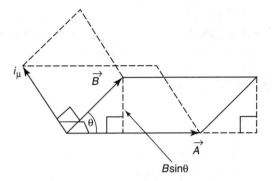

I FIGURE 1.3

The cross product is not commutative:

$$\vec{B} \times \vec{A} = -\vec{A} \times \vec{B} \tag{1.6}$$

The vector product of **A** and **B** can be written in terms of the Cartesian components of these vectors,

$$\vec{A} \times \vec{B} = (A_1 \hat{e}_1 + A_2 \hat{e}_2 + A_3 \hat{e}_3) \times (B_1 \hat{e}_1 + B_2 \hat{e}_2 + B_3 \hat{e}_3)$$

using the relations

$$\hat{e}_i \times \hat{e}_i = 0, \quad i = 1, 2, 3; \quad \hat{e}_1 \times \hat{e}_2 = \hat{e}_3, \quad \hat{e}_2 \times \hat{e}_3 = \hat{e}_1, \quad \hat{e}_3 \times \hat{e}_1 = \hat{e}_2, \tag{1.7}$$

It follows after some computation that

$$\vec{A} \times \vec{B} = (A_2 B_3 - A_3 B_2) \hat{e}_1 + (A_3 B_1 - A_1 B_3) \hat{e}_2 + (A_1 B_2 - A_2 B_1) \hat{e}_3 \tag{1.8}$$

or, in determinant,

$$\vec{A} \times \vec{B} = \begin{vmatrix} \hat{e}_1 & \hat{e}_2 & \hat{e}_3 \\ A_1 & A_2 & A_3 \\ B_1 & B_2 & B_3 \end{vmatrix} \tag{1.9}$$

c. The Triple Scalar Product

The triple scalar product $A \cdot B \times C$ is a useful operation but of less importance. Its value equals the volume of the parallelepiped formed by **A**, **B**, and **C**. Note that the scalar result from the triple scalar product changes sign on an inversion of the coordinates axes. For this reason, it is not a true scalar and is called a pseudo-scalar.

d. The Triple Vector Product

The triple product $A \times (B \times C)$ is a vector, because it is the vector product of two vectors, A and $B \times C$. It is easy to show that

$$A \times (B \times C) = B(A \cdot C) - C(A \cdot B) \tag{1.10}$$

1.3 Coordinate Systems

In addition to the Cartesian coordinate system that is the basic type, we also need cylindrical and spherical polar coordinates in the following chapters. Other coordinate systems may have occasional use, but they are not discussed here.

a. Cylindrical Polar Coordinates (ρ, ϕ, z)

The cylindrical coordinates ρ, ϕ, and z are defined by their relationships to the Cartesian coordinates x_1, x_2, and x_3:

$$\left.\begin{array}{l} x_1 = \rho \cos\phi \\ x_2 = \rho \sin\phi \\ x_3 = z\text{........} \end{array}\right\} \quad \text{or} \quad \left.\begin{array}{l} \rho^2 = x_2^2 + x_1^2 \\ \phi = \tan^{-1}(x_2/x_1) \\ z = x_3 \end{array}\right\} \tag{1.11}$$

As shown in **Figure 1.4**, ρ is the projection of the radius vector r onto the x_1x_2-plane, and ϕ is the angle that meets with the x_1-axis. The unit vectors $\hat{e}_\rho$ and $\hat{e}_z$ are in the directions of increasing ρ and z, respectively. The unit vector $\hat{e}_\phi$ is parallel to the x_1x_2-plane and perpendicular to $\hat{e}_\rho$ in the direction of increasing ϕ. Both $\hat{e}_\rho$ and $\hat{e}_\phi$, but not $\hat{e}_z$, change orientations in space as P moves on the cylindrical surface.

We can express the unit vectors $\hat{e}_\rho, \hat{e}_\phi$, and $\hat{e}_z$ in terms of rectangular unit vectors x_1, x_2, x_3. We first find the vectors that are tangent to the ρ-, ϕ-, and z-curves. They are given, respectively, by $\partial\vec{r}/\partial\rho$, $\partial\vec{r}/\partial\phi$, and $\partial\vec{r}/\partial z$. Hence, the unit vectors $\hat{e}_\rho, \hat{e}_\phi$, and $\hat{e}_z$ are given by

$$\hat{e}_\rho = \frac{\partial\vec{r}/\partial\rho}{|\partial\vec{r}/\partial\rho|} = \frac{\cos\phi\,\hat{e}_1 + \sin\phi\,\hat{e}_2}{[\cos^2\phi + \sin^2\phi]^{1/2}} = \cos\phi\,\hat{e}_1 + \sin\phi\,\hat{e}_2 \tag{1.12}$$

$$\hat{e}_\phi = \frac{\partial\vec{r}/\partial\phi}{|\partial\vec{r}/\partial\phi|} = -\sin\phi\,\hat{e}_1 + \cos\phi\,\hat{e}_2 \tag{1.13}$$

$$\hat{e}_z = \hat{e}_3 \tag{1.14}$$

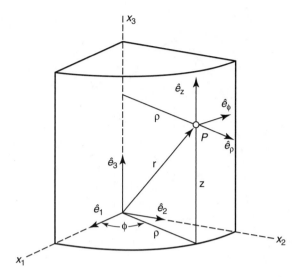

I FIGURE 1.4

b. Spherical Polar Coordinates (r, θ, ϕ)

The spherical coordinates are defined by the relations (**Figure 1.5**)

$$\left.\begin{array}{l} x_1 = r\sin\theta\cos\phi \\ x_2 = r\sin\theta\sin\phi \\ x_3 = r\cos\theta \end{array}\right\} \quad \text{or} \quad \left.\begin{array}{l} r = (x_1^2 + x_2^2 + x_3^2)^{1/2} \\ \theta = \tan^{-1}(x_1^2 + x_2^2)^{1/2}/x_3 \\ \phi = \tan^{-1}(x_2/x_1) \end{array}\right\} \tag{1.15}$$

The three unit vectors we use in spherical coordinates are defined by the equations

$$\hat{e}_r = \frac{\partial\vec{r}/\partial r}{|\partial\vec{r}/\partial r|} = \hat{e}_1\sin\theta\cos\phi + \hat{e}_2\sin\theta\sin\phi + \hat{e}_3\cos\theta,$$

$$\hat{e}_\phi = \frac{\partial\vec{r}/\partial\phi}{|\partial\vec{r}/\partial\phi|} = \hat{e}_1\cos\theta\cos\phi + \hat{e}_2\cos\theta\sin\phi - \hat{e}_3\sin\theta, \tag{1.16}$$

$$\hat{e}_\theta = \frac{\partial\vec{r}/\partial\theta}{|\partial\vec{r}/\partial\theta|} = -\hat{e}_1\sin\phi + \hat{e}_2\cos\phi$$

and their directions are defined in the following manner:

$\hat{e}_r$ is directed along increasing r

$\hat{e}_\theta$ is perpendicular to r in the direction of increasing θ in the plane containing r and z

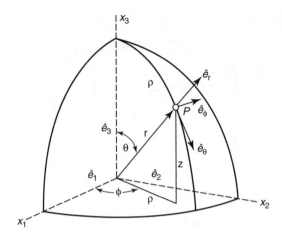

I FIGURE 1.5

$\hat{e}_\phi$ completes the right-handed triad. That is, $\hat{e}_r$ turned into $\hat{e}_\theta$ according to the right-hand rule goes in the direction of $\hat{e}_\theta$.

Thus, $\hat{e}_r$ and $\hat{e}_\theta$ are in the same vertical plane with $\hat{e}_3$, whereas $\hat{e}_\phi$ is parallel to the horizontal $x_1 x_2$-plane.

1.4 Vector Differentiation

Most physical quantities discussed in this book fall into two classes: (1) scalar point functions and (2) vector point functions. A scalar point function, also known as a scalar field, is scalar function of the vector variable r. Similarly, a vector point function, also known as a vector field, is vector function of the vector variable r. We will see later that the electrostatic potential function $V(r)$ is a scalar field and the electric and the magnetic fields $E(r)$ and $B(r)$ are vector fields.

It is straightforward to develop a calculus involving vectors, scalar functions of vectors, and vector functions of vectors. For example, consider a vector $\vec{A} = A\hat{A}$, where $\hat{A}$ is a unit vector pointing in the direction of the vector $\vec{A}$, and it is a function of some parameter s. The derivative of $\vec{A}$ with respect to s is defined as

$$\frac{d\vec{A}}{ds} = \frac{dA}{ds}\hat{A} + A\frac{d\hat{A}}{ds}$$

where dA/ds is the rate of change in the magnitude of $\vec{A}$ with respect to s and $d\hat{A}/ds$ is the rate of change in the direction of $\vec{A}$ with respect to s.

We next consider the evaluation of the derivative of an arbitrary scalar function of position $\varphi(r)$. Because φ is a function of three independent variables, the three components of r, any derivative with respect to an explicit variable is a partial derivative of the kind $\partial\varphi/\partial x_1$, $\partial\varphi/\partial x_2$, $\partial\varphi/\partial x_3$.

We can evaluate the total derivative of φ with respect to a variable contained only implicitly in φ. For example, $\varphi(r)$ may not depend explicitly on the time t but may change with t because of the variation of r with time. The total derivative of φ with respect to t is then

$$\frac{d\varphi}{dt} = \frac{\partial\varphi}{\partial x_1}\frac{dx_1}{dt} + \frac{\partial\varphi}{\partial x_2}\frac{dx_2}{dt} + \frac{\partial\varphi}{\partial x_3}\frac{dx_3}{dt}$$

If φ depends explicitly on t as well as on x_i ($i = 1, 2, 3$), an additional term $\partial\varphi/\partial t$ appears on the right-hand side of the above equation.

We now define four quantities known as the gradient, the divergence, the curl, and the Laplacian; they arise in practical applications.

a. The Gradient

Gradient is an operation performed on a scalar function that results in a vector function. The magnitude of this vector function at any point in the region of the scalar field is the maximum rate of increase of the scalar function at that point. The direction of the vector function at that point is the direction in which this maximum rate of increase occurs. To illustrate these mathematically, consider a scalar field described by a scalar function $\psi(x, y, z)$ that is defined and differentiable at each point with respect to the position coordinates (x, y, z). The total differential corresponding to an infinitesimal change $dr = (dx, dy, dz)$ is

$$d\psi = \frac{\partial\psi}{\partial x}dx + \frac{\partial\psi}{\partial y}dy + \frac{\partial\psi}{\partial z}dz$$

We can express $d\psi$ as a scalar product of two vectors,

$$d\psi = \left(\frac{\partial\psi}{\partial x}\hat{e}_x + \frac{\partial\psi}{\partial y}\hat{e}_y + \frac{\partial\psi}{\partial z}\hat{e}_z\right) \cdot (dx\hat{e}_x + dy\hat{e}_y + dz\hat{e}_z) = \nabla\psi \cdot d\vec{r} \qquad (1.17)$$

where the symbol ∇ stands for "del" and is vector operator defined as

$$\nabla = \frac{\partial}{\partial x}\hat{e}_x + \frac{\partial}{\partial y}\hat{e}_y + \frac{\partial}{\partial z}\hat{e}_z \qquad (1.18)$$

When the ∇ operator operates on a scalar function, the operation is known as evaluating the gradient of the scalar function; that is, $\nabla\psi$ is the gradient of ψ:

$$\nabla\psi = \frac{\partial\psi}{\partial x}\,\hat{e}_x + \frac{\partial\psi}{\partial y}\,\hat{e}_y + \frac{\partial\psi}{\partial z}\,\hat{e}_z \tag{1.19}$$

To discuss the physical significance of $\nabla\psi$, we first note that $\psi(x, y, z) = c$, where c is a constant, represents a surface. Let $\vec{r} = x\hat{e}_x + y\hat{e}_y + z\hat{e}_z$ be the position vector to a point $P(x, y, z)$ on the surface. When we move along the surface to a nearby point $Q(\vec{r} + d\vec{r})$, then

$$d\vec{r} = dx\hat{e}_x + dy\hat{e}_y + dz\hat{e}_z$$

lies in the tangent plane to the surface at P. But as we move along the surface $d\psi = 0$. Consequently, from equation (1.17) we have

$$\nabla\psi \cdot d\vec{r} = 0$$

which states that $\nabla\psi$ is perpendicular to dr and, therefore, to the surface (Figure 1.6).

Now consider two adjacent surfaces of constant ψ equal to c_1 and c_2, respectively, and let dl be the displacement vector drawn from point P on surface $\psi = c_1$ to point Q on surface $\psi = c_2$. Then, because the two surfaces are infinitesimally close to each other, we have according to

$$d\psi = (\nabla\psi)_{at\,P} \cdot dl = |\nabla\psi|dl\cos\alpha$$

or

$$\left(\frac{d\psi}{dl}\right)_{at\,P} = |\nabla\psi|\cos\alpha$$

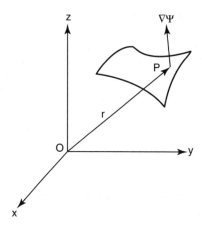

I **FIGURE 1.6**

where α is the angle between the two vectors $\nabla\psi$ and dl. $(d\psi/dl)_{\text{at P}}$ is maximum when $\cos\alpha = 1$ or $\alpha = 0$. Thus, $\nabla\psi$ is in the direction of maximum increase of $\psi(x, y, z)$.

The del operator ∇, as defined by equation (1.18) in Cartesian coordinates, is a very useful vector operator, which has properties analogous to those of ordinary vectors. It will help us in the future to keep in mind that ∇ acts both as a differential operator and as a vector. For example,

$$\nabla(uv) = \left(\hat{e}_x\frac{\partial}{\partial x} + \hat{e}_y\frac{\partial}{\partial y} + \hat{e}_z\frac{\partial}{\partial z}\right)(uv)$$

$$= \left(\hat{e}_x\frac{\partial u}{\partial x} + \hat{e}_y\frac{\partial u}{\partial y} + \hat{e}_z\frac{\partial u}{\partial z}\right)v + \left(\hat{e}_x\frac{\partial v}{\partial x} + \hat{e}_y\frac{\partial v}{\partial y} + \hat{e}_z\frac{\partial v}{\partial z}\right)u$$

$$= u\nabla v + v\nabla u$$

This result is easily remembered if we keep in mind that ∇ is a differential operator, so that we can apply the ordinary rule of calculus.

In other coordinate systems, the gradient ψ takes different forms. The defining equation (1.19) can be transformed term by term to any other representation. In cylindrical coordinates, $(\hat{e}_x, \hat{e}_y, \hat{e}_z)$ can be expressed via the following equations in cylindrical coordinates:

$$\hat{e}_x = \hat{e}_\rho\cos\phi - \hat{e}_\phi\sin\phi, \quad \hat{e}_y = \hat{e}_\rho\sin\phi + \hat{e}_\phi\cos\phi, \quad \hat{e}_z = \hat{e}_z;$$

they are the inverse transformations of equations (1.12), (1.13), and (1.14). In addition, terms such as $\partial\psi/\partial x$ may be calculated from the rules for implicit differentiation of a function of several variables:

$$\frac{\partial\psi(\rho,\phi,z)}{\partial x} = \frac{\partial\psi}{\partial\rho}\frac{\partial\rho}{\partial x} + \frac{\partial\psi}{\partial\phi}\frac{\partial\phi}{\partial x} + \frac{\partial\psi}{\partial z}\frac{\partial z}{\partial x}$$

Using the fact that

$$\rho = \sqrt{x^2 + y^2}, \text{ and } \phi = \cos^{-1}(x/\rho)$$

It is straightforward to show that

$$\nabla\psi(\rho,\phi,z) = \hat{e}_\rho\frac{\partial\psi}{\partial\rho} + \hat{e}_\phi\frac{1}{\rho}\frac{\partial\psi}{\partial\phi} + \hat{e}_z\frac{\partial\psi}{\partial z} \qquad (1.20)$$

Similarly, in spherical coordinates

$$\nabla\psi(r,\theta,\phi) = \hat{e}_r\frac{\partial\psi}{\partial r} + \frac{\hat{e}_\theta}{r}\frac{\partial\psi}{\partial\theta} + \frac{\hat{e}_\phi}{r\sin\theta}\frac{\partial\psi}{\partial\phi} \qquad (1.21)$$

We next apply the differential operator ∇ on a vector field. There are two types of products involving two vectors, namely the scalar and vector products. Vector differential operations on vector fields can also be separated into two types, called the divergence and the curl.

b. The Divergence of a Vector

If $\vec{V}(x, y, z) = V_1\hat{e}_1 + V_2\hat{e}_2 + V_3\hat{e}_3$ is a differentiable vector field (i.e., it is defined and differentiable at each point (x, y, z) in a region of space), the divergence of V, written $\nabla \cdot V$ or $divV$, is defined by the scalar product

$$\nabla \cdot \vec{V} = \left(\frac{\partial}{\partial x}\hat{e}_1 + \frac{\partial}{\partial y}\hat{e}_2 + \frac{\partial}{\partial z}\hat{e}_3\right) \cdot (V_1\hat{e}_1 + V_2\hat{e}_2 + V_3\hat{e}_3) = \frac{\partial V_1}{\partial x} + \frac{\partial V_2}{\partial y} + \frac{\partial V_3}{\partial z} \tag{1.22}$$

It is obvious that the result is a scalar field. Note the analogy with $A \cdot B = A_1B_1 + A_2B_2 + A_3B_3$. Also note that $\nabla \cdot V \neq V \cdot \nabla$, and $V \cdot \nabla$ is a scalar differential operator:

$$\vec{V} \cdot \nabla = V_1\frac{\partial}{\partial x} + V_2\frac{\partial}{\partial y} + V_3\frac{\partial}{\partial z}$$

The physical meaning of the divergence is best seen in connection with flow of a fluid. Consider a steady flow of fluid of density $\rho(x, y, z)$, the velocity field is given by $\vec{v}(x, y, z) = v_1(x, y, z)\hat{e}_x + v_2(x, y, z)\hat{e}_y + v_3(x, y, z)\hat{e}_z$ **(Figure 1.7)**.

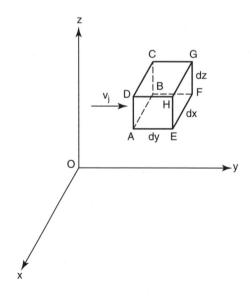

I FIGURE 1.7

We first concentrate on the flow passing through a small parallelepiped ABCDEFGH of dimensions $dxdydz$. The x and y components of the velocity v contribute nothing to the flow through the face ABCD. The mass of fluid entering ABCD per unit time is $\rho v_2 dxdz$. The amount leaving the face EFGH per unit time is

$$\left[\rho v_2 + \frac{\partial(\rho v_2)}{\partial y}\, dy\right] dxdz$$

so the loss of mass per unit time is

$$\frac{\partial(\rho v_2)}{\partial y}\, dxdydz$$

Adding the net rate of flow-out of all three pairs of surfaces of our parallelepiped, we get the total mass loss per unit time

$$\left[\frac{\partial(\rho v_1)}{\partial x} + \frac{\partial(\rho v_2)}{\partial y} + \frac{\partial(\rho v_3)}{\partial z}\right] dxdydz = \nabla \cdot (\rho \bar{v})\, dxdydz$$

The mass loss per unit time per unit volume is $\nabla \cdot (\rho \bar{v})$, and this quantity is called the divergence of the vector $\rho \bar{v}$. Similarly, we can calculate the divergence of $f(x, y, z)V$, where f is a scalar:

$$\nabla \cdot (f\vec{V}) = \frac{\partial}{\partial x}(f\,V_1) + \frac{\partial}{\partial y}(f\,V_2) + \frac{\partial}{\partial z}(f\,V_3)$$

$$= f\,(\partial V_1/\partial x + \partial V_2/\partial y + \partial V_3/\partial z) + (V_1 \partial f/\partial x + V_2 \partial f/\partial y + V_3 \partial f/\partial z)$$

or

$$\nabla \cdot (f\vec{V}) = f\nabla \cdot \vec{V} + \vec{V} \cdot \nabla f \tag{1.23}$$

This result can be remembered easily if we remember that ∇ acts both as a differential operator and as a vector. Thus, when operating on $f V$, we first keep f fixed and let ∇ operate on V. We then keep V fixed and let ∇ operate on f (remember that $\nabla \cdot f$ is nonsense); V and ∇f are vectors and we complete their multiplication by taking their dot product.

A vector whose divergence is zero is said to be solenoidal. It is straightforward to show that in cylindrical coordinates, $\nabla \cdot V$ is

$$\nabla \cdot \vec{V}(\rho, \phi, z) = \frac{1}{\rho}\frac{\partial}{\partial \rho}(\rho V_\rho) + \frac{1}{\rho}\frac{\partial}{\partial \phi} V_\phi + \frac{\partial}{\partial z} V_z \tag{1.24}$$

and in spherical coordinates

$$\nabla \cdot \vec{V}(r, \theta, \phi) = \frac{1}{r^2}\frac{\partial}{\partial r}(r^2 V_r) + \frac{1}{r\sin\theta}\left[\frac{\partial}{\partial \theta}(\sin\theta V_\theta) + \frac{\partial}{\partial \phi} V_\phi\right] \tag{1.25}$$

c. The Laplacian Operator ∇^2

The divergence of a vector field is defined by the scalar product of the operator ∇ with the vector field. What is the scalar product of ∇ with itself?

$$\nabla^2 = \nabla \cdot \nabla = \left(\frac{\partial}{\partial x} \hat{e}_x + \frac{\partial}{\partial y} \hat{e}_y + \frac{\partial}{\partial z} \hat{e}_z \right) \cdot \left(\frac{\partial}{\partial x} \hat{e}_x + \frac{\partial}{\partial y} \hat{e}_y + \frac{\partial}{\partial z} \hat{e}_z \right)$$

$$= \frac{\partial^2}{\partial x^2} + \frac{\partial^2}{\partial y^2} + \frac{\partial^2}{\partial z^2}$$

This important quantity

$$\nabla^2 = \frac{\partial^2}{\partial x^2} + \frac{\partial^2}{\partial y^2} + \frac{\partial^2}{\partial z^2} \tag{1.26}$$

is a scalar differential operator called the Laplacian.

Because the Laplacian is a scalar differential operator, it does not change the vector character of the field on which it operates. Thus, $\nabla^2 \psi(r)$ is a scalar field as $\psi(r)$ is a scalar field, and $\nabla^2(\nabla \psi(r))$ is a vector field because the gradient $\nabla \psi(r)$ is a vector field. The equation $\nabla^2 \psi = 0$ is called Laplace's equation.

d. The Curl of a Vector

If $V(x, y, z)$ is a differentiable vector field, the curl of V is defined by the vector product $\nabla \times V$:

$$\nabla \times \vec{V} = \begin{vmatrix} \hat{e}_x & \hat{e}_y & \hat{e}_z \\ \dfrac{\partial}{\partial x} & \dfrac{\partial}{\partial y} & \dfrac{\partial}{\partial z} \\ V_1 & V_2 & V_3 \end{vmatrix} \tag{1.27}$$

$$= \hat{e}_x \left(\frac{\partial V_3}{\partial y} - \frac{\partial V_2}{\partial z} \right) + \hat{e}_y \left(\frac{\partial V_1}{\partial z} - \frac{\partial V_3}{\partial x} \right) + \hat{e}_z \left(\frac{\partial V_2}{\partial x} - \frac{\partial V_1}{\partial y} \right)$$

A vector is said to be irrotational if its curl is zero. We see that the gradient of any scalar field $\psi(r)$ is irrotational:

$$\nabla \times (\nabla \psi) = \begin{vmatrix} \hat{e}_x & \hat{e}_y & \hat{e}_z \\ \partial/\partial x & \partial/\partial y & \partial/\partial z \\ \partial\psi/\partial x & \partial\psi/\partial y & \partial\psi/\partial z \end{vmatrix} = \begin{vmatrix} \hat{e}_x & \hat{e}_y & \hat{e}_z \\ \partial/\partial x & \partial/\partial y & \partial/\partial z \\ \partial/\partial x & \partial/\partial y & \partial/\partial z \end{vmatrix} \psi(x, y, z) = 0$$

A vector V is solenoidal (or divergence free) if $\nabla \cdot V = 0$. From this we see that the curl of any vector field V is solenoidal:

$$\nabla \cdot (\nabla \times \vec{V}) = 0.$$

The proof is straightforward but is very tedious. We leave it to the reader.

If $\psi(r)$ is a scalar field and $V(r)$ a vector field, then one can show that

$$\nabla \times (\psi V) = \psi (\nabla \times V) + \nabla \psi \times V$$

To prove it, we first write

$$\nabla \times (\psi \vec{V}) = \begin{vmatrix} \hat{e}_x & \hat{e}_y & \hat{e}_z \\ \dfrac{\partial}{\partial x} & \dfrac{\partial}{\partial y} & \dfrac{\partial}{\partial z} \\ \psi V_1 & \psi V_2 & \psi V_3 \end{vmatrix}$$

Then notice that $\partial(\psi V_2)/\partial x = \psi(\partial V_2/\partial x) + (\partial \psi/\partial x)V_2$, and so on, we can expand the determinant as a sum of two determinants:

$$\nabla \times (\psi \vec{V}) = \psi \begin{vmatrix} \hat{e}_x & \hat{e}_y & \hat{e}_z \\ \dfrac{\partial}{\partial x} & \dfrac{\partial}{\partial y} & \dfrac{\partial}{\partial z} \\ V_1 & V_2 & V_3 \end{vmatrix} + \begin{vmatrix} \hat{e}_x & \hat{e}_y & \hat{e}_z \\ \dfrac{\partial \psi}{\partial x} & \dfrac{\partial \psi}{\partial y} & \dfrac{\partial \psi}{\partial z} \\ V_1 & V_2 & V_3 \end{vmatrix} = \psi(\nabla \times \vec{V}) + \nabla \psi \times \vec{V}$$

It is too long to discuss the physical meaning of the curl of a vector here, which is not quite as transparent as that for the divergence. We refer the reader to *Mathematical Methods for Physicists*, by Tai L. Chow (Cambridge University Press, 2000).

The $\nabla \times V$ takes different forms in cylindrical and spherical coordinates:

Cylindrical

$$\nabla \times \vec{V} = \left(\frac{1}{\rho} \frac{\partial V_z}{\partial \phi} - \frac{\partial V_\phi}{\partial z} \right) \hat{e}_\rho + \left(\frac{\partial V_\rho}{\partial z} - \frac{\partial V_z}{\partial \rho} \right) \hat{e}_\phi + \left(\frac{1}{\rho} \frac{\partial (\rho V_\phi)}{\partial \rho} - \frac{1}{\rho} \frac{\partial V_\rho}{\partial \phi} \right) \hat{e}_\rho \tag{1.28}$$

$$\nabla \times \vec{V} = \frac{1}{\rho} \begin{vmatrix} \hat{e}_\rho & \rho \hat{e}_\phi & \hat{e}_z \\ \dfrac{\partial}{\partial \rho} & \dfrac{\partial}{\partial \phi} & \dfrac{\partial}{\partial z} \\ V_\rho & \rho V_\phi & V_z \end{vmatrix} \tag{1.28a}$$

Spherical

$$\nabla \times \vec{V} = \frac{1}{r \sin \theta} \left[\frac{\partial (\sin \theta V_\phi)}{\partial \theta} - \frac{\partial V_\theta}{\partial \phi} \right] \hat{e}_r$$

$$+ \frac{1}{r} \left[\frac{1}{\sin \theta} \frac{\partial V_r}{\partial \phi} - \frac{\partial (r V_\phi)}{\partial r} \right] \hat{e}_\theta + \frac{1}{r} \left[\frac{\partial (r V_\theta)}{\partial r} - \frac{\partial V_r}{\partial \theta} \right] \hat{e}_\phi \tag{1.29}$$

$$\nabla \times \vec{V} = \frac{1}{r^2 \sin\theta} \begin{vmatrix} \hat{e}_r & r\hat{e}_\theta & r\sin\theta\,\hat{e}_\phi \\ \frac{\partial}{\partial r} & \frac{\partial}{\partial \theta} & \frac{\partial}{\partial \phi} \\ V_r & rV_\theta & r\sin\theta\,V_\phi \end{vmatrix} \qquad (1.29a)$$

A vector field that has nonvanishing curl is called a vortex field, and the curl of the field vector is a measure of the vorticity of the vector field.

We now list formulas involving the operator ∇, which are used in the chapters that follow. *A* and *B* are differentiable vector functions, and ϕ and ψ are differentiable scalar functions of position (x, y, z):

1. $\nabla(\phi + \psi) = \nabla\phi + \nabla\psi$ or $\text{grad}(\phi + \psi) = \text{grad }\phi + \text{grad }\psi$

2. $\nabla \cdot (A + B) = \nabla \cdot A + \nabla \cdot B$ or $\text{div}(A + B) = \text{div }A + \text{div }B$

3. $\nabla \times (A + B) = \nabla \times A + \nabla \times B$ or $\text{curl}(A + B) = \text{curl }A + \text{curl }B$

4. $\nabla \cdot (\phi A) = (\nabla\phi) \cdot A + \phi(\nabla \cdot A)$

5. $\nabla \times (\phi A) = (\nabla\phi) \times A + \phi(\nabla \times A)$

6. $\nabla \cdot (A \times B) = B \cdot (\nabla \times A) - A \cdot (\nabla \times B)$

7. $\nabla \times (A \times B) = (B \cdot \nabla)A - B(\nabla \cdot A) - (A \cdot \nabla)B + A(\nabla \cdot B)$

8. $\nabla(A \cdot B) = (B \cdot \nabla)A + (A \cdot \nabla)B + B \times (\nabla \times A) + A \times (\nabla \times B)$

9. $\nabla \cdot (\nabla\phi) \equiv \nabla^2\phi \equiv \dfrac{\partial^2 \phi}{\partial x^2} + \dfrac{\partial^2 \phi}{\partial y^2} + \dfrac{\partial^2 \phi}{\partial z^2}$

where $\nabla^2 \equiv \dfrac{\partial^2}{\partial x^2} + \dfrac{\partial^2}{\partial y^2} + \dfrac{\partial^2}{\partial z^2}$ is called the *Laplacian operator*.

10. $\nabla \times (\nabla\phi) = 0$. The curl of the gradient of ϕ is zero.

11. $\nabla \cdot (\nabla \times A) = 0$. The divergence of the curl of **A** is zero.

12. $\nabla \times (\nabla \times A) = \nabla(\nabla \cdot A) - \nabla^2 A$

In formulas 9 through 12, ϕ and *A* are supposed to have continuous second partial derivatives.

1.5 Vector Integration and Integral Theorems

a. Volume Integrals and Line Integrals

The volume integral of a function $f(r)$ is defined by

$$\int_V f(\bar{r})\, dV = \lim_{\delta V_i \to 0}\left(\sum_i f(\bar{r}_i)\,\delta V_i\right), \tag{1.30}$$

where the volume V is dissected into elements δV_i, and r_i lies in δV_i, which goes to zero in the limit. In Cartesian coordinates, $dV = dxdydz$, and in spherical polar coordinates, $dV = r^2\sin\theta\, dr d\theta d\phi$. We may use any other coordinate system that is convenient for the problem under consideration.

To define an integral along a line C in space joining points A and B, we dissect the line into elements that, if C is a sufficiently smooth curve, can be approximated by straight line elements of length δl pointing in the direction $\hat{t}$, $\hat{t}$ being a unit vector tangent to the curve. A vector line element $\delta l = \hat{t}\delta l$ can then be defined. The direction of $\hat{t}$ is taken in the sense we choose to go along the curve C, from A to B say.

The total length of the curve (**Figure 1.8**) is defined to be

$$length = \int_C dl = \lim_{\delta l \to 0}\left(\sum_i \delta l_i\right) \tag{1.31}$$

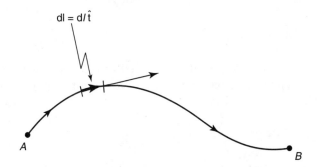

dl = dl$\hat{t}$

A

B

I FIGURE 1.8

and we can define line integrals

$$\int_C f(\vec{r})\,dl = \lim_{\delta l \to 0}\left(\sum_i f(\vec{r_i})\,\delta l_i\right)$$

(1.32)

where r_i is a point in the element δl_i.

Line integrals with $f(r) = F(r)\cdot\hat{t}$ often occurs in physics, and we write

$$\int_C \vec{F}\cdot\hat{t}\,dl = \int_C \vec{F}\cdot d\vec{l}$$

(1.33)

If $\vec{F}$ can be expressed as $\vec{F} = \hat{i}F_1 + \hat{j}F_2 + \hat{k}F_3$, then equation (1.13) becomes

$$\int_C \vec{F}\cdot\hat{t}\,dl = \int_C \vec{F}\cdot d\vec{l} = \int_C (\hat{i}F_1 + \hat{j}F_2 + \hat{k}F_3)\cdot(\hat{i}dx + \hat{j}dy + \hat{k}dz)$$

$$\int_C \vec{F}\cdot\hat{t}\,dl = \int_C F_1(x,y,z)\,dx + \int_C F_2(x,y,z)\,dy + \int_C F_3(x,y,z)\,dz$$

(1.34)

Each integral on the right-hand side requires for its execution more than knowledge of the limits. In fact, each integral on the right-hand side is not completely defined because in the first integral, for example, we do not know the value of y and z in F_1. What is needed is a statement such as

$$y = g(x),\ z = h(x)$$

that specifies y, z for each value of x (and so specifies the path of integration). The integrand now reduces to $F_1(x, y, z) = F_1(x, g(x), h(x))$ so that the first integral on the right-hand side becomes well defined. Thus, in general, a line integral is a path-dependent integral. If the scalar product $\vec{F}\cdot d\vec{l}$ is equal to an exact differential, $\vec{F}\cdot d\vec{l} = d\varphi = -\nabla\varphi\cdot d\vec{l}$, the integration depends only on the limits and is therefore path independent:

$$\int_A^B \vec{F}\cdot d\vec{l} = \int_A^B d\varphi = \varphi_B - \varphi_A$$

It follows that the line integral of the gradient of any scalar function of position φ around a closed curve vanishes. We denote an integral around a closed curve by the symbol $\oint$:

$$\oint \vec{F}\cdot d\vec{l} \quad \text{or} \quad \oint_\Gamma \vec{F}\cdot d\vec{l}$$

where Γ specifies the closed path. A vector field F that has the above property (path independent) is called conservative:

$$\oint \vec{F} \cdot \vec{dl} = 0 \tag{1.35}$$

The curl of a conservative vector field is zero: $\nabla \times F = \nabla \times (-\nabla\varphi) = 0$. Both gravitational force and Coulomb force are conservative forces.

b. Surfaces and Surface Integrals

Surfaces can be represented by vectors. A bounded plane surface has a magnitude represented by its area and a direction specified by the direction of its normal. If the surface is part of a closed surface, the outward drawn normal is taken as positive. If the surface is not part of a closed surface, we specify the positive direction of the normal by the rule that a right-handed screw rotated in the plane of the surface in the positive sense of the periphery advances along the positive normal.

For a nonplane surface, we may dissect it into a number of elementary surfaces, δS_i. If the surface is smooth and the elements are sufficiently small, each element can be approximated by an element of a plane. It is often useful to consider the surface element as itself a vector, $\vec{\delta S} = \hat{n}\delta S$, of magnitude δS, pointing in the direction of the unit vector $\hat{n}$. The vector representative of the entire surface is the sum of the vectors representing its elements.

Integral of function $f(r)$ over a surface S is defined to be

$$\int_S f(\vec{r})\, dS = \lim_{\delta S_i \to 0} \left(\sum_i f(\vec{r}_i)\, \delta S_i \right) \tag{1.36}$$

The case when $f(r) = F(r) \cdot \hat{n}$ often occurs in physics problems, for example, when we consider the flow of electric charge across a surface. We may conveniently write

$$\int_S \vec{F} \cdot \hat{n}\, dS = \int_S \vec{F} \cdot \vec{dS}$$

c. The Divergence Theorem and Stokes' Theorem

We use these two important theorems over and over again in Chapter 2. We state them here without proof. Readers interested in the proofs may consult books on vector analysis or on mathematical methods for physicists.

The divergence theorem was introduced by Joseph L. Lagrange and was first used in the modern sense by George Green. It states that the volume integral of the divergence of a vector function $A(r)$ taken over any volume τ is equal to the surface integral of A taken over the surface Σ enclosing the volume τ (**Figure 1.9**):

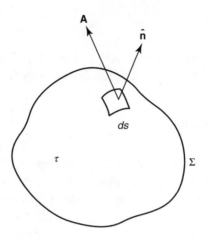

$$\int_\tau \nabla \cdot \vec{A}\, d\tau = \int_\Sigma \vec{A} \cdot \hat{n}\, ds = \int_\Sigma \vec{A} \cdot d\vec{s} \tag{1.37}$$

Stokes' theorem was first discovered by Lord Kelvin in 1850 and was rediscovered by George Gabriel Stokes 4 year later. It states that the surface integral of the curl of a vector function $A(r)$ taken over any surface Σ is equal to the line integral of $A(r)$ around the periphery Γ of the surface **(Figure 1.10):**

$$\int_S (\nabla \times \vec{A}) \cdot \hat{n}\, ds = \oint_\Gamma \vec{A} \cdot d\vec{l} \tag{1.38}$$

Along with these theorems are similar, and essentially equivalent, theorems:

$$\int_\tau \nabla \times \vec{A}\, d\tau = -\int_\Sigma \vec{A} \times d\vec{s} \tag{1.39}$$

$$\int_\tau \nabla u\, d\tau = \int_\Sigma u\, d\vec{s} \tag{1.40}$$

$$\int_\Sigma \nabla u \times d\vec{s} = -\oint_\Gamma u\, d\vec{l} \tag{1.41}$$

We occasionally find these formulations useful.

The ability to relate physics inside a volume to physics on the enclosing surface or physics on a surface to physics on the bounding curve is of paramount importance in a field theory.

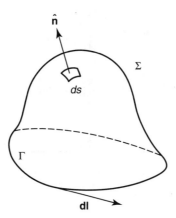

I FIGURE 1.10

1.6 Dirac's Delta Function

The δ-function, introduced by Paul A. M. Dirac, is a very useful tool in physics; the one-dimensional δ-function can be defined by the relations

$$\delta(x) = \begin{cases} 0, & \text{if } x \neq 0 \\ \infty, & \text{if } x = 0 \end{cases} \tag{1.42}$$

and

$$\int_{-\infty}^{\infty} \delta(x)\, dx = 1 \tag{1.43}$$

Thus, the δ-function is not a function in the usual mathematical sense, because its value is not finite at $x = 0$. Mathematicians call it a generalized function or distribution. We can picture the δ-function as an infinitely high and infinitesimally narrow spike, as shown by area 1 in **Figure 1.11**.

If $f(x)$ is an arbitrary function that is continuous at $x = 0$, then the product $f(x)\delta(x)$ is zero everywhere except at $x = 0$. It follows that

$$f(x)\delta(x) = f(0)\delta(x) \tag{1.44}$$

That is, we replace $f(x)$ by the value it assumes at $x = 0$, and

$$\int_{-\infty}^{\infty} f(x)\,\delta(x)\, dx = f(0) \int_{-\infty}^{\infty} \delta(x)\, dx = f(0) \tag{1.45}$$

If we shift the spike from the origin $x = 0$ to some other point $x = a$, equations (1.42) to (1.44) correspondingly become

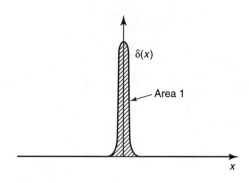

I FIGURE 1.11

$$\delta(x-a) = \begin{cases} 0, & if\ x \neq a \\ \infty, & if\ x = a \end{cases} \tag{1.42a}$$

with

$$\int_{-\infty}^{\infty} \delta(x-a)\,dx = 1 \tag{1.43a}$$

and

$$f(x)\delta(x-a) = f(a)\delta(x-a) \tag{1.44a}$$

$$\int_{-\infty}^{\infty} f(x)\,\delta(x-a)\,dx = f(a) \tag{1.45a}$$

We can easily verify the following most frequently required properties of the δ-function:

1. $\int_a^b f(x)\,\delta(x-x')\,dx = \begin{cases} f(x'), & if\ a < x' < b \\ 0, & if\ x' < a\ or\ x' > b \end{cases}$

2. $\delta(-x) = \delta(x)$

3. $\delta'(x) = -\delta'(-x)$, where $\delta'(x) = d\delta(x)/dx$

4. $x\delta(x) = 0$

5. $\delta(ax) = a^{-1}\delta(x),\ a > 0$

6. $\delta(x^2 - a^2) = (2a)^{-1}[\delta(x-a) + \delta(x+a)],\ a > 0$

7. $\int \delta(a-x)\delta(x-b)\,dx = \delta(a-b)$

8. $f(x)\delta(x-a) = f(a)\delta(x-a)$

Each of the first six listed properties can be established by multiplying both sides by a continuous differentiable function $f(x)$ and then integrating over x.

It is easy to generalize the δ-function to three dimensions:

$$\delta^3(\vec{r}) = \delta(x)\delta(y)\delta(z) \tag{1.46}$$

with the volume integral equal to 1

$$\int_{all\ space} \delta^3(\vec{r})\,d\vec{r} = \int_{-\infty}^{\infty}\int_{-\infty}^{\infty}\int_{-\infty}^{\infty} \delta(x)\delta(y)\delta(z)\,dx\,dy\,dz = 1 \tag{1.47}$$

and

$$\int_{all\ space} f(\vec{r})\,\delta^3(\vec{r}-\vec{r_0})\,d\vec{r} = f(\vec{r_0}) \tag{1.48}$$

where $\vec{r_0}$ is any fixed point.

As an example of the application of the δ-function, let us consider Poisson's equation:

$$\nabla^2 \Phi = -\rho/\varepsilon_0 \tag{1.49}$$

We shall see in Chapter 2 that Φ is the electric scalar potential, ρ the charge density, and ε_0 the permittivity of free space.

We may verify by direct differentiation that $1/r$ is a solution of Laplace's equation for $r > 0$:

$$\nabla^2(1/r) = 0, \quad r > 0 \tag{1.50}$$

Recall from general physics the potential Φ of a point charge q is

$$\Phi = q/r, \quad r > 0 \tag{1.51}$$

Because q is a point charge, the density ρ can be represented by a δ-function:

$$\int_{all\ space} \rho\,dV = \int_{all\ space} q\delta(\vec{r})\,dV = q \tag{1.52}$$

Now we may write the integral of Poisson's equation as

$$\int_{all\ space} \nabla^2 \Phi\,dV = q\int_{all\ space} \nabla^2\left(\frac{1}{r}\right)dV = -\varepsilon_0^{-1}\int_{all\ space} \rho\,dV \tag{1.53}$$

Combining this with equation (1.51), we have

$$\int_{all\ space} \nabla^2\left(\frac{1}{r}\right)dV = -\varepsilon_0^{-1} \tag{1.54}$$

Because $\nabla^2(1/r)$ vanishes for $r > 0$ and has an integral over all space of $1/\varepsilon_0$, we may then write

$$\nabla^2\left(\frac{1}{r}\right) = -\varepsilon_0^{-1}\delta(\vec{r}) \tag{1.55}$$

In Chapter 3, we shall see that the δ-function is of considerable value in the formal development of solutions to Poisson's equation by the method of Green's function.

1.7 Analytical Mechanics

The essential physics of analytical mechanics is contained in the three laws of motion. The second law describes how particles and objects behave under the application of specified forces. Specifically, it relates force F to the time rate of change of the linear momentum of the particle designed by $p = mv$:

$$\vec{F} = \frac{d}{dt}(m\vec{v}) \tag{1.56}$$

When m is constant, equation (1.56) reduces to the familiar form

$$\vec{F} = m\vec{a}$$

where $\vec{a}$ is the acceleration of the particle produced by the applied force $\vec{F}(r)$. If the velocity $\vec{v}$ of the particle is much less than the velocity of light, then we have

$$\vec{F} = m\frac{d^2\vec{r}}{dt^2} \tag{1.57}$$

Equation (1.57) is a second-order differential equation whose solution in terms of the initial position and velocity of the particle specifies unambiguously the motion of the particle for all future time. Classical mechanics is therefore often called deterministic.

One important consequence of Newton's second law is the introduction of the concept of energy. This concept arises naturally through a consideration of the work done on the particle by the force during the motion from position a to position b along its path:

$$W = \int_a^b \vec{F} \cdot d\vec{r} = \int_a^b \vec{F} \cdot \dot{\vec{r}}dt \tag{1.58}$$

where a dot placed over a symbol denotes derivative with time. Substituting equation (1.57) into equation (1.58) and after simple mathematical manipulation, we find

$$W = \int_a^b d\left(m\dot{\vec{r}} \cdot \dot{\vec{r}}/2\right) = \int_a^b d\left(mv^2/2\right) = T_b - T_a \tag{1.59}$$

where $T = mv^2/2$ is the kinetic energy of the particle.

Equation (1.59) states that the work done on the particle is equal to the increase in its kinetic energy; this is known as the work principle. In general, the work done by a force depends on the path along which the work is performed; thus the work principle is really of not much use to us unless the path followed by the particle is known. But if the inner product $\vec{F} \cdot d\vec{r}$ is an exact differential of certain function of the integration variable r, say, $V(r)$

$$\vec{F} \cdot d\vec{r} = -dV(r) \quad \text{and} \quad \vec{F} = -\nabla V(r) \tag{1.60}$$

the work integral becomes independent of the path of integration—it depends only on the initial and final position of the particle. Forces of this type are called *conservative forces*. The work integral now becomes

$$W = \int_a^b \vec{F} \cdot d\vec{r} = \int_a^b (-dV) = V_a - V_b$$

where V_a and V_b are called the potential energy of the particle at points a and b, respectively. Combining this with the work principle of equation (1.59), we obtain

$$T_a + V_a = T_b + V_b$$

or

$$E = T + V = \tfrac{1}{2}mv^2 + V(r) = \text{const.} \tag{1.61}$$

i.e., the total (mechanical) energy of the particle is conserved (constant in time) in a conservative force field. From equation (1.60), we see that, for a conservative force, $\nabla \times \vec{F} = 0$.

Energy conservation is one of the three most important conservation laws in physics. The other two conservation laws are momentum and angular momentum conservation. Most problems in mechanics cannot be solved completely in terms of known functions or it is too tedious to solve them, but very often certain information about the system that is of greater interest and importance than a complete knowledge of the path can be obtained easily from the conservation laws.

The momentum conservation follows directly from the second law of motion. When there is no external applied force, equation (1.56) reduces to

$$\vec{p} = m\vec{v} = \text{const.}$$

The angular momentum conservation also follows from the second law of motion. The angular momentum $\vec{L}$ of a particle moving with velocity $\vec{v}$ (or with a linear momentum $\vec{p} = m\vec{v}$) is

$$\vec{L} = \vec{r} \times \vec{p} \tag{1.62}$$

where $\vec{r}$ is the position vector of the particle from the origin O, and the angular momentum $\vec{L}$ is measured with respect to O. The rate of change of angular momentum is obtained by differentiating equation (1.62) with respect to time:

$$\frac{d\vec{L}}{dt} = \frac{d}{dt}\left(m\vec{r} \times \vec{v}\right) = m\dot{\vec{v}} \times \vec{v} + m\vec{r} \times \dot{\vec{v}} = m\vec{r} \times \dot{\vec{v}} = \vec{r} \times \vec{F}$$

where $\vec{F}$ is the force acting on the particle m. The vector product $\vec{r} \times \vec{F}$ is the moment of the force about the origin O and is called the torque, $\vec{N}$, of the force about O. And we have

$$\vec{N} = d\vec{L}/dt \tag{1.63}$$

which says the rate of change of angular momentum of a particle about a fixed point O is equal to the applied torque about O.

Equation (1.63) is Newton's second law for rotational motion. From (1.63) it follows that if there is no component of torque in a given direction, the component of angular momentum in that direction is constant.

For a central force field, $\vec{F} = F(r)\hat{r}$ and, consequently, $\vec{N} = \vec{r} \times \vec{F} = 0$. Therefore $\vec{L}$ is a constant of the motion.

The second law of motion is an assertion about the workings of nature, and the force is supposed to have some independent properties that Newton did not describe completely. But he did give one rule about the force, which is the third law of motion: Action and reaction are equal in magnitude but opposite in direction. Newton recognized that there is no such thing as a single force in the universe. Instead, forces are properties of the interaction of objects. Consider the force on particle 1 due to particle 2 ($\vec{F}_{12}$); this force is related to the force exerted on particle 2 by particle 1 ($\vec{F}_{21}$) by the equation

$$\vec{F}_{12} + \vec{F}_{21} = 0 \qquad\qquad (1.64)$$

The Coulomb or electrostatic force between two charged particles satisfies equation (1.64). However, it is important to realize that not all physical forces satisfy equation (1.64). If the forces under consideration do satisfy equation (1.64), then it is straightforward to show that the two-body system with internal forces can be replaced by an equivalent one-body system. First, the equations of motion for the two particles are

$$\vec{F}_{12} = m_1 \frac{d\vec{v}_1}{dt}, \quad \vec{F}_{21} = m_2 \frac{d\vec{v}_2}{dt}$$

Next, we introduce a new variable, $\vec{R}$, the coordinate of the center of mass, and $\vec{r}$, the relative coordinate:

$$\vec{R} \equiv \frac{m_1 \vec{r}_1 + m_2 \vec{r}_2}{m_1 + m_2}, \quad \vec{r} = \vec{r}_1 - \vec{r}_2$$

We further introduce $\vec{F} = \vec{F}_{12} = -\vec{F}_{21}$.

In terms of the new variables and making use of equation (1.64), we obtain after some mathematical manipulations

$$\frac{d^2 \vec{R}}{dt^2} = 0 \qquad\qquad (1.65)$$

$$\vec{F} = \mu \frac{d^2 \vec{r}}{dt^2} \qquad\qquad (1.66)$$

where $\mu = m_1 m_2 / (m_1 + m_2)$ is the reduced of the system. Equation (1.65) states that the center of mass of the system moves with uniform velocity. Equation (1.66) states that the relative motion of the two particles is that of a particle of mass μ (the reduced mass of the system) in the force field acting between the particles. The treatment of many-particle systems follows along similar lines; we do not pursue it here.

Problems

1. Find the angle between the face diagonals of a cube. (For simplicity you may use a cube of side 1 and place one corner at the origin.)
2. Find the projection of the vector $\vec{A} = \hat{i} - 2\hat{j} + \hat{k}$ on the vector $\vec{B} = 4\hat{i} - 4\hat{j} + 7\hat{k}$.
3. Find a unit vector normal to the plane of $\vec{A} = 2\hat{i} - 6\hat{j} - 3\hat{k}$ and $\vec{B} = 4\hat{i} + 3\hat{j} - \hat{k}$.

4. Find an equation for the plane perpendicular to the vector $\vec{A} = 2\hat{i} + 3\hat{j} + 6\hat{k}$ and passing through the terminal point of the vector $\vec{B} = \hat{i} + 5\hat{j} + 3\hat{k}$.

5. Prove the law of sines for plane geometry.

6. Find a unit normal to the surface $x^2y + 2xz = 4$ at the point $(2, -2, 3)$.

7. Show that $\nabla(FG) = F\nabla(G) + G\nabla(F)$.

8. Find the angle between the surface $x^2 + y^2 + z^2 = 9$ and $z = x^2 + y^2 - 3$ at the point $(2, -1, 2)$.

9. Show that (a) $\nabla r^n = nr^{n-2}\vec{r}$, (b) $\nabla \cdot (\vec{r}/r^3) = 0$, (c) $\nabla^2(1/r) = 0$.

10. Let P be any point on an ellipse whose foci are at points A and B. Prove that lines AP and BP make equal angles with the tangent to the ellipse at P. (The problem has a physical application. Light rays or sound waves originating at focus A, for example, are reflected from the ellipse to focus B.) **(Figure 1.12)**.

11. If $\vec{A} = (3x^2 + 6y)\hat{i} - 14yz\hat{j} + 20xz^2\hat{k}$, evaluate the line integral $\int \vec{A} \cdot d\vec{r}$ from $(0, 0, 0)$ to $(1, 1, 1)$ along the following paths:
    (a) $x = t$, $y = t^2$, $z = t^3$;
    (b) The straight line joining $(0, 0, 0)$ and $(1, 1, 1)$;
    (c) The straight line from $(0, 0, 0)$ to $(1, 0, 0)$, then to $(1, 1, 0)$, and then to $(1, 1, 1)$.

12. If $\vec{A} = 4xz\hat{i} - y^2\hat{j} + yz\hat{k}$, evaluate $\iint_S \vec{A} \cdot \hat{n}\,dS$ where S is the surface of the cube bounded by $x = 0$, $x = 1$, $y = 0$, $y = 1$, $z = 0$, $z = 1$.

13. Express the divergence theorem and Stokes' theorem in rectangular form.

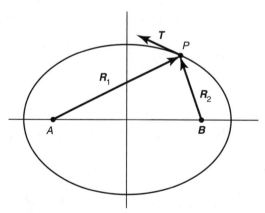

I FIGURE 1.12

14. Prove equation (1.39). (Use the divergence theorem.)
15. Prove equation (1.40). (Use the divergence theorem.)
16. Prove equation (1.41). (Use Stokes' theorem.)
17. Prove that a necessary and sufficient condition that $\oint_C \vec{A} \cdot d\vec{r} = 0$ for every closed curve C is that $\nabla \times \vec{A} = 0$ identically.

2 Electrostatics

Electrostatics covers the special case when all electric charges are stationary. Electric charge is a fundamental and characteristic property of the microscopic particles that make up matter. We can summarize its observed properties in three statements:

1. There are two kinds of electric charges: positive and negative. Like charges repel and unlike charges attract.
2. Charge can be neither created nor destroyed, and the total electric charge in an isolated system never changes. This is often referred to as the charge conservation law.
3. Millikan's oil-drop experiment and many other experiments have shown that in nature, electric charge comes in units of one magnitude only, the electronic charge carried by an electron or a proton. That is, electric charge is quantized. Of course, the fact of charge quantization lies outside the scope of classical electromagnetism. We thus usually ignore it and act as if charges q could have any strength whatsoever. High-energy physicists have demonstrated the existence of quarks, which are microscopic particles with electric charge $\pm 1/3\ e$ and $\pm 2/3\ e$. Quarks, however, are always confined inside nucleus.

Electrons and atomic nuclei, the basic constituents, all carry electric charge. An electron carries a negative charge $-e$ and an atomic nucleus a positive charge Ze, where Z is the atomic number, ranging from 1 to 92 for uranium and higher for some unstable nuclei. At the present limits of experimental resolution, electrons seem to be the only point particles, in the sense that no intrinsic size or structure has yet been discerned for them. On an atomic scale, we can regard the atomic nuclei as structureless point particles. The charge density $\rho(r)$ of a point charge at the point r_1 can be expressed in terms of Dirac's δ-function:

$$\rho(\vec{r}) = e\delta(\vec{r} - \vec{r_1})$$

We can regard a charged body as a point charge when its dimension is small compared with values of *r* of interest, and its internal structure may be neglected.

2.1 Coulomb's Law

Toward the end of the 18th century, technique in experimental science achieved sufficient sophistication. Coulomb refined observations of the forces between electric charges with a torsion balance. A small metal sphere *A* is fixed at the end of an insulated rod (**Figure 2.1**). A similar sphere *B* is fixed at one end of an ebonite arm that is suspended at its midpoint by a fine thread from a torsion head. When *A* and *B* are similarly charged, they repel each other and the ebonite arm rotates. The twist in the thread brings *B* to rest at a definite distance *r* from *A*. Coulomb found that

1. Two point charges exert on each other forces that are along the line joining them and are inversely proportional to the square of the distance between them.
2. These forces are also proportional to the product of the charges and are repulsive for like charges and attractive for unlike charges.

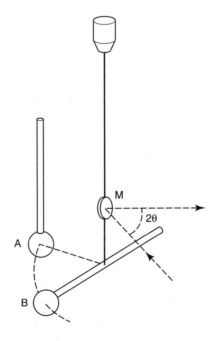

I FIGURE 2.1

These two statements are known as Coulomb's law, after Charles Augustin de Coulomb (1736–1806):

$$F \propto \frac{q^2}{r^2}$$

We can rewrite the above expression as an equation:

$$F = p \frac{q^2}{r^2}$$

where p is a constant of proportionality.

This equation leads to a system of units for measuring charges. If the Gaussian system is used, F is measured in dynes, r in cm, p put arbitrarily to unity, and q is numerically equal to $r(F)^{1/2}$ electrostatic units of charge. The unit is thus defined: A charge of 1 electrostatic unit when placed 1 cm from an exactly similar charge in air repels it with a force of 1 dyne.

A more convenient unit of charge has found common usage: coulomb (C). All measuring instruments such as ammeters and voltmeters are based on this unit that is defined by reference to magnetic effects of currents and is equal to 2.9978×10^9 electrostatic units of charge. One coulomb charge placed 1 cm away from a similar charge in air experiences a force of

$$F = \frac{2.9978 \times 10^9 \times 2.9978 \times 10^9}{1} = 8.98755 \times 10^{18} \, dynes.$$

When the coulomb unit of charge is used, we use mks units of force and length—newton and meter. We also have to choose a new value for p. Because two equal charges of 1 coulomb placed 1 cm apart in air experience a force of 8.98755×10^{18} dynes, i.e., 8.98755×10^{13} newtons, and because 1 cm = 10^{-2} m, the required value of p is given by

$$8.98755 \times 10^{13} = p \frac{1 \times 1}{(10^{-2})^2}$$

from which we find

$$p = 8.98755 \times 10^9 \, \text{mks units} \, (N \cdot m^2 / C^2).$$

It is common to write $p = 1/4\pi\varepsilon_0$; thus

$$4\pi\varepsilon_0 = 1.113 \times 10^{-10} \, C^2 / N \cdot m^2$$

where ε_0 is regarded as an intrinsic constant of vacuum and called the permittivity of free space. The practical units discussed here are called rationalized mks units or SI units, and these are used throughout the book. To go from the Gaussian system to the SI system, it is necessary only to multiply e^2 in a formula with $1/4\pi\varepsilon_0$.

The inclusion of the factor 4π is to simplify the form of some important relations occurring in the electromagnetic theory. We often deal with spherical shapes and hence a factor of 4π is bound to appear. It is therefore advantageous to use a constant containing 4π explicitly.

Experiment and theory support the more general form of Coulomb's law for two "point" charges q and Q (**Figure 2.2**):

$$\vec{F} = \frac{qQ}{4\pi\varepsilon_0 r^2}\,\hat{r} \tag{2.1}$$

Coulomb's torsion balance experiments did not yield high accuracy, but more recent experiments prove that the law is true to within very narrow limits. For all practical purposes, Coulomb's law is accepted as an exact law.

a. Principle of Superposition

Electrostatic forces are two-body forces, which means that the force between any pair is unaltered by the presence of other charges in their neighborhood. If there are more than two particles present with charges $q_1, q_2, \ldots,$ the total forces on any one particle is the vector sum of forces it experiences due to all other particles separately. This is known as the principle of superposition. In **Figure 2.3** there are three charges; force on q_3 is given by

$$\vec{F} = \vec{F}_{13} + \vec{F}_{23} = \frac{q_1 q_3}{4\pi\varepsilon_0 |\vec{r}_{13}|^2}\,\hat{r}_{13} + \frac{q_2 q_3}{4\pi\varepsilon_0 |\vec{r}_{23}|^2}\,\hat{r}_{23}$$

The force on a charge q_j due to a number of other charges present is

$$\vec{F}_j = \frac{1}{4\pi\varepsilon_0} \sum_{i \neq j} \frac{q_i q_j}{|\vec{r}_{ij}|^2}\,\hat{r}_{ij} \tag{2.2}$$

If r_i and r_j are the position vectors of the charges q_i and q_j, respectively, then

$$\vec{F}_j = \frac{1}{4\pi\varepsilon_0} \sum_{i \neq j} \frac{q_i q_j}{|\vec{r}_j - \vec{r}_i|^3}\left(\vec{r}_j - \vec{r}_i\right) \tag{2.2a}$$

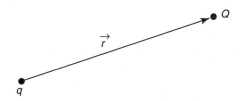

I FIGURE 2.2

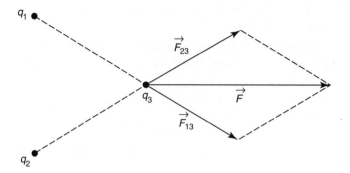

A simple extension of the idea of N interacting point charges is the interaction of a point charge with a continuous charge distribution. Now what do we mean by a continuous charge distribution? Millikan's oil-drop experiment and many other experiments have shown that in nature, electric charge is quantized: Any charged object in nature contains an integral multiple of the charge that has a magnitude of 1.6×10^{-19} C that is extremely small. Any macroscopic charged body contains a large number of electronic charges. Thus, the discreteness of charge causes no difficulties in establishing the concept of continuous charge distribution.

If charge is distributed through a volume V with a volume charge density ρ and on the surface S that bounds V with a surface charge density σ, then the force exerted by this charge distribution on a point charge q at r is obtained from equation (2.2a) by replacing q_i with $\rho_i \, dv_i$ or $\sigma_i da_i$:

$$\vec{F}_q = \frac{q}{4\pi\varepsilon_0} \int_V \frac{\vec{r}-\vec{r}'}{|\vec{r}-\vec{r}'|^3} \rho\,(\vec{r}')\,dv' + \frac{q}{4\pi\varepsilon_0} \int_s \frac{\vec{r}-\vec{r}'}{|\vec{r}-\vec{r}'|^3}\,\sigma\,(\vec{r}')\,da' \tag{2.3}$$

where

$$\rho = \lim_{\Delta v \to 0} \frac{\Delta q}{\Delta v} \quad \text{and} \quad \sigma = \lim_{\Delta s \to 0} \frac{\Delta q}{\Delta s}$$

In **Figure 2.4**, the point charge q is outside the charge distribution. What happens if the point r falls inside the charge distribution? It may appear at first sight that the first integral of equation (2.3) diverges, but this is not true. The region of integration in the vicinity of r contributes a negligible amount.

The interaction of two continuous charge distributions is too much for us to handle, so we avoid it.

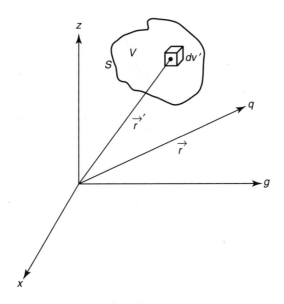

I FIGURE 2.4

Example 1

Two particles are suspended by a string of the same length l from the same point. Each has a mass m and a charge q. Show that the angle θ, which each string makes with the vertical, is given by

$$16\pi\varepsilon_0 mgl^2 \sin^3\theta = q^2\cos\theta$$

Solution

Let T be the tension in the string and F the force between the two charges; then we have

$$F = \frac{q^2}{4\pi\varepsilon_0(2l\sin\theta)^2}, \quad T\sin\theta = F, \quad \text{and} \quad T\cos\theta = mg$$

Elimination of T and F gives the answer (**Figure 2.5**).

2.2 The Electric Field

We are accustomed with direct contact forces, such as push or pull. But electric forces, like gravitational forces, act between objects that may be widely separated. Coulomb's law implies that the force on object A emanates directly and instantaneously from object B. This contradicts the special theory of relativity. An appropriate way to regard such forces involves the concept of force fields. When a charge is present somewhere, the properties of

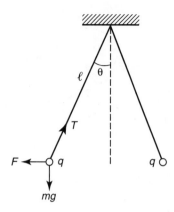

I FIGURE 2.5

space in its vicinity is so altered that another charge brought to this region experiences a force there. The alternation in space caused by a charge is called the charge's electric field, and any other charge is thought of interacting with the field and not directly with the charge responsible for it. You may believe that the field language is somewhat redundant, merely offering a purely semantic change of view. But there are reasons why we consider a field to be a real entity, not just a semantic trick. We will see later that fields do carry energy, momentum, and angular momentum.

We may explore the electric field by carrying a unit positive charge around the field and by carefully keeping all the charges producing the field fixed in position so that the field is not changed by the attractions or repulsions exerted by the test charge. The force experienced by the unit test charge when at rest relative to the observer at any point in the field is then the electric field E at that point. Accordingly, we operationally define the electric field at a point as the limit of the force on a test charge placed at that point to the charge of the test charge:

$$\vec{E} = \lim_{q \to 0} \frac{\vec{F}}{q} \tag{2.4}$$

The limiting process included in equation (2.4) is to ensure that the test charge does not affect the charge distributions that produce the field $\vec{E}$. Obviously, the electric field has units of newton/coulomb. By convention, the electric field leaves a positive charge and enters into a negative charge.

If the force given by equation (2.2a) is written as

$$\vec{F}_q = \frac{q}{4\pi\varepsilon_0} \sum_i \frac{q_i}{|\vec{r} - \vec{r}_i|^3} (\vec{r} - \vec{r}_i)$$

in which q_i is taken as the test charge q and so r_i is replaced by r the position vector of the test charge q, then equation (2.4) gives

$$\vec{E} = \frac{\vec{F_q}}{q} = \frac{1}{4\pi\varepsilon_0}\sum_i \frac{q_i}{|\vec{r} - \vec{r_i}|^3}(\vec{r} - \vec{r_i}) \tag{2.5}$$

For a charge distribution, the force is given by equation (2.3), and then equation (2.4) gives

$$\vec{E} = \frac{1}{4\pi\varepsilon_0}\int_V \frac{\vec{r} - \vec{r}'}{|\vec{r} - \vec{r}'|^3}\rho(\vec{r}')\,dv' + \frac{1}{4\pi\varepsilon_0}\int_S \frac{\vec{r} - \vec{r}'}{|\vec{r} - \vec{r}'|^3}\sigma(\vec{r}')\,da' \tag{2.6}$$

Note that the above expression for the electric field is quite general and includes equation (2.5). Using Dirac's δ-function, we can represent a distribution of N charges q_i located at points r_i by

$$\rho(\vec{r}) = \sum_{i=1}^{N} q_i\delta(\vec{r} - \vec{r}')$$

and equation (2.6), with $\sigma = 0$, reduces to equation (2.5)

$$\vec{E}(\vec{r}) = \frac{1}{4\pi\varepsilon_0}\int_V \sum_{i=1}^{N} q_i\delta(\vec{r}' - \vec{r_i}) \cdot \frac{(\vec{r} - \vec{r}')}{|\vec{r} - \vec{r}'|^3}\,d\vec{r}' = \frac{1}{4\pi\varepsilon_0}\sum_{i=1}^{N} q_i \frac{(\vec{r} - \vec{r}')}{|\vec{r} - \vec{r}'|^3}$$

Knowing the distributions ρ and σ, equation (2.6) allows us, at least in principle, to determine the electric field. The problem is that in practice these distributions are rarely known.

Example 2

Charge is distributed uniformly along an infinite straight line with density ρ_l. Find the electric field at the general point P, as shown in **Figure 2.6**.

Solution

We use cylindrical coordinates, with the line charge as the z-axis. At P we have dE

$$d\vec{E} = \frac{dQ}{4\pi\varepsilon_0 R^2}\left(\frac{r}{R}\hat{e}_r + \frac{-z}{R}\hat{e}_z\right), \quad R = \sqrt{r^2 + z^2}$$

where $\hat{e}_r$ and $\hat{e}_z$ are the unit vectors in the direction of r and z, respectively, and $dQ = \rho_l dz$. Now for every dQ at z there is another charge dQ at $-z$, and the z components cancel. Then

$$\vec{E} = \int_{-\infty}^{\infty} \frac{\rho_l r\,dz}{4\pi\varepsilon_0(r^2 + z^2)^{3/2}}\hat{e}_r = \frac{\rho_l r}{4\pi\varepsilon_0}\left[\frac{z}{r^2\sqrt{r^2 + z^2}}\right]_{-\infty}^{\infty} = \frac{\rho_l}{2\pi\varepsilon_0 r}\hat{e}_r$$

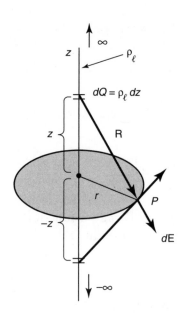

I FIGURE 2.6

Example 3

Find the electric field due to charge uniformly distributed over an infinite plane with charge density ρ_S.

Solution

We use the cylindrical coordinate system, with the plane of charge in the $z = 0$ plane, as shown in **Figure 2.7**. At point P, dE is given by

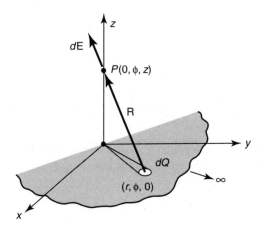

I FIGURE 2.7

$$d\vec{E} = \frac{\rho_s r \, dr \, d\phi}{4\pi\varepsilon_0 R^2}\left(\frac{-r}{R}\hat{e}_r + \frac{z}{R}\hat{e}_z\right), \quad R = \sqrt{r^2 + z^2}$$

Now symmetry about the z-axis dictates the cancellation of the radial components. Then

$$\vec{E} = \int_0^{2\pi}\int_0^\infty \frac{\rho_s r z \, dr \, d\phi}{4\pi\varepsilon_0(r^2 + z^2)^{3/2}}\hat{e}_z = \frac{\rho_s z}{2\varepsilon_0}\left[\frac{-1}{\sqrt{r^2 + z^2}}\right]_0^\infty \hat{e}_z = \frac{\rho_s}{2\varepsilon_0}\hat{e}_z$$

The electric field may be calculated at each point in space in the vicinity of a system of charges or of a charge distribution, as shown in the above two simple examples. As an aid to visualizing the electric field structure associated with a particular distribution of charge, Michael Faraday introduced the concept of lines of force (also known as field lines). Lines of force are drawn in such a way that everywhere they are parallel to the direction of the field. Thus, a line of force is a continuous curve in the electric field, such that the tangent at every point is in the direction of the electric field at that point. As there is a single direction for the electric field at every point of the field, there is just one field line through any given point and two lines can never intersect (**Figure 2.8**).

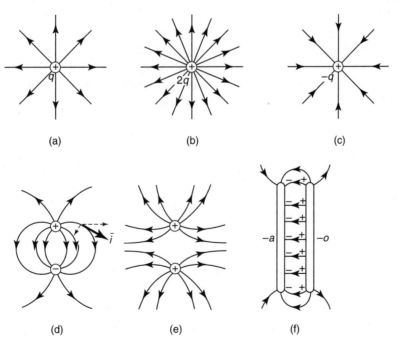

(a)   (b)   (c)

(d)   (e)   (f)

I FIGURE 2.8

Because the direction of the field line is that of **E**, the lines must be defined by the differential equations

$$d\vec{r} = \lambda \vec{E}$$

or, in Cartesian coordinates,

$$\frac{dx}{E_x} = \frac{dy}{E_y} = \frac{dz}{E_z} \tag{2.7}$$

and in spherical coordinates.

$$\frac{dr}{E_\theta} = \frac{r d\theta}{E_\theta} = \frac{r \sin\theta d\theta}{E_\phi} \tag{2.8}$$

It is obvious that free positive charges tend to move along lines of forces in the forward sense and negative charges in the backward sense.

2.3 Earnshaw Theorem

A theorem due to Earnshaw states that no charge can be in stable equilibrium in an electrostatic field under the influence of electric forces alone. The proof of this theorem is very simple. Assume that point O (**Figure 2.9**) is a point at which a positive test charge would find itself in stable equilibrium in the field under consideration. Then the test charge is displaced from O in whatever direction, and the force due to the field must urge it back toward O. This means that lines of force must converge at O. However, this would require the presence of negative charge(s). Therefore, there can be no point of stable equilibrium in an electrostatic field.

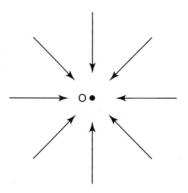

I FIGURE 2.9

2.4 The Electrostatic Potential

The electric field has a number of interesting mathematical properties, which we now explore. As noted in Chapter 1, if the curl of a vector vanishes, then the vector may be expressed as the gradient of a scalar function, and the field is said to be conservative. The electric field given by equation (2.6) meets this criterion. We note that taking the curl of equation (2.5) or (2.6) involves differentiating with respect to r. This variable appears in the equation only in function of the $(\vec{r} - \vec{r}')/|\vec{r} - \vec{r}'|^3$; hence, it suffices to show that the function of this form has zero curl. Using the vector identity $\nabla \times (\varphi \vec{A}) = \nabla \varphi \times \vec{A} + \varphi \nabla \times \vec{A}$, we obtain

$$\nabla \times \frac{\vec{r} - \vec{r}'}{|\vec{r} - \vec{r}'|^3} = \frac{1}{|\vec{r} - \vec{r}'|^3} \nabla \times (\vec{r} - \vec{r}') + \left(\nabla \frac{1}{|\vec{r} - \vec{r}'|^3} \right) \times (\vec{r} - \vec{r}')$$

A direct calculation shows that

$$\nabla \times (\vec{r} - \vec{r}') = 0, \quad \nabla \frac{1}{|\vec{r} - \vec{r}'|^3} = -3 \frac{\vec{r} - \vec{r}'}{|\vec{r} - \vec{r}'|^5}$$

Then

$$\nabla \times \frac{\vec{r} - \vec{r}'}{|\vec{r} - \vec{r}'|^3} = -3 \frac{\vec{r} - \vec{r}'}{|\vec{r} - \vec{r}'|^5} \times (\vec{r} - \vec{r}') = 0$$

These results show that the curl of the electric field vanishes:

$$\nabla \times \vec{E} = 0 \tag{2.9}$$

Hence, a scalar function exists whose gradient is the electric field:

$$\vec{E}(\vec{r}) = -\nabla V(\vec{r}) \tag{2.10}$$

For convenience, a minus sign has been included in equation (2.9). The function $V(r)$ is called the electrostatic potential. Because V is a scalar field, it is much simpler to handle than the vector field $E(r)$. To the superposition principle of the vector field $E(r)$, we now have the additivity of potentials. Note that if we add a constant to V in equation (2.10), the electric field is unchanged so that to this extent the potential V is arbitrary.

If the potential V is expressed as a function of the Cartesian coordinates (x, y, z), then equation (2.10) becomes, in component form,

$$E_x = -\frac{\partial V}{\partial x}, \quad E_y = -\frac{\partial V}{\partial y}, \quad E_z = -\frac{\partial V}{\partial z} \tag{2.10a}$$

Similarly, in spherical coordinates r, θ, ϕ, we have

$$E_r = -\frac{\partial V}{\partial r}, \quad E_\theta = -\frac{\partial V}{r\partial\theta}, \quad E_\phi = -\frac{\partial V}{r\sin\theta\partial\phi} \qquad (2.10b)$$

and in cylindrical coordinates r, θ, z

$$E_r = -\frac{\partial V}{\partial r}, \quad E_\theta = -\frac{\partial V}{r\partial\theta}, \quad E_z = -\frac{\partial V}{\partial z} \qquad (2.10c)$$

The SI unit of potential is N · m/C (newton · meters per coulomb) or J/C (joules per coulomb). A joule per coulomb is often called a volt. The electric field is then measured in volts per meter.

The potential has an important physical interpretation. Let us suppose that a unit test charge is moved along a curve Γ under the influence of an electric field $E(r)$. The work done against the field in moving the charge (which is just (-1) times the work done by the field) is (**Figure 2.10**)

$$W = -\int_\Gamma \vec{F} \cdot d\vec{l} = -q\int_\Gamma \vec{E} \cdot d\vec{l}, \quad (\vec{F} = q\vec{E})$$

Now $\vec{E}(\vec{r}) = -\nabla V(\vec{r})$, and W becomes

$$\begin{aligned}
W &= -q\int_\Gamma \vec{E} \cdot d\vec{l} = q\int_\Gamma \nabla V \cdot d\vec{l} \\
&= q\int_\Gamma \left(\frac{\partial V}{\partial x}\hat{i} + \frac{\partial V}{\partial y}\hat{j} + \frac{\partial V}{\partial z}\hat{k}\right) \cdot (dx\hat{i} + dy\hat{j} + dz\hat{k}) \\
&= q\int_\Gamma \left[(\partial V/\partial x)\,dx + (\partial V/\partial y)\,dy + (\partial V/\partial z)\,dz\right] \\
&= q\int_\Gamma dV = q\left[V(\vec{r}_2) - V(\vec{r}_1)\right]
\end{aligned} \qquad (2.11)$$

The quantity qV may be interpreted as the potential energy of the test charge in the electric field. Then the potential V can be interpreted as potential energy per unit charge.

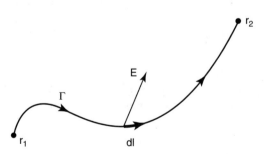

I FIGURE 2.10

Note that in equation (2.11) the work done does not depend on the particular path Γ and is given by the difference in electric potential between two points. And so for a closed path the line integral vanishes, a result that follows directly from equation (2.9) and Stokes' theorem:

$$\int_S \nabla \times \vec{E} \cdot d\vec{S} = \oint_\Gamma \vec{E} \cdot d\vec{l} = 0$$

The electrostatic potential V can be obtained directly as soon as its existence is established. For example, the electric field due to point charges q_i is given by equation (2.5):

$$\vec{E} = \frac{1}{4\pi\varepsilon_0} \sum_i \frac{q_i}{|\vec{r} - \vec{r}_i|^3} (\vec{r} - \vec{r}_i)$$

Now

$$\frac{\vec{r} - \vec{r}'}{|\vec{r} - \vec{r}'|^3} = -\nabla \frac{1}{|\vec{r} - \vec{r}'|}$$

Using this result, we can rewrite the expression for E as

$$\vec{E} = \frac{1}{4\pi\varepsilon_0} \sum_i \frac{q_i}{|\vec{r} - \vec{r}'|^3} (\vec{r} - \vec{r}_i) = -\nabla \left(\frac{1}{4\pi\varepsilon_0} \sum_i \frac{q_i}{|\vec{r} - \vec{r}_i|} \right)$$

from which it follows that

$$V(\vec{r}) = \frac{1}{4\pi\varepsilon_0} \sum_i \frac{q_i}{|\vec{r} - \vec{r}_i|} \tag{2.12}$$

Similarly, the potential that corresponds to the electric field of equation (2.6) is

$$V(\vec{r}) = \frac{1}{4\pi\varepsilon_0} \int_V \frac{\rho(\vec{r}')}{|\vec{r} - \vec{r}'|} \, dv' + \frac{1}{4\pi\varepsilon_0} \int_S \frac{\sigma(\vec{r}')}{|\vec{r} - \vec{r}'|} \, da' \tag{2.13}$$

which is also easily verified by direct differentiation.

Example 4

A thin glass rod of length l placed along the x-axis with one end at the origin is charged uniformly along its length with a total charge Q. Find the potential and the field intensity at any point on the x-axis beyond the end of the rod.

Solution

$$V(x) = \frac{1}{4\pi\varepsilon_0} \int_0^l \frac{Q/l}{x - \alpha}\, d\alpha = \frac{Q}{4\pi\varepsilon_0 l}\Big[-\log(x - \alpha)\Big]_0^l = \frac{Q}{4\pi\varepsilon_0 l} \log \frac{x}{x - l}$$

and

$$E(x) = -\frac{dV}{dx} = \frac{Q}{4\pi\varepsilon_0 l}\frac{d}{dx}\left(\log \frac{x - l}{x}\right) = \frac{Q}{4\pi\varepsilon_0 x (x - l)}$$

as shown in **Figure 2.11**.

I FIGURE 2.11

Example 5

A charge Q is distributed uniformly over the surface of a sphere of radius a. Find the potential and the field intensity everywhere.

Solution

Let P be a point distant r from the center of the sphere. By symmetry V is a function of r only. Let A, A' be two points on the line OP and the surface of the sphere that are, respectively, nearer to and further from P. Then

$$AP = |r - a|, \quad A'P = r + a$$

For convenience, **Figure 2.12** has been drawn with $r > a$. Consider an elementary zone of the sphere with OP subtending an angle $\delta\theta$ at the center of the sphere, its area given by $2\pi a^2 \sin\theta\delta\theta$ and its distance $R = (a^2 + r^2 - 2ar\cos\theta)^{1/2}$ from P. Thus

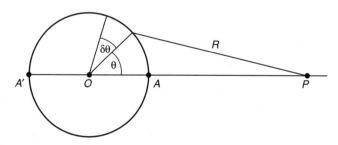

I FIGURE 2.12

$$\delta V = \frac{\sigma}{4\pi\varepsilon_o R}\, 2\pi a^2 \sin\theta\delta\theta, \quad \text{where } \sigma = Q/4\pi a^2$$

Now $R^2 = a^2 + r^2 - 2ar\cos\theta$, and so $R\delta R = ar\sin\theta\delta\theta$. δV then becomes

$$\delta V = \frac{\sigma 2\pi a^2}{4\pi\varepsilon_0}\, \frac{\delta R}{ar}$$

As θ ranges from 0 to π, R ranges from AP to $A'P$. Thus

$$V(r) = \frac{\sigma 2\pi a^2}{4\pi\varepsilon_0 ar}\int_{|r-a|}^{r+a} dR = \frac{Q}{4\pi\varepsilon_0 ar}\frac{1}{2}\left\{(r+a) - |r-a|\right\} = \begin{cases} Q/4\pi\varepsilon_0 r & r > a \\ Q/4\pi\varepsilon_0 a & r < a \end{cases}$$

It follows that

$$\vec{E} = \begin{cases} q\hat{r}/4\pi\varepsilon_0 r^2 & r > a \\ 0 & r < a \end{cases}$$

2.5 Equipotentials and Field Lines

Because fields and potentials are often described by fairly complicated functions, it is helpful to have some means of depicting them graphically. A good picture of field behavior can often be obtained by drawing a sampling of the field lines and the traces of equipotential surfaces in some appropriately chosen plane.

An equipotential surface is a surface at every point of which the potential has the same value:

$$V(r) = C \text{ (constant)}$$

By choosing different numerical values of the constant, we generate a family of such surfaces filling the space. Because the potential at each point is unique, there is one and only one equipotential through any point.

The direction of E is normal to the equipotential surface through that point. To see this, consider a displacement dr on an equipotential surface; we have

$$V(\vec{r}) - V(\vec{r} + d\vec{r}) = 0$$

or

$$V(\vec{r}) - \left[V(\vec{r}) + \frac{\partial V}{\partial r}\cdot d\vec{r}\right] = 0$$

From which it follows that

$$\nabla V \cdot d\vec{r} = 0 \quad \text{or} \quad \vec{E} \cdot d\vec{r} = 0$$

Because dr is on the surface, E must be normal to the surface, that is, the field lines are normal to the equipotential surface. This provides a way of mapping the electric field. **Figure 2.13** shows field lines and equipotential surfaces for a point charge and for two equal and opposite point charges.

A closed equipotential surface (Figure 2.13) with no charge inside encloses an equipotential region, that is, a region where all points are at the same potential as the surface. Suppose that an equipotential surface of different potential V' lies inside the closed equipotential surface of potential V; if V' is greater than V, lines of force are directed from V' toward V at all points of the equipotential surface V'. Therefore, the flux through this closed surface is positive. But a positive flux requires a positive charge inside the surface V'. Consequently, V' cannot be greater than V. A similar line of reasoning shows that V' cannot be less than V. V' is therefore equal to V, and the same is true of the potential at all points inside the original surface. As the potential is uniform throughout the volume inside this surface, no field can be present there. Conversely, a region throughout which no field exists is equipotential.

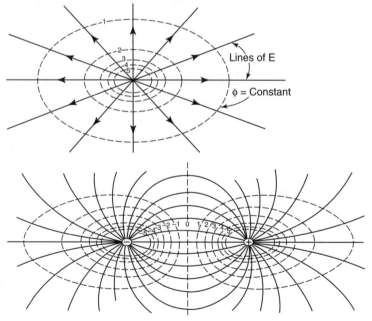

I FIGURE 2.13

Consider a charged conductor (**Figure 2.14**) with a cavity inside. While the conductor is being charged, the free electrons in its interior move under the influence of electrical forces until they so distribute themselves that the electric intensity is zero everywhere inside the conductor. The number of free electrons in any metallic conductor is so great that this condition can always be realized. Therefore, in the equilibrium state, there is no electric field in the body of the conductor. Consequently, there can be no charge in the interior of the conductor, for the presence of charge requires diverging or converging lines of force and therefore the existence of an electric field. So all the charge must be located on the surface of the conductor. Finally, the absence of an electric field in the body of the conductor means that the potential is the same everywhere. The outer surface and the surface of the cavity are equipotential surfaces of the same potential.

If the cavity is empty, it is, of course, an equipotential region at the same potential as the conductor. On the other hand, if it contains a charge, a field is present, and the cavity is no longer an equipotential region. Because the lines of force originating on the charge must end on the surface of the cavity, a charge of equal magnitude and opposite sign is induced there. This releases to the outer surface of the conductor an equal charge of the same sign as that introduced into the cavity.

If the force between charges did not vary inversely with the square of the distance, the field inside a cavity in a charged conductor would not vanish. Cavendish and, later, Maxwell attempted to detect the presence of electrical forces in such a cavity with null results. So sensitive is this method of showing that the force varies inversely with the square of the distance that Maxwell calculated that if the exponent differed from 2 by as much as one part in 21,600, evidence of a field would have been found.

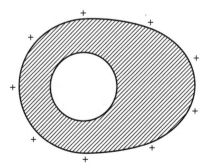

I FIGURE 2.14

2.6 Gauss' Law: Integral and Differential Forms

Gauss's law relates the electric flux through a closed surface to the net amount of charge enclosed by that surface. What do we mean by electric flux? We talk of the "number of field lines" crossing a given area. If we have a surface S bounded by a closed curve C we may write flux of E across $S = \int E \cdot dS$, where the product dot in $E \cdot dS$ picks out the area normal to the field lines. If S is a closed surface, we speak of the flux of E out of, or into, S. If the direction of dS is that of the outward normal, then

$$\text{flux of } \vec{E} \text{ out of } S = \oint \vec{E} \cdot \vec{dS} = \oint \vec{E} \cdot \hat{n} dS$$

Let us take the first case of a point charge q inside the closed surface S. Then

$$\vec{E} \cdot \hat{n} dS = \frac{q}{4\pi\varepsilon_0} \frac{\hat{r} \cdot \hat{n}}{r^2} dS$$

Now $dA = \hat{r} \cdot \hat{n} dS$ is the projection of dS on a plane perpendicular to r, and $d\Omega = dA/r^2$ is the solid angle subtended by dS at q (**Figure 2.15**) and so

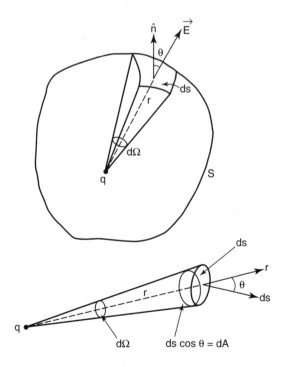

I FIGURE 2.15

$$\vec{E} \cdot \hat{n} dS = \frac{q}{4\pi\varepsilon_0} \frac{\hat{r} \cdot \hat{n}}{r^2} dS = \frac{q}{4\pi\varepsilon_0} d\Omega$$

and

$$\oint \vec{E} \cdot d\vec{S} = \oint \vec{E} \cdot \hat{n} dS = \frac{q}{4\pi\varepsilon_0} \oint d\Omega = \frac{q}{4\pi\varepsilon_0} 4\pi = \frac{q}{\varepsilon_0}$$

$$\oint \vec{E} \cdot d\vec{S} = \oint \vec{E} \cdot \hat{n} dS = \frac{q}{\varepsilon_0} \tag{2.14}$$

This result states that the flux of E out of any closed surface containing a charge q is q/ε_0. And it is known as Gauss' law.

If the charge q lies outside the closed surface, we can show that the total flux out of the surface is zero. As shown in **Figure 2.16**, with q as vertex describe a solid angle $d\Omega$. Let ds_1 be the area of the surface intercepted at P_1 and ds_2 at P_2. The projections of these surfaces perpendicular to the radius vector are $ds_1 \cos\theta_1$ and $ds_2 \cos\theta_2$, and

$$d\Omega = \frac{ds_1 \cos\theta_1}{r_1^2} = \frac{ds_2 \cos\theta_2}{r_2^2}, \quad \text{or} \quad d\Omega = \frac{\hat{r}_1 \cdot \hat{n}_1 ds_1}{r_1^2} = \frac{\hat{r}_2 \cdot \hat{n}_2 ds_2}{r_2^2}$$

Because the angle $\pi - \theta_1$ between the directions of E_1 and n_1 is obtuse, the flux through ds_1 is negative, signifying that it is directed inward through the closed surface instead of outward. Taking ds_1 and ds_2 together, the outward flux is

$$-\vec{E}_1 \cdot \hat{n}_1 ds_1 + \vec{E}_2 \cdot \hat{n}_2 ds_2 = -\frac{q\hat{r}_1 \cdot \hat{n}_1 ds_1}{4\pi\varepsilon_0 r_1^2} + \frac{q\hat{r}_2 \cdot \hat{n}_2 ds_2}{4\pi\varepsilon_0 r_2^2} = -\frac{q}{4\pi\varepsilon_0} d\Omega + \frac{q}{4\pi\varepsilon_0} d\Omega = 0$$

Hence, because the whole surface S can be divided into pairs of elements subtending the same solid angle at q such that the inward flux through one annuls outward flux through the other, the net outward flux through the entire surface due to a charge outside is zero.

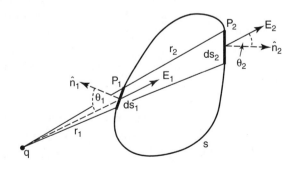

I **FIGURE 2.16**

If a set of N charges $\{q_i\}$ are present inside S, then by the principle of superposition the total electric field E is the sum of the contributions from each individual charge. The same is true for the flux of E out of S. This leads to the relation

$$\oint \vec{E} \cdot \vec{dS} = \oint \vec{E} \cdot \hat{n} dS = \frac{1}{\varepsilon_0} \sum_{i=1}^{N} q_i \tag{2.15}$$

This is Gauss' law in integral form: The total electric flux out of any closed surface S is equal to $1/\varepsilon_0$ times the total charge enclosed by S.

We have proved Gauss' law on the hypothesis that the charges are discrete, but the result may be easily extended to a continuous charge distribution with a charge density ρ. Regarding the volume distribution as a set of discrete charges $\rho d\vec{r}$, the enclosed charge is then simply $\int_V \rho d\vec{r}$, where V is the volume of space enclosed by S, and Gauss' law now takes the form

$$\oint \vec{E} \cdot \vec{dS} = \oint \vec{E} \cdot \hat{n} dS = \frac{1}{\varepsilon_0} \int_V \rho d\vec{r} \tag{2.16}$$

Gauss' law can as well be expressed in a differential form. Using divergence theorem (see equation 1.18)

$$\int_V \nabla \cdot \vec{E} d\vec{r} = \int_S \vec{E} \cdot \vec{dS}$$

Gauss' law (2.16) can be written as

$$\int_V \left(\nabla \cdot \vec{E} - \frac{\rho}{\varepsilon_0} \right) d\vec{r} = 0$$

Because this result holds for arbitrary V, it follows that

$$\nabla \cdot \vec{E} = \rho/\varepsilon_0 \tag{2.17}$$

This is the differential form of Gauss' law. It is a field equation, relating the derivative of $\vec{E}$ at a point to the charge density ρ at that point.

We have fully specified the electrostatic field because earlier we showed that $\nabla \times \vec{E} = 0$. The fact that $\nabla \times \vec{E} = 0$ in no way depends on the inverse square nature of electrostatic interaction but requires only that the force be central. On the other hand, Gauss's law demands an inverse square law as well as invoking the principle of superposition and so is simply an alternative expression of Coulomb's law. We may combine the two differential equations for the electric field. The vanishing curl of $\vec{E}$ means that $\vec{E} = -\Delta V$ that is equation (2.8). Substituting this into equation (2.17) gives

$\nabla \cdot (-\nabla V) = \rho/\varepsilon_0$

i.e., $\nabla^2 V = -\rho/\varepsilon_0$ (2.18)

This is known as Poisson's equation. Our formula (2.13) for the potential

$$V(\vec{r}) = \frac{1}{4\pi\varepsilon_0} \int_V \frac{\rho(\vec{r}')}{|\vec{r}-\vec{r}'|}\, dv'$$

may be regarded as the solution of the differential equation (2.18).

In the particular case of $\rho = 0$, Poisson's equation reduces to Laplace's equation

$\nabla^2 V = 0$ (2.19)

Equations (2.18) and (2.19) are alternative expressions of Gauss' law; our method of deriving this result shows that it depended only on the validity of the inverse square law and the principle of superposition. It follows therefore that (2.18) and (2.19) are equivalent to a statement that the law of force between charges is the inverse square and that electric fields and potentials are additive.

a. Applications of Gauss' Law

Gauss' law is a powerful tool for finding the electric intensity in fields of such symmetry that we can draw Gaussian surfaces everywhere normal to $\vec{E}$ at all points at which the magnitude of the electric field intensity is the same. The most common kinds of symmetry are plane, cylindrical, and spherical, as illustrated by the following simple examples.

Example 6

This example examines the field due to a uniformly charged plane *MN* of very great extent (**Figure 2.17**).

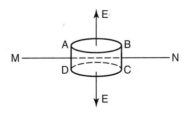

I FIGURE 2.17

Solution

Technically, we should consider an infinitely extended plane. But approximate answers can be provided with large planes. Let σ be the charge per unit area of the surface. From symmetry it is clear that the electric field is perpendicular to the plane, being directed upward above the plane and downward below if the charge is positive. Describe a pillbox-shaped surface $ABCD$, the flat bases AB and DC of the box lying parallel to and equidistant from the plane MN. If Δs is the area of one of the bases, then the flux through the surface of the pillbox is $2E\Delta s$ and the charge enclosed is $\sigma\Delta s$. Gauss' law now takes the form $2E\Delta s = \sigma\Delta s/\varepsilon_0$, giving

$$E = \frac{\sigma}{2\varepsilon_0}$$

Consequently, the field is uniform on each side of the plane, the magnitude of E being independent of the distance from the plane.

The potential can be evaluated easily. In general, for a uniform field in the x direction, $dV/dx = -E$, and because the field is constant,

$$V = C - Ex = C - \sigma x/2\varepsilon_0$$

where C is a constant determined by the zero of V and x is the distance from the plane.

Example 7

This example examines the field due to two parallel conducting plates AB and CD (**Figure 2.18**) of very great extent. The lower plate CD has a uniform positive charge density σ and the upper plate AB has an equal negative charge density.

Solution

We first consider the field between the plates. At a point P between the plates, the electric field intensity due to AB is $\sigma/2\varepsilon_0$ upward and that due to

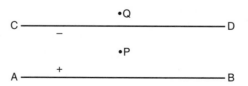

I FIGURE 2.18

CD is also $\sigma/2\varepsilon_0$ upward. Therefore, the total field is

$$E = \frac{\sigma}{\varepsilon_0}$$

upward, everywhere between the plates. The electric intensity at a point Q outside the plates, on the other hand, is $\sigma/2\varepsilon_0$ upward due to AB and $\sigma/2\varepsilon_0$ downward due to CD. So the field outside the plates vanishes. These results apply, of course, regardless of the way in which the charge densities divide between the inside and outside surfaces of their respective plates. There can be no field within the plates themselves, however, and this requires that the charge densities be confined to the inside surfaces.

Example 8: The Electric Field Just Outside a Conductor

Because the surface of the conductor is an equipotential, the field just outside a conductor must be normal to the surface of the conductor, and its value may be calculated in terms of the surface charge density σ by applying Gauss' law to the Gaussian surface in **Figure 2.19**. The total charge contained within this surface is σdS. No flux of $\vec{E}$ comes across dS_1 or out of the sides of the Gaussian surface so that the total flux of $\vec{E}$ out of the volume is EdS_2. Hence by Gauss' law

$$EdS_2 = EdS = \sigma dS/\varepsilon_0$$

we find the field just outside of a conductor is

$$E = \sigma/\varepsilon_0$$

We see from this that a sheet of charge on a conductor gives rise to an electric field twice as big as that due to a sheet of charge on its own (see Example 6); it is instructive to see why this is so. The charges in the neighborhood of a point P on the surface of a conductor do give rise to a field $\sigma/2\varepsilon_0$ both

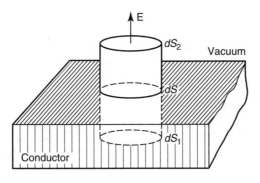

I FIGURE 2.19

inside and outside the surface, but because $\vec{E} = 0$ inside a conductor, we conclude that the residual charges (i.e., those other than charges in the neighborhood of P) are the source of an additional field at P, equal to $\sigma/2\varepsilon_0$ (in magnitude). The resultant field inside the conductor vanishes, whereas that just outside becomes σ/ε_0. When we considered a sheet of charge on its own in Example 6, no other charges are present.

Example 9

The field due to a very long uniformly charged rod of radius a.

Solution

Consider a very long, uniformly charged rod of radius a. This system has cylindrical symmetry. The Gaussian surface is a coaxial cylinder of radius r (**Figure 2.20**). Technically, we need to consider an infinitely long cylinder, but we can get approximate answers for very long cylinders. By the cylindrical symmetry, the field is normal to the axis directed away from the axis and is the same at equal distances from the axis. If l is the length of the cylindrical surface, the flux through this surface is $2\pi rlE$. Note that the end faces do not contribute to the field because E is tangential to these surfaces. Gauss's law gives

$$2\pi rlE = \frac{\lambda l}{\varepsilon_0}, \quad \text{or} \quad E = \frac{\lambda}{2\pi r\varepsilon_0}$$

where λ is the charge per unit length.

The potential V cannot be evaluated with infinity as a zero because the integral $\int_{\infty}^{r} (-\vec{E}) \cdot d\vec{l}$ becomes infinite. Any other point may be chosen and if

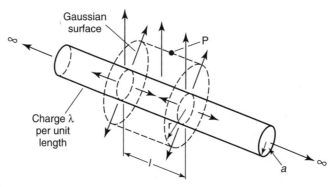

I FIGURE 2.20

we take the cylinder itself, using x as the variable distance:

$$V = \int_a^r -\frac{\lambda dx}{2\pi\varepsilon_0 x} = \frac{\lambda}{2\pi\varepsilon_0} \ln (a/r) \quad \text{(zero at cylinder)}$$

or, in general,

$$V = C - \frac{\lambda}{2\pi\varepsilon_0} \ln r$$

where C is a constant determined by the zero of V.

Example 10

The field due to a charged sphere inside of which electric charge is so distributed that the charge density ρ is a function of the radius vector r alone.

Solution

Obviously, the Gaussian surface is a concentric sphere. For the field outside the sphere, we describe a concentric spherical surface s_1 of radius r_1 greater than a (**Figure 2.21**), where a is the radius of the charged sphere. It is clear from symmetry that the electric field intensity is everywhere normal to s_1 and has the same magnitude E_1 at all points of this surface. Therefore, the total flux through s_1 is $4\pi r_1^2 E_1$, and Gauss' law requires that $4\pi r_1^2 E_1 = Q/\varepsilon_0$, where Q is the total charge inside s. Solving for E_1 we obtain

$$E_1 = \frac{Q}{4\pi\varepsilon_0 r_1^2}$$

This is just the expression we should have found if all the charge had been concentrated at the center 0 of the sphere s. Therefore, any distribution of charge in which the density is a function of the radius vector only, such

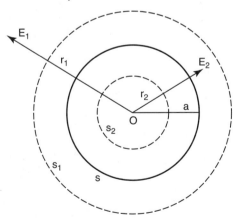

I FIGURE 2.21

as that on an isolated charged conducting sphere, produces the same field at exterior points as if the entire charge were located at the central point.

If we apply Gauss' law to the spherical surface s_2 of radius less than a, we find in the same way

$$E_2 = \frac{Q_2}{4\pi\varepsilon_0 r_2^2}$$

where Q_2 is the portion of the charge inside s_2. In this case the electric intensity is that which would be produced by a charge Q_2 located at 0. The charge lying between the spherical surfaces s_2 and s is without effect. If all the charge lay between these two surfaces, there would be no field at points on the surface s_2 or at points inside this surface. Thus, a charge spread uniformly over the surface of a sphere produces no field at points in its interior, although the field at exterior points is the same as if the entire charge were concentrated at the center of the sphere. On the other hand, if the charge is distributed uniformly through the volume of the sphere,

$$Q_2 = (r_2^3/a^3)Q$$

then

$$E_2 = \frac{Q}{4\pi\varepsilon_0 a^3} r_2 = \frac{\rho r}{3\varepsilon_0}, \quad \rho = \frac{Q}{(4\pi/3)a^3}$$

in the interior of the sphere.

The two forms of E (i.e., E_1 and E_2) match, as they should, at the surface of the sphere ($r_1 = r_2 = a$). The form of the electric field is shown in **Figure 2.22.**

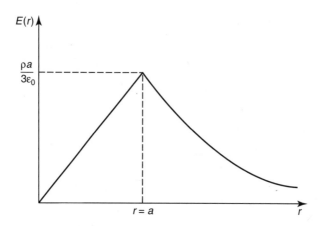

I FIGURE 2.22

To evaluate V, we use the general equation

$$V_B - V_A = \int_A^B -E \cos\theta dl = \int_A^B -\vec{E} \cdot \vec{dl}$$

which reduces to $V_B = \int_A^B -\vec{E} \cdot \vec{dl}$ when A is the zero of potential. If taking zero at infinity and using x as the variable distance from the center, we obtain

$$V = \int_\infty^a -\frac{Qdx}{4\pi\varepsilon_0 x^2} + \int_a^r -\frac{Qxdx}{4\pi\varepsilon_0 a^3} = \frac{Q}{8\pi\varepsilon_0 a}\left(3 - \frac{r^2}{a^3}\right), \quad \text{(zero at } \infty\text{)}$$

Suppose we wish to find the force between two spherical charges Q_1 and Q_2 in each of which the charge density ρ is a function of the distance from the center only. Denote the centers of the two spheres by 0_1 and 0_2 and the distance between centers by R. Replace the second spherical distribution by a point charge Q_2 located at 0_2. The electric field at 0_2 due to the first sphere is

$$E_1 = \frac{Q_1}{4\pi\varepsilon_0 R^2}$$

And, therefore, the force on the point charge Q_2 at 0_2 due to Q_1 is

$$F = Q_1 Q_2 / 4\pi\varepsilon_0 R^2 = Q_1 E_2$$

where E_2 is the electric field intensity at 0_1 due to Q_2.

But the law of action and reaction requires that this expression should also represent the force exerted by Q_2 on Q_1. If now the point charge at 0_2 is replaced by the spherical distribution originally assumed, the field due to Q_2 and therefore the force exerted by it on Q_1 remain unchanged. Consequently, the force between the two spherical charges is the same as if each were a point charge located at its geometric center. It is of interest to note that to deduce this result, which follows so simply from Gauss' law, for the corresponding case of gravitational attraction, Newton delayed publication of the law of gravitation for 20 years.

2.7 Is the Field of a Point Charge Exactly $1/r^2$?

As shown above, the field inside a uniform spherical shell of charge is zero. If we look in detail at how the field inside the shell gets to be zero, we can see more clearly why it is that Gauss' law is true only because the Coulomb force depends exactly on the square of the distance. Imagine a cone with apex at P and extending on either side to cut out surface elements dS_1 and

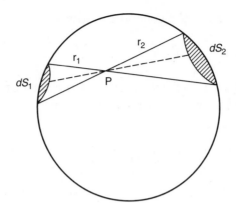

I FIGURE 2.23

dS_2 (**Figure 2.23**). Let r_1, r_2 be the distances of these elements from P. If σ is the surface density of charge, the fields at P due to elements are $\sigma dS_1/r_1^2$ and $\sigma dS_2/r_2^2$ and act in opposite directions. But $dS_1/r_1^2 = dS_2/r_2^2 = d\Omega$, and $d\Omega$ is the solid angle subtended at P by the two elements of area. Hence, the contributions to the field due to the two elements being equal and opposite, so cancel exactly.

We can proceed to cover the entire sphere in this fashion by balancing off field contributions of opposite differential areas. Because each pair of differential areas gives a zero contribution, we conclude that the field at P is zero.

2.8 Energy of Electrostatic Systems

Any system of interacting objects has an internal energy associated with each configuration of the system. The energy of an electrostatic system is obviously the potential energy arising from the interaction between the charges. In this section we calculate the work that must be done to assemble a charge distribution. The work done is most easily calculated for the case of a distribution of discrete point charges, which we perform here.

We know that electric potential is the potential energy per unit charge and the potential difference between points **1** and **2** in an electrostatic field is equal to the work per unit charge to carry a charged particle q from **1** to **2**, as given by equation (2.11):

$$V(\vec{r}_2) - V(\vec{r}_1) = W/q$$

In particular, if we bring the charge q to point $\vec{r}$ from far away and set the reference point at infinity, then

$$V(\vec{r}) = W/q, \quad \text{or} \quad W = qV(\vec{r})$$

We now apply this to calculate the work that must be done to assemble a distribution of discrete point charges, which is bringing in from far away, one by one. Let the first charge q_1 be placed at $\vec{r}_1$, and this takes no work, because there is no field yet. Next bring in q_2 from infinity and place it at a distance r_{12} from q_1, where $r_{12} = |\vec{r}_2 - \vec{r}_1|$. The work required to bring in q_2 is $q_2 V_1(\vec{r}_2)$, where $V_1(\vec{r}_2)$ is the potential due to q_1:

$$W_{12} = q_2 \frac{1}{4\pi\varepsilon_0} \frac{q_1}{r_{12}} = \frac{q_1 q_2}{4\pi\varepsilon_0 r_{12}}$$

Now a third charge q_3 is added to the system at $\vec{r}_3$, and work needs to be done against the field of q_1 and q_2:

$$W_{123} = q_3 \left[V_1(\vec{r}_3) + V_2(\vec{r}_3) \right] = \frac{q_3}{4\pi\varepsilon_0} \left(\frac{q_1}{r_{13}} + \frac{q_2}{r_{23}} \right)$$

Continuing to build up the assembly in this way, the total energy of the assembly is

$$W = \frac{q_2}{4\pi\varepsilon_0} \left(\frac{q_1}{r_{12}} \right) + \frac{q_3}{4\pi\varepsilon_0} \left(\frac{q_1}{r_{13}} + \frac{q_2}{r_{23}} \right) + \frac{q_4}{4\pi\varepsilon_0} \left(\frac{q_1}{r_{14}} + \frac{q_2}{r_{24}} + \frac{q_3}{r_{34}} \right) + \cdots$$

$$= \frac{1}{4\pi\varepsilon_0} \sum_i q_i \sum_{j<i} \frac{q_j}{r_{ij}} \tag{2.20}$$

The restriction $j < i$ ensures that the interaction between every pair is counted only once. We can write equation (2.20) in a different form:

$$W = \frac{1}{8\pi\varepsilon_0} \sum_{i=1} \sum_{j=1} \frac{q_i q_j}{r_{ij}} = \frac{1}{2} \sum_{i=1} q_i \left(\sum_{j=1} \frac{1}{4\pi\varepsilon_0} \frac{q_j}{r_{ij}} \right), \quad i \neq j$$

The factor 1/2 appears in the expression because each pair is counted twice. The term in parentheses is the potential at point $\vec{r}_i$ (the position of q_i) produced by all the charges except q_i. Thus

$$W = \frac{1}{2} \sum_i q_i V(\vec{r}_i) \tag{2.21}$$

This is the amount of work required to assemble the system.

If the charges are not localized as point charges but are distributed with volume density ρ and surface density σ, the equivalent statement to equation (2.21) is

$$W = \frac{1}{2} \int_V \rho V d\vec{r} + \frac{1}{2} \int_{S+S'} \sigma V dS \tag{2.22}$$

where the first term is a volume integral and the second is a surface integral: S' is the surface of all conductors in the system and S is a surface that bounds our system from the outside and that we may choose to locate at infinity. Replacing ρ using Gauss' law in differential form $\nabla \cdot \vec{E} = \rho/\varepsilon_0$, the first integral becomes

$$\frac{1}{2} \int_V \rho V d\vec{r} = \frac{\varepsilon_0}{2} \int_V V(\nabla \cdot \vec{E}) d\vec{r} = \frac{\varepsilon_0}{2} \int_V \left[\nabla \cdot (V\vec{E}) - \vec{E} \cdot \nabla V \right] d\vec{r}$$

$$= \frac{\varepsilon_0}{2} \int_{S+S'} V\vec{E} \cdot d\vec{S} + \frac{\varepsilon_0}{2} \int_V \vec{E} \cdot \vec{E} d\vec{r}$$

where we have used the formula $\nabla \cdot (V\vec{E}) = V\nabla \cdot \vec{E} + \vec{E} \cdot \nabla V$.

A consideration of orders of magnitude shows that the surface integral over S vanishes for any finite set of charges (at large distances from the charge, E goes like $1/r^2$ and potential V like $1/r$, but S grows like r^2). Substituting these results into equation (2.22), we obtain

$$W = \frac{\varepsilon_0}{2} \int_{S'} V\vec{E} \cdot d\vec{S} + \frac{\varepsilon_0}{2} \int_V \vec{E} \cdot \vec{E} d\vec{r} + \frac{1}{2} \int_{S'} \sigma V dS \tag{2.23}$$

Now

$$\frac{\varepsilon_0}{2} \int_{S'} V\vec{E} \cdot d\vec{S} = \frac{\varepsilon_0}{2} \int_{S'} V E_n dS$$

E_n is the normal component of $\vec{E}$ out of volume V and hence into each conductor. Applying the result of Example 8 and allowing for the change in the direction of E_n, we have

$$E_n = -\sigma/\varepsilon_0$$

and

$$\frac{\varepsilon_0}{2} \int_{S'} V\vec{E} \cdot d\vec{S} = \frac{\varepsilon_0}{2} \int_{S'} V E_n dS = -\frac{1}{2} \int_{S'} \sigma V dS$$

Hence, equation (2.23) becomes

$$W = \frac{\varepsilon_0}{2} \int_V \vec{E} \cdot \vec{E} d\vec{r} = \frac{\varepsilon_0}{2} \int_V E^2 d\vec{r} \tag{2.24}$$

This integration is to be performed over all space except that occupied by conductors. But because $\vec{E} = 0$ inside any conductor, it may equally well be extended over all space.

Where is the electrostatic energy located? Equations (2.22) and (2.24) offer two different methods of computing the electrostatic energy: Equation (2.22) is an integral over the charge distribution, and equation (2.24) is an integral over the field. Is the energy stored in the charge? Or is it stored in the field? As Feynman said, "If we restrict ourselves to electrostatics there is really no way to tell where the energy is located." The complete Maxwell equations of electrodynamics give us much more information (although even then the answer is, strictly speaking, not unique.) We therefore revisit this question in detail again in Chapter 6. We now give you only the result for the particular case of electrostatics. The energy is located in space, where the electric field is. This seems reasonable because we know that when charges are accelerated, they radiate electric fields. When light or radio waves travel from one point to another, they carry their energy with them. But there are no charges in the waves. So we want to locate the energy where the electromagnetic field is and not at the charges from which it came. We thus describe the energy not in terms of the charges, but in terms of the fields they produce. We interpret equation (2.24) as saying that when an electric field is present, there is located in space an energy whose density (energy per unit volume) is

$$u = \frac{1}{2} \varepsilon_0 E^2$$

that is, each volume element dV in an electric field contains the energy $(\varepsilon_0 E^2/2)dV$.

Strictly speaking we cannot apply equation (2.24) to a point charge, for it shows that the energy of a point charge is infinite!

$$W = \frac{\varepsilon_0}{2} \int_0^r \int_0^{\pi/2} \int_0^{2\pi} \left(\frac{1}{4\pi\varepsilon_0} \frac{q}{r^2} \right)^2 r^2 \sin^2\theta \, dr d\theta d\phi = \frac{q^2}{8\pi\varepsilon_0} \int_0^\infty \frac{1}{r^2} \, dr \to \infty$$

The energy of a system as given by equation (2.24) is the work required to assemble it; starting from one specified configuration, it does not include any self-energy (interaction energy of a charge with itself). The electron cannot be considered capable of assembly by bringing up smaller charges to form a submicroscopic charge density, because no smaller quantities of charge are known experimentally than the total charge of the electron. Therefore, if we apply equation (2.24) to a supposed structure for the electron, there is serious question whether the result means anything at all.

The idea of locating energy in the field is inconsistent with the assumption of the existence of point charges. One way out of this "divergence" problem would be to think that elementary charges, such as electrons, are not points but are really small distributions of charge. On this note, let us apply our energy formulas to such a simple model of the electron and to equate the electrostatic self-energy to the Einstein self-energy m_0c^2. Here m_0 is the rest mass of the electron and c the velocity of light in vacuum. If, for instance, we suppose the charge e of the electron to be distributed uniformly over the surface of a sphere of radius r, the potential at all points of this surface would be $e/4\pi\varepsilon_0r$ and the formal value for the electrostatic self-energy would be $W = e^2/(8\pi\varepsilon_0r)$. Then equating this to m_0c^2 we obtain

$$r = \frac{e^2}{8\pi\varepsilon_0 m_0 c^2} \tag{2.25}$$

Now $e = 4.8 \times 10^{-10}$ esu, $m_0 = 9 \times 10^{-28}$ g, $c = 3 \times 10^{10}$ cm/s, we find $r = 1.41 \times 10^{-13}$ cm, which is not inconsistent with other knowledge. Equation (2.25) shows that the heavier the particle, the smaller its classical radius. Because a proton is about 1838 times more massive than an electron, its classical radius is 1836 times smaller.

Does the classical radius of an electron have any meaning? This is a serious question. We do not intend to discuss it here, because it demands more physics and so it really falls off the scope of this introductory text. But we point out that the classical radius of the electron is expressed through universal constants e, c, and m_0. The classical radius of the electron also determines, for example, the so-called effective cross-section of scattering of light or long-wave x radiation on free electron (Thomson scattering).

Example 11

Find the energy of a uniformly charged sphere of radius R.

Solution

The energy we are going to calculate is equal to the work done in bringing the charges from infinity to the sphere. We may imagine that the charged sphere is formed by assembling various thin shells of charge. Consider a small sphere of charge of radius r (**Figure 2.24**). If ρ is the density of charge, the total charge on the sphere is $4/3\ \pi r^3\rho$. Suppose a small layer of charge dq in the form of thin shell of thickness dr is deposited on the sphere, where $dq = \rho 4\pi r^2 dr$. The work done in bringing this charge from infinity is

$$dW = \text{potential at } r \times dq$$

$$= \frac{1}{4\pi\varepsilon_0}\frac{4/3\pi r^3\rho}{r}\rho 4\pi r^2 dr = \frac{4\pi\rho^2}{3\varepsilon_0}r^4 dr$$

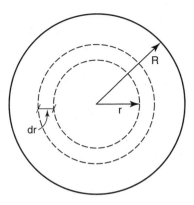

I FIGURE 2.24

The total energy required assembling a charged sphere of radius R is

$$W = \frac{4\pi\rho^2}{3\varepsilon_0} \int_0^R r^4 \, dr = \frac{4\pi\rho^2}{15\varepsilon_0} R^5$$

2.9 Conductors in the Electrostatic Field

An electric conductor, a solid that contains many "free" electrons, has many obvious basic properties in electrostatic fields:

1. Under static conditions, the electric field *inside* a conductor vanishes. Otherwise, free electrons would move.
2. Because the electric field vanishes in a conductor, the potential is the same at all points in the conducting material. Thus, the surface of the conductor is an equipotential surface and the whole conductor is an equipotential region.
3. The charge of a conductor resides on its surface, where there are strong forces to keep them from leaving.
4. The electric field just outside a conductor must be perpendicular to the surface of the conductor; otherwise, charge flows around the surface. The magnitude of the field just outside the conductor is equal to $E = \sigma/\varepsilon_0$, where σ is the local surface charge density (see Example 8).

a. Electrical Shielding

The fact that the electric field inside a conductor is zero may be used to shield sensitive instruments from the effect of the electrostatic field. To see this, let us consider a conductor with a cavity in it (**Figure 2.25**). Everywhere on

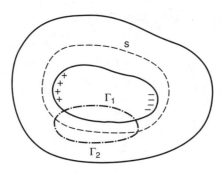

the Gaussian surface S the field is zero, so there is no flux through S and the total charge inside S is zero. But Gauss' law cannot rule out equal amounts of positive and negative charges on the inner surface of the conductor. We now consider a second path Γ that consists of two parts: Γ_1 inside the cavity and Γ_2 in the conductor. The line integral of $\vec{E}$ around any closed path in an electrostatic field is zero, so

$$\oint_{\Gamma} \vec{E} \cdot \vec{dl} = \int_{\Gamma_1} \vec{E} \cdot \vec{dl} + \int_{\Gamma_2} \vec{E} \cdot \vec{dl} = 0$$

Now $\int_{\Gamma_2} \vec{E} \cdot \vec{dl} = 0$, as it is inside the conductor, so we are left with

$$\int_{\Gamma_1} \vec{E} \cdot \vec{dl} = 0$$

that is, there is neither a field inside the empty cavity nor any charges on the inner surface. Sensitive instruments placed inside such empty cavity are shielded from outside electric fields. The enclosure does not even need to be a solid conductor; chicken wire is often enough.

Obviously, if there is a charge q inside the cavity, the electric field inside is not zero. The charge q induces an opposite charge $-q$ on the wall of the cavity, which in turn induces an equal amount of opposite charge $+q$ on the surface of the conductor (**Figure 2.26**). Still, no external fields penetrate the conductor.

b. Force on a Charged Conducting Surface

Because an electric field exists at the surface of a charged conductor, the charge on the surface experiences a force; this force, like field itself, is everywhere normal to the surface. Strictly speaking, the charge density σ does not lie on a mathematical surface but occupies a very thin layer of thickness t that is of the order of a few atomic diameters. Assume that at distance x from the

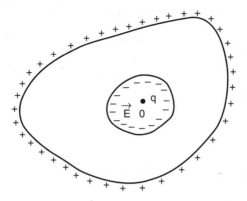

I FIGURE 2.26

surface of the conductor, the charge density is ρ and the field is $\vec{E}$. From Poisson's equation, because $\vec{E}$ is normal to the conductor

$$\frac{d^2V}{dx^2} = -\frac{d}{dx}\left(-\frac{dV}{dx}\right) = -\rho/\varepsilon_0$$

we obtain $dE/dx = \rho/\varepsilon_0$.

Consider the charge at different distances inside a small cylinder of unit cross-section, as shown in **Figure 2.27**. The amount of charge between the

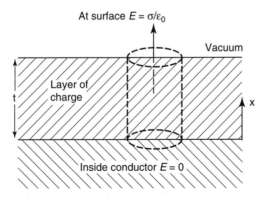

I FIGURE 2.27

planes x and $x + dx$ is ρdx, and the force on this charge is

$$E\rho dx = E\left(\varepsilon_0 \frac{dE}{dx}\right)dx$$

The total force on the charge in the whole cylinder is then

$$\vec{F} = \int_0^t \varepsilon_0 E \frac{dE}{dx}\,dx = \varepsilon_0 \int_0^{\sigma/\varepsilon_0} E\,dE = \frac{\sigma^2}{2\varepsilon_0} \qquad (2.26)$$

Hence, the required force is $\sigma^2/2\varepsilon_0$ per unit area of the surface. As σ appears as a square, the force on the charges is always a tension. The charges cannot normally escape from the conductor; they communicate this force to the surface itself. In terms of the electric field $E = \sigma/\varepsilon_0$ just outside the surface of the conductor, the tension is

$$\vec{F} = \frac{\varepsilon_0 E^2}{2} \qquad (2.27)$$

c. High-Voltage Breakdown

The charges on the surface of a conductor repel other like charges. For a charged sphere, this mutual repulsion causes the charges to spread out over the surface of the sphere uniformly; in this way the charges are as far possible from each other. However, on an irregularly shaped object, the charges tend to distribute over the more outwardly curved or pointed regions. We can see qualitatively why this is so. The repulsive forces are predominantly directed parallel to the surface (**Figure 2.28a**). Thus, the charges move apart until repulsive forces from other nearby charges create an equilibrium situation. But at the sharp end (Figure 2.28b), the repulsive forces between two charges are directed predominantly away from the surface. Thus, there is less tendency for the charges to move apart along the surface here and the

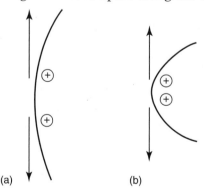

(a)   (b)

I FIGURE 2.28

amount of charge per unit area is greater than at the flat end. The electric field just outside the surface of these regions is also greater (**Figure 2.29**). Because the field is higher at the sharp end, a charge can be more easily pulled on or off the conductor in these regions. Air breaks down on passing these high field regions.

A better way to see the high field at a pointed object is to connect a big conducting sphere to a small sphere by a conducting wire; the wire keeps the two spheres at the same potential. It is a somewhat idealized version of the conductor of Figure 2.29. If the big sphere has the radius a and charge Q, then its potential is approximately (because the presence of one sphere changes somewhat the charge distribution on the other) given by

$$V_1 = \frac{1}{4\pi\varepsilon_0} \frac{Q}{a}$$

Similarly, if the small sphere has radius b and charge q, we have

$$V_2 = \frac{1}{4\pi\varepsilon_0} \frac{q}{b}$$

Because $V_1 = V_2$, $Q/a = q/b$.

For the two fields we have

$$\frac{E_1}{E_2} = \frac{Q/a^2}{q/b^2} = \frac{b}{a}$$

Thus, the surface having the smallest radius of curvature has the highest surface charge density and accordingly the highest electric field. This result is technically very important, because air breaks down if the electric field is too great. What happens is that a loose charge (electron or ion) somewhere in the air is accelerated by the field, and if the field is very great, the charge can

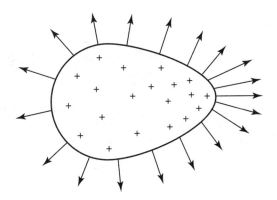

I FIGURE 2.29

pick up enough speed before it hits another atom to be able to knock an electron off that atom. As a result, more and more charges are produced. Their motion constitutes a discharge or spark. If you want to charge an object to a high potential and not have it discharge itself by sparks in the air, you must be sure that the surface of the object is smooth, so that there is no place where the electric field is abnormally large. But high fields at a pointed object find many practical applications. For example, the field-emission microscope depends for its operation on the higher fields produced at a sharp metal point.

2.10 Capacitors

A capacitor (also known as a condenser) is a device for storing charge and therefore energy. Let us see how it is introduced.

We have seen a definite relation between charge and potential in any electrostatic system. As an example, consider a conductor with charge Q at potential V. As the charge is located entirely on the surface of the conductor, we have

$$V = \frac{1}{4\pi\varepsilon_0} \int_S \frac{\sigma dS}{r}$$

for the potential at any point in space, including, in particular, any point P on the surface of the conductor.

Evidently multiplying every element of charge by any factor multiplies V by the same factor. The surface of the conductor remains equipotential and equilibrium is not disturbed. Hence, V is proportional to the total charge Q, and we may write

$$Q = CV \tag{2.28}$$

The constant of proportionality C is called the *capacitance* of the conductor. It is the charge that can be placed on the conductor per unit potential. Capacitance is measured in coulombs per volt, which is called farad. In practice, capacitances are usually measured in microfarads, a microfarad being 10^{-6} farad. In terms of the fundamental units, the farad is 1 meter^{-1}sec^2.

The capacitance depends on the geometry of the conductors. In the case of a uniformly charged sphere of radius a, the electric field intensity outside the sphere is given by

$$E = \frac{Q}{4\pi\varepsilon_0 r^2}$$

and the potential at the surface of the sphere is

$$V = -\int_{\infty}^{a} E \, dr = \frac{Q}{4\pi\varepsilon_0} \int_{a}^{\infty} \frac{dr}{r^2} = \frac{Q}{4\pi\varepsilon_0 a}$$

Therefore, from equation (2.28)

$$C = Q/V = 4\pi\varepsilon_0 a \tag{2.29}$$

A very common and convenient type of capacitor consists of two equal parallel plates of area A separated by a small distance d. If we place charges $+Q$ on one plate and $-Q$ on the other, these will be attracted by one another and will spread uniformly over the inner surfaces of the plates. Except the fringe effects near the edges of the two plates, the lines of force go straight from the positive plate to the negative plate. The electric field intensity is constant between the plates and zero outside, except near the edges. In the central portion of the plates, electric field E is perpendicular to the plates and is equal to

$$E = \sigma/\varepsilon_0$$

where σ is the surface charge density on the plate. Then the potential V is

$$V = Ed = \sigma/\varepsilon_0 d$$

and the capacitance per unit area is

$$C = \varepsilon_0 /d \tag{2.30}$$

Evidently, the smaller we make d for a given area A, the larger capacitance C becomes. But in practice there is a limit on d imposed by the need to avoid electrical breakdown. High-voltage breakdown occurs when the electric field strength is such that a stray electron in the gap between the two plates picks up enough energy to enable it to ionize a gas molecule, thus creating another electron. A cascade of electrons then very quickly gives rise to a spark across the gap.

Concentric spheres form another simple capacitor. Let the outer sphere be grounded, that is, be at the same potential as the earth. Because we are concerned with potential difference only, it is usually convenient to take the invariable potential of the earth as zero. The electric field intensity at a point P between the spheres is $Q/4\pi\varepsilon_0 r$, as the $-Q$ on the outer sphere does not produce an electric field inside. The difference in potential between the spheres is (**Figure 2.30**)

$$V_1 - V_2 = \frac{Q}{4\pi\varepsilon_0}\int_a^b \frac{dr}{r^2} = \frac{Q(b-a)}{4\pi\varepsilon_0 ab}$$

Therefore

$$C = \frac{Q}{V_1 - V_2} = 4\pi\varepsilon_0 \frac{ab}{b-a}.$$ (2.31)

a. The Energy of a Capacitor

We know already that charges reside only on the surface of the conductor, and the potential energy of a conductor can be found from the following equation

$$W = \frac{1}{2}\int \sigma V dS$$ (2.32)

Where the integral is taken over the surface of the conductor and also, if it is necessary, over a surface at "infinity" enclosing the system, where the potential is zero.

Applying equation (2.40) to capacitors we have

$$W = \frac{1}{2}\int \sigma_+ V_+ dS + \frac{1}{2}\int \sigma_- V_- dS$$

where V_+, V_- are the potentials and σ_+, σ_- the surface densities of charges on the two conductors. In terms of total charges $Q = \int \sigma_+ dS$ and $-Q = \int \sigma_- dS$, we have

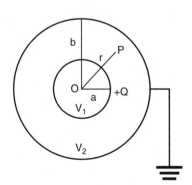

I FIGURE 2.30

$$W = \frac{1}{2} Q \left(V_+ - V_- \right) = \frac{1}{2} Q V \tag{2.33}$$

This energy can as well be expressed in terms of the electric field. Now the charge density $\sigma = \varepsilon_0 E_n = \varepsilon_0 \vec{E} \cdot \hat{n}$, we can write equation (2.32) as

$$W = \frac{1}{2} \varepsilon_0 \int_S V \vec{E} \cdot d\vec{S}$$

We now use Gauss' divergence theorem to transform this expression into the volume integral:

$$W = -\frac{\varepsilon_0}{2} \int_{vol} \nabla \cdot (V\vec{E}) \, d\tau$$

You may have already noticed the sign change in front of the integral on the right. This occurs because the sense of $d\vec{S}$ is outward from the conductors, whereas the convention of the divergence theorem requires it to be outward from the volume of integration.

Now

$$\nabla \cdot (V\vec{E}) = \vec{E} \cdot \nabla V + V\nabla \cdot \vec{E}, \ \ \vec{E} = -\nabla V$$

and in empty space

$$\nabla \cdot \vec{E} = 0$$

Substituting these results in the last expression, we obtain

$$W = \frac{\varepsilon_0}{2} \int_{vol} E^2 \, d\tau$$

It is in agreement with equation (2.24).

2.11 Electric Dipole

We have seen that the electric flux vector of a point charge is directed radially outward and that its magnitude can be found by application of Gauss' law. If, however, more than one charge is present, the field pattern becomes more complicated. Consider, for example, two equal and opposite charges $\pm q$ separated by a small distance (**Figure 2.31**); such a system is known as a dipole. We often come across such a system in physics. For example, when an atom or a molecule is placed in an electric field, the positive and negative charges are affected by the opposite forces and are displaced slightly, forming a dipole.

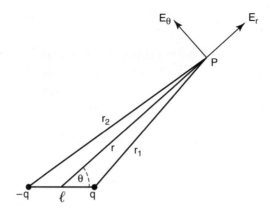

I FIGURE 2.31

Let l be the distance between the two charges q and $-q$. The potential at P due to the dipole is

$$V = \frac{q}{4\pi\varepsilon_0 r_1} + \frac{-q}{4\pi\varepsilon_0 r_2} = \frac{1}{4\pi\varepsilon_0}\left[\frac{q}{r-(l/2)\cos\theta} - \frac{q}{r+(l/2)\cos\theta}\right]$$

$$= \frac{1}{4\pi\varepsilon_0}\left[\frac{ql\cos\theta}{r^2-(l^2/4)\cos^2\theta}\right] = \frac{ql\cos\theta}{4\pi\varepsilon_0 r^2},$$

as l^2 is negligible compared with r^2. The product ql is known as the dipole moment and is designated by p. The dipole moment is a vector having the direction of the axis of the dipole. We shall take its positive sense to be that of the line from $-q$ to q. Then the numerator of the expression for the potential is just the component of the dipole moment in the direction of the radius vector $\vec{r}$ and we can write

$$V(\vec{r}) = \frac{p\cos\theta}{4\pi\varepsilon_0 r^2} = \frac{\vec{p}\cdot\vec{r}}{4\pi\varepsilon_0 r^3} \tag{2.34}$$

The dipole potential varies as $1/r^2$ and hence falls off more rapidly than that around an isolated point charge (its potential varies as $1/r$).

We can express dipole potential as a gradient:

$$V(\vec{r}) = -\frac{1}{4\pi\varepsilon_0}\vec{p}\cdot\nabla(1/r) = -\vec{p}\cdot\nabla\Phi \tag{2.35}$$

where $\Phi = \dfrac{1}{4\pi\varepsilon_0 r}$ is the potential of a unit charge and $\nabla(1/r) = -\vec{r}/r^3$.

Once the potential is found, the electrostatic field can be obtained by taking the gradient of V:

$$E_r = -\frac{\partial V}{\partial r} = -\frac{\partial}{\partial r}\left(\frac{p\cos\theta}{4\pi\varepsilon_0 r^2}\right) = \frac{2p\cos\theta}{4\pi\varepsilon_0 r^3} \qquad (2.36a)$$

$$E_\theta = -\frac{\partial}{r\partial\theta}\left(\frac{p\cos\theta}{4\pi\varepsilon_0 r^2}\right) = \frac{p\sin\theta}{4\pi\varepsilon_0 r^3} \qquad (2.36b)$$

Evidently,

$$E_\phi = -\frac{1}{r\sin\theta}\frac{\partial V}{\partial\phi} = 0 \qquad (2.36c)$$

Some electric field lines (with arrows) and sections of equipotential surfaces (dotted lines) of a dipole (in the plane of paper) are shown in **Figure 2.32**. The equation for the lines of force can be found by considering **Figure 2.33**, which shows that the slope of the tangent to the line of force at P is $rd\theta/dr$. Hence,

$$rd\theta/dr = E_\theta/E_r = \sin\theta/2\cos\theta$$

Integration results in $r = A\sin^2\theta$, which determines the family of lines of forces for an electric dipole. The integration constant A is a parameter that varies from one line of force to another.

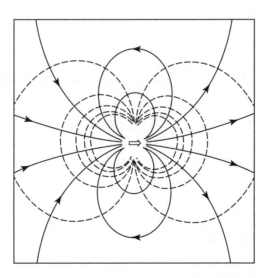

I FIGURE 2.32

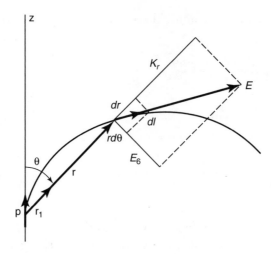

In rectangular coordinates the field of an electric dipole has the following form:

$$E_z = -\frac{\partial}{\partial z}\left(\frac{p}{4\pi\varepsilon_0}\,\frac{z}{r^3}\right) = -\frac{p}{4\pi\varepsilon_0}\left(\frac{1}{r^3} - \frac{3z^2}{r^5}\right), \quad z = r\cos\theta$$

or

$$E_z = \frac{p}{4\pi\varepsilon_0}\,\frac{3\cos^2\theta - 1}{r^3} \tag{2.37a}$$

The x- and y-components are

$$E_x = \frac{p}{4\pi\varepsilon_0}\,\frac{3zx}{r^5}, \quad E_y = \frac{p}{4\pi\varepsilon_0}\,\frac{3zy}{r^5} \tag{2.37b}$$

These two can be combined to give a component directed perpendicular to the z-axis, $E_\perp$:

$$E_\perp = \sqrt{E_x^2 + E_y^2} = \frac{p}{4\pi\varepsilon_0}\,\frac{3z}{r^5}\sqrt{x^2 + y^2} = \frac{p}{4\pi\varepsilon_0}\,\frac{3\cos\theta\sin\theta}{r^5} \tag{2.37c}$$

The field of an electric dipole may be expressed in the following way:

$$\vec{E} = -\nabla V = -\frac{1}{4\pi\varepsilon_0}\,\nabla\left(\frac{\vec{p}\cdot\vec{r}}{r^3}\right)$$

$$= -\frac{1}{4\pi\varepsilon_0}\left[\frac{1}{r^3}\,\nabla(\vec{p}\cdot\vec{r}) + (\vec{p}\cdot\vec{r})\,\nabla\left(\frac{1}{r^3}\right)\right] \tag{2.38}$$

Now $\nabla(\vec{p}\cdot\vec{r}) = \vec{p}, \quad \nabla(1/r^3) = -3\vec{r}/r^3$

The expression for $\vec{E}$ becomes

$$\vec{E} = \frac{1}{4\pi\varepsilon_0}\left[\frac{3\,(\vec{p}\cdot\vec{r})\,\vec{r}}{r^5} - \frac{\vec{p}}{r^3}\right] = \frac{1}{4\pi\varepsilon_0 r^3}\left[\frac{3\,(\vec{p}\cdot\vec{r})\,\vec{r}}{r^2} - \vec{p}\right] \qquad (2.39)$$

a. Electric Dipole in an External Electric Field

What happens when a dipole is placed in an electric field? We now consider a uniform field; the field moves from left to right and exerts a force qE on $+q$ and $-qE$ on $-q$. There is net torque acting on the dipole and non-zero energy of orientation of the dipole. The magnitude of the torque is $qEl\sin\theta$ = $pE\sin\theta$. Because this is about an axis perpendicular to $\vec{p}$ and $\vec{E}$, it may be written as $\vec{p}\times\vec{E}$ **(Figure 2.34)**:

$$\vec{\tau} = \vec{p}\times\vec{E} \qquad (2.40)$$

The torque tends to turn the dipole into a position parallel to the field. The work performed in this case is

$$dW = \vec{\tau}\cdot d\vec{\theta} = pE\sin\theta\,d\theta$$

where $d\theta$ is the differential angular displacement. This work is done at the expense of the electrostatic energy U because the system is isolated. So $dW = -dU$ and

$$dU = -pE\sin\theta\,d\theta$$

If we choose the zero of potential energy at $\theta = \pi/2$, then

$$U = -pE\cos\theta = -\vec{p}\cdot\vec{E} \qquad (2.41)$$

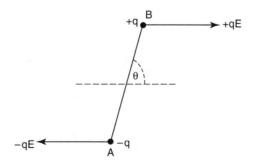

I FIGURE 2.34

U is zero when p is transverse to E and it has a minimum or maximum value when p is parallel or antiparallel to E.

If the field in which the dipole is placed is nonuniform, then the potential energy U varies with position as well as the orientation of the dipole. The variation with position results a translational force acting on the dipole. To see this, let us choose the origin of the coordinate system at the charge $-q$ as shown in **Figure 2.35**. The force acting on the dipole in the negative x-direction is $F_x^- = qE_x$ and in the positive direction is

$$F_x^+ = q(E_x + dE_x)$$

The net component of the force in the positive x-direction is $F_x = qdE_x$.

Because the field is nonuniform,

$$dE_x = \frac{\partial E_x}{\partial x}\,dx + \frac{\partial E_x}{\partial y}\,dy + \frac{\partial E_x}{\partial z}\,dz$$

Hence,

$$F_x = q\left(\frac{\partial E_x}{\partial x}\,dx + \frac{\partial E_x}{\partial y}\,dy + \frac{\partial E_x}{\partial z}\,dz\right)$$

$$= q\left(\frac{\partial}{\partial x}\,dx + \frac{\partial}{\partial y}\,dy + \frac{\partial}{\partial z}\,dz\right)E_x$$

$$= q\,(\vec{dl}\cdot\nabla)\,E_x = (\vec{p}\cdot\nabla)\,E_x$$

where $\vec{dl} = \hat{i}dx + \hat{j}dy + \hat{k}dz$, and $\nabla = \hat{i}\frac{\partial}{\partial x} + \hat{j}\frac{\partial}{\partial y} + \hat{k}\frac{\partial}{\partial z}$.

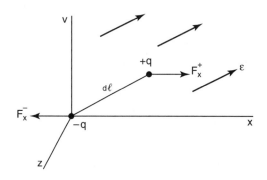

I FIGURE 2.35

Similarly, we can obtain other components. The force on the dipole is, in general,

$$\vec{F} = (\vec{p} \cdot \nabla)\vec{E} \tag{2.42}$$

where $p \cdot \nabla$ is a scalar differential operator whose form in Cartesian coordinates is

$$\vec{p} \cdot \nabla = p_x \frac{\partial}{\partial x} + p_y \frac{\partial}{\partial y} + p_z \frac{\partial}{\partial z}$$

One application of equation (2.41) of some interest concerns the case when the external field acting on some dipole $\vec{p}_1$ is produced by a second point-like dipole of moment $\vec{p}_2$. The mutual potential energy of interaction of this pair U_{12} is given by equation (2.41):

$$U_{12} = -\vec{p}_1 \cdot \vec{E}$$

where $\vec{E}$ is the field produced by the second dipole and it is given by equation (2.39). Hence,

$$W = -\vec{p}_1 \cdot \frac{1}{4\pi\varepsilon_0 r^3}\left[\frac{3\,(\vec{p}_2 \cdot \vec{r})}{r^2}\,\vec{r} - \vec{p}_2 \right] = \frac{1}{4\pi\varepsilon_0 r^3}\left[\vec{p}_1 \cdot \vec{p}_2 - \frac{3\,(\vec{p}_1 \cdot \vec{r})(\vec{p}_2 \cdot \vec{r})}{r^2} \right] \tag{2.43}$$

b. Electric Double Layers

An electric double layer is a surface that has two layers of charges of opposite sign, one just outside the other (**Figure 2.36**). Such double layers have been observed in many biological and colloid problems. The strength p of the layer is defined as the product of the surface charge density σ and the separation l between the layers:

$$p = l\sigma$$

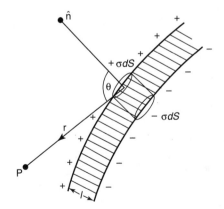

I FIGURE 2.36

The two charges $\pm\sigma dS$ are equivalent to a dipole of moment $(\sigma dS)l = p dS$. We may therefore picture an electric double layer as a layer of dipoles, all pointing normal to the surface. The dipole moment per unit area is p. The potential of the dipole at a point P is

$$\Phi = \frac{p}{4\pi\varepsilon_0}\int \frac{\vec{r}\cdot \vec{dS}}{r^3} = \frac{p}{4\pi\varepsilon_0}\int \frac{\hat{r}\cdot \vec{dS}}{r^2} = \frac{p\,d\Omega}{4\pi\varepsilon_0}$$

where $d\Omega$ is the solid angle subtended at P by the elementary area dS.

2.12 Multipole Expansion of Potentials

It is apparent from the discussion of dipole moments given above that the potential distribution produced by a specified distribution of charge might also possess higher order moments. Suppose that we have a finite number of charges grouped about the origin and that the distance $\vec{R_i}$ of any charge q_i from the origin is much smaller than the distance r from the origin to the point $P(x, y, z)$ at which we wish to compute V (**Figure 2.37**).

Let the coordinates of q_i be ξ_i, η_i, ζ_i; we then have the distance r_i from q_i to P

$$r_i = \sqrt{(x - \xi_i)^2 + (y - \eta_i)^2 + (z - \zeta_i)^2}$$

then the potential at P is

$$V = \frac{1}{4\pi\varepsilon_0}\sum_i \frac{q_i}{r_i} \tag{2.44}$$

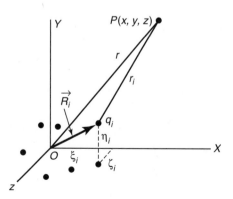

I FIGURE 2.37

As ξ_i, η_i, ζ_l are small compared with x, y, z, we may expand equation (2.44) for V in a three-dimensional Taylor series about the origin, as follows:

$$V = \frac{1}{4\pi\varepsilon_0}\sum_i q_i \left\{ \left[\frac{1}{r_i}\right]_0 + \left[\frac{\partial}{\partial\xi_i}\left(\frac{1}{r_i}\right)\right]_0 \xi_i + \left[\frac{\partial}{\partial\eta_i}\left(\frac{1}{r_i}\right)\right]_0 \eta_i + \left[\frac{\partial}{\partial\varsigma_i}\left(\frac{1}{r_i}\right)\right]_0 \varsigma_i \right.$$

$$+ \left[\frac{1}{2}\frac{\partial^2}{\partial\xi_i^2}\left(\frac{1}{r_i}\right)\right]_0 \xi_i^2 + \left[\frac{1}{2}\frac{\partial^2}{\partial\eta_i^2}\left(\frac{1}{r_i}\right)\right]_0 \eta_i^2 + \left[\frac{1}{2}\frac{\partial^2}{\partial\varsigma_i^2}\left(\frac{1}{r_i}\right)\right]_0 \varsigma_i^2$$

$$\left. + \left[\frac{\partial^2}{\partial\xi_i\partial\eta_i}\left(\frac{1}{r_i}\right)\right]_0 \xi_i\eta_i + \left[\frac{\partial^2}{\partial\eta_i\partial\varsigma_i}\left(\frac{1}{r_i}\right)\right]_0 \eta_i\varsigma_i + \left[\frac{\partial^2}{\partial\varsigma_i\partial\xi_i}\left(\frac{1}{r_i}\right)\right]_0 \varsigma_i\xi_i + \cdots \right\}$$

The subscript 0 means that following the indicated differentiations, all terms are evaluated at the origin ($\xi_i = \eta_i = \zeta_i = 0$); we get for a few of the terms:

$$\left[\frac{1}{r_i}\right]_0 = \frac{1}{r}, \quad \left[\frac{\partial}{\partial\xi_i}\left(\frac{1}{r_i}\right)\right]_0 = \frac{x}{r^3} \left[\frac{\partial^2}{\partial\xi_i^2}\left(\frac{1}{r_i}\right)\right]_0 = -\frac{1}{r^3} + \frac{3x^2}{r^5}, \quad \left[\frac{\partial^2}{\partial\xi_i\partial\eta_i}\left(\frac{1}{r_i}\right)\right]_0 = \frac{3xy}{r^5}$$

We can get all other terms in the same manner, and the last expression for V becomes

$$V = \frac{1}{4\pi\varepsilon_0}\left\{ \frac{1}{r}\sum_i q_i + \frac{1}{r^2}\left[l\sum_i q_i\xi_i + m\sum_i q_i\eta_i + n\sum_i q_i\varsigma_i\right] + \frac{1}{r^2}\left[\frac{1}{2}(3l^2 - 1)\sum_i q_i\xi_i^2 \right.\right.$$

$$+ \frac{1}{2}(3m^2 - 1)\sum_i q_i\eta_i^2 + \frac{1}{2}(3n^2 - 1)\sum_i q_i\varsigma_i^2 + 3lm\sum_i q_i\xi_i\eta_i + 3mn\sum_i q_i\eta_i\varsigma_i \tag{2.45}$$

$$\left.\left. + 3nl\sum_i q_i\varsigma_i\xi_i\right] + \cdots\cdots \right\}$$

where $l = x/r$, $m = y/r$, $n = z/r$ are the direction cosines of the vector $\bar{r}$. Evidently, the quantities indicated by the summation signs depend on the distribution of the charges only and not on the location of P; these quantities are properties of the charge distribution.

The first summation, $\sum q_i$, is the total net charge; it is called the monopole moment of the distribution and is a scalar quantity. The next three summation terms within the first square bracket are the three components of the dipole moment of the distribution; this is a vector with the components

$$p_x = \sum_i q_i\xi_i, \quad p_y = \sum_i q_i\eta_i, \quad p_z = \sum_i q_i\varsigma_i$$

Similarly, the second square bracket contains six summations that are the components of the quadrupole moment of the distribution; this is a tensor of the second rank. Note that the potential due to the monopole varies as

$1/r$, that due to the dipole moment as $1/r^2$, and that due to the quadrupole moment as $1/r^3$.

We do not expound further on multipole expansion but make two general remarks:

1. The potential due to an arbitrary distribution of charge may always be expressed in a multipole expansion. In general, all terms of such an expansion are present, and each higher order term decreases with distance by an additional factor of $1/r$.

2. Multipole moments are in general dependent on the choice of origin for their calculation. But the dipole moment is independent of the choice of origin if the distribution has zero net charge. In addition to the general dependence on the choice of origin, the quadrupole and higher order moments also depend on the orientation of the axis.

2.13 Minimum Energy Theorem

It is possible to establish various general theorems concerning the energy of an electrostatic system. We introduce one only, Kelvin's minimum energy theorem. Suppose that we are given certain conductors and each receives a given fixed total charge. Kelvin's minimum energy theorem states that the charges distribute themselves over the surfaces of the various conductors in such a way that the energy W is a minimum. The actual charge distribution, of course, makes all the surfaces of the conductors equipotentials and the interior of the conductors correspondingly field free.

To prove the theorem, let $\vec{E}$ and $\vec{E}'$ be the respective fields of the actual charge distribution and of any different distribution. Then the corresponding energies are

$$W = \frac{1}{2}\varepsilon_0 \int_V E^2 \, d\tau, \quad W' = \frac{1}{2}\varepsilon_0 \int_V E'^2 \, d\tau \tag{2.46}$$

where V is the region exterior to the conducting material. Thus,

$$W' - W > \frac{1}{2}\varepsilon_0 \int_V (E'^2 - E^2) \, d\tau = \varepsilon_0 \int_V \left[\frac{1}{2}(E' - E)^2 - \vec{E} \cdot (\vec{E} - \vec{E}') \right] d\tau \tag{2.47}$$

If Φ is the potential of the actual charge distribution, then

$$-\int_V -\vec{E} \cdot (\vec{E} - \vec{E'}) = d\tau \int_V (\vec{E'} - \vec{E}) \Delta \cdot \Phi d\tau = \int_V \Delta \cdot [\Phi (\vec{E} - \vec{E'})] d\tau \qquad (2.48)$$

because div $\vec{E}$ and div $\vec{E'}$ vanish throughout V. Now the right-hand side of equation (2.48) can be written as the surface integral

$$\int \Phi (\vec{E} \cdot \vec{E'}) \cdot d\vec{S}$$

It is zero: Over the sphere at infinity $\vec{E}$ and $\vec{E'}$ are $O(1/r^2)$ and Φ is $O(1/r)$ and over the surface of each conductor Φ is constant, and $\int \vec{E} \cdot d\vec{S} = \int \vec{E'} \cdot d\vec{S}$ because each integral gives $-1/\varepsilon_0$ times the net charges on the conductor.

Equation (2.47) now reduces to

$$W' - W > \frac{1}{2} \varepsilon_0 \int_V (E' - E)^2 d\tau$$

from which we find

$$W < W' - \varepsilon_0 \int_V (E' - E)^2 d\tau < W' \qquad (2.49)$$

and the theorem is proved.

Kelvin's minimum energy theorem provides a means of getting approximate solutions to electrostatic problems: It suggests that a "trial solution" can be optimized by allocating to adjustable parameters those values that minimize the energy.

2.14 Applications of Electrostatic Fields

Electrostatic fields find wide applications. We briefly discuss some of them in this section.

a. Electrostatic Particle Precipitators

F. G. Cottrell, a professor of Chemistry at the University of California at Berkeley, developed the first electrostatic particle precipitator in the early 1900s and was first used in a coal-fired power plant by the Detroit Edison Company in 1923. We now use them as air-pollution control devices. Tiny particles (ranging in size from about 10^{-3} to 10^{-5} cm) of soot, ash, and dust are major components of the air-borne emissions from fossil fuel-burning power plants and from many industrial processing plants. Electrostatic precipitators can remove nearly all these particles from the emissions (up to 99% efficient).

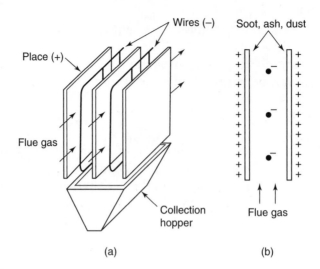

I FIGURE 2.38

Most of the particles in flue gases are neutral, however. Thus, we must first place charges on the particles and then collect them with electrostatic force. In practice, the charging and collection occur in the same region. As shown in **Figure 2.38**, the flue gases containing the soot, ash, or dust particles are passed between a series of positively charged metal plates and negatively charged wires. The flue gases are ionized by the strong electric field around the wires, attracted toward, and collected on the positive plates. Periodically, the plates are shaken so the collected soot, ash, and dust slide down into a collection hopper. Figure 2.37b is a top view of one channel, where the three black dots are the negatively charged wire hanging between the two positively charged plates.

b. Photoduplication (Xerography)

Xerography is the dry-copying method used in the Xerox copier and in similar electronic copying machines. **Figure 2.39** shows the sequence of steps used in the xerography:

(a)   The surface of a large drum in the machine is coated with a thin film of the photoconductive material, such as selenium or some compound of selenium, and the surface is then given a positive charge uniformly in the dark. A photoconductive material is a poor electric conductor in the dark but becomes a reasonably good conductor when exposed to light.

(b)   The page to be copied is projected onto the charged photoconductive surface. The surface becomes conducting in areas

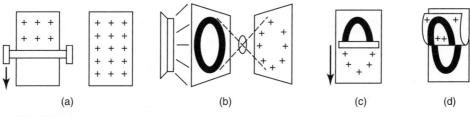

I FIGURE 2.39

where light strikes: The light produces electrons that neutralize the positive charges placed there. In areas not stroked by light, positive charges remain, leaving a hidden image of the original in the form of a positive surface charge distribution.

(c)   The drum now passes the black toner, whose particles are negatively charged and so attracted to the image. The image now becomes visible. A sheet of paper is brought over the drum, and its backside is charged so that the toner adheres to the paper.

(d)   Heat is applied to melt the plastic toner particles onto the paper, producing a copy of the original.

A laser printer operates on a similar basis. A laser beam charges those parts of the drum that the beam strikes, by ejection of electrons. The laser beam is turned on and off while scanning back and forth to charge the images to be printed. The remaining processes of picking up toner and fusing it onto paper are similar to those in the copy machine.

c. Electrostatic Lenses

In oscilloscopes or other electronic instruments for visual displays, we need to focus electron beams in a vacuum. **Figure 2.40** shows an actual electron gun in an oscilloscope. The electrons leaving the cathode are constricted into a narrow beam when they pass through the aperture in the electrode W, which is held at a negative potential with respect to the cathode. The remaining three electrodes are used to focus the beam onto a fine spot on the screen. We do not plan to give a detailed discussion on how they work to focus the beam onto a fine spot on the screen. Instead, we outline the general principles of electrostatic lenses for nonrelativistic electrons.

As shown in **Figure 2.41**, we have a narrow beam of electrons moving at velocity v_1 in an equipotential region of V_1 volts into another equipotential region of V_2 volts. At the boundary between the two equipotential regions, there is a potential gradient and so an electric field that accelerates

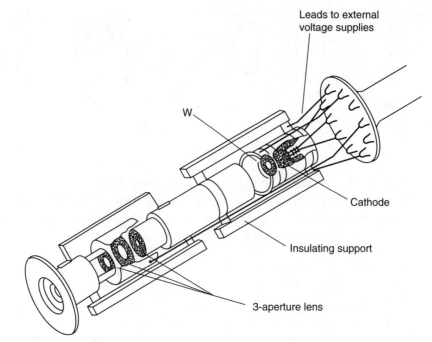

I FIGURE 2.40

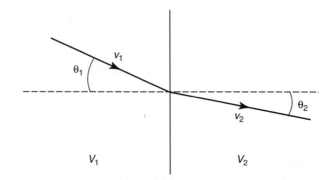

I FIGURE 2.41

the electrons normal to the boundary but does not change their velocity parallel to the boundary. Thus, the beam travels in the second region at a different direction with a different velocity v_2, and the following relation relates it to v_1:

$$v_1 \sin \theta_1 = v_2 \sin \theta_2.$$

We also have

$$\tfrac{1}{2}mv_1^2 = eV_1, \quad \tfrac{1}{2}mv_2^2 = eV_2$$

From these three equations we obtain

$$\frac{\sin\theta_1}{\sin\theta_2} = \frac{v_2}{v_1} = \sqrt{\frac{V_2}{V_1}}$$

We see that this is analogous to Snell's law in optics, $\sin\theta_1/\sin\theta_2 = n_2/n_1$, with the refractive index of the medium replaced by electron velocity.

A convex lens focuses light; we expect, in a similar way, a curved equipotential surface will focus electrons. That is, we need nonuniform electric fields to focus electrons. **Figure 2.42** shows two simple ways to produce focusing: Figure 2.42a is a pair of apertures in a plane conductor, and Figure 2.42b is a gap between two cylindrical electrodes. The dashed lines are equipotentials and the electric fields are indicated by solid lines. We leave the detailed analysis to Chapter 14.

Electrostatic fields also find application in painting. The liquid paint is fed onto a spinning conducting disk or cone that is at a high voltage. The body to be painted is nearby and kept grounded (at potential zero). The electric field between the charged disk and the body exerts a force on the spin-off charged droplets toward the body and deposits them there. Essentially, all the paint is deposited on the object uniformly.

There are many other commercial applications, but here we just mention two. Ink jet printers use electrostatic fields to achieve remarkable control for depositing charged ink droplets on paper. Electrostatic fields are used to deposit charged fiber particles as flocking onto tee shirts, greeting cards, wallpaper, and so forth.

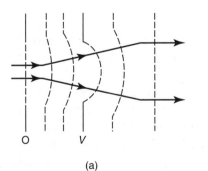

(a)

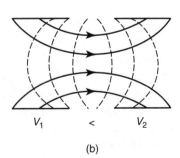

(b)

I FIGURE 2.42

Problems

1. A small positively charged conducting sphere is brought into contact with an identical uncharged sphere. When separated by 10 cm, the repulsion between them is 9×10^{-3} newton. Find the original charge on the first sphere.

2. In Rutherford's scattering experiment, gold nuclei were bombarded with α particles energetic enough to approach within 2×10^{-14} meters of the nucleus. Find the electrostatic force of repulsion experienced by the α particle.

3. Point charges having 0.1, 0.2, –0.3, and –0.2 microcoulombs (μC) are placed at the corners A, B, C, and D, respectively, of a square of side 1 meter. Calculate the force on a charge of 1 μC placed at the center of the square.

4. Two point charges, $q_1 = 50\ \mu C$ and $q_2 = 10\ \mu C$, are placed at (−1, 1, −3)m and (3, 1, 0)m, respectively. Find the force on q_1.

5. Two infinite uniform sheets of charge, each with density ρ_S, are placed at $x = \pm 1$. Determine E in all regions.

6. Show that (a) $\nabla \times (\vec{r} - \vec{r}')$ and (b) $\nabla 1/(\vec{r} - \vec{r}')^3 = -3\ \vec{r} - \vec{r}'/(\vec{r} - \vec{r}')^5$. Note that ∇ only operates on r.

7. Point charges $1C$ and $2C$ are located at the origin and $z = 3$ m, respectively. Find the potential and the field at (1m, 1m, 1m).

8. Using Laplace's equation, argue that a charged particle cannot be held in a stable equilibrium by electrostatic forces alone (Earnshaw theorem).

9. Show that the average value of the electrostatic field over the volume V of a sphere due to a point charge q somewhere within the sphere is

$$\langle E \rangle_{av} = - q\vec{r}_0/3\varepsilon_0 V$$

10. What is the potential due to a uniformly charged spherical shell of radius a at a point (a) outside, (b) on the surface, and (c) in the interior of the shell.

11. Given a slab of thickness t, the charge is uniformly distributed throughout its volume. Find the electric field intensity at a distance y from the median plane of the slab less than one-half t.

12. Find the field at a distance r greater than a from an axis of an infinitely long straight rod of radius a whose charge per unit length is λ. Find the field intensity inside the rod (a) if the charge is distrib-

uted uniformly over the surface of the cylinder, and (b) if the charge is distributed uniformly throughout its volume.

13. Find the electrostatic energy of a uniformly charged spherical shell of total charge q and radius a.

14. Of two coplanar dipoles, the first is fixed and the second is able to rotate about its center that is fixed. If the position vector of the second relative to the first makes angles θ and ϕ with the first and the second dipoles, such that $\theta + \phi$ is the angle between them, show that $\tan \theta = 2\tan \phi$ for equilibrium.

15. An uncharged spherical conductor has a cavity of some weird shape. Somewhere within the cavity is a charge $+q$. Find the electric field outside the sphere.

16. Consider the linear electric quadrupole as shown in **Figure 2.43**. Find its potential and electric field at a point P, where $r \gg s$.

17. Consider the capacitor made of two concentric spheres discussed in the text. Find the capacitance when the inner sphere instead of the outer is grounded, keeping the potential difference $V_1 - V_2$ unchanged.

18. Consider two infinitely long concentric cylinders, of radii a and b $(b > a)$. Find the capacitance per unit length when the outer cylinder is ground.

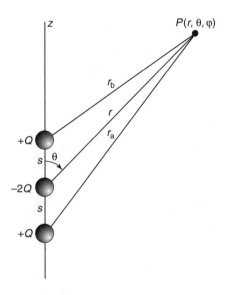

I FIGURE 2.43

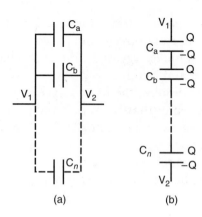

(a)   (b)

I FIGURE 2.44

19. The capacitance of combinations of capacitors can be calculated easily. Suppose we have several capacitances C_a, C_b, . . . , C_n connected in parallel (**Figure 2.44a**). Find the resultant capacitance. If they are connected in series (**Figure 2.44b**), what is the resultant capacitance?

3 Electrostatic Boundary Value Problems

3.1 Introduction

We learned in Chapters 1 and 2 that whenever the charge distribution is completely specified, the potential and electric field can be calculated directly as integrals over this charge distribution:

$$V(\vec{r}) = \frac{1}{4\pi\varepsilon_0} \int \frac{dq'}{|\vec{r} - \vec{r}'|} \tag{3.1}$$

$$\vec{E}(\vec{r}) = \frac{1}{4\pi\varepsilon_0} \int \frac{(\vec{r} - \vec{r}')\, dq'}{|\vec{r} - \vec{r}'|^3} \tag{3.2}$$

Later, this idea will be extended to include the bound charge density in dielectrics. There are, however, many problems encountered in practice in which the charge density is not specified in advance or is only incompletely specified. In these cases, it is fruitful to recast the problem in differential equations, known as Poisson's equation

$$\nabla^2 V = -\rho/\epsilon_0 \tag{3.3}$$

This, together with appropriate boundary conditions, is equivalent to equation (3.1). In a region where $\rho = 0$, Poisson's equation reduces to Laplace's equation:

$$\nabla^2 V = 0 \tag{3.4}$$

The Laplacian ∇^2 is a scalar operator; it takes different forms in different coordinate systems. The choice of the particular set of coordinates is arbitrary, but substantial simplification of the problem can be achieved by choosing a set compatible with the symmetry of the problem. The three common coordinate systems generally using ∇^2 have the following forms:

1. Rectangular coordinates *(x, y, z)*

$$\nabla^2 = \frac{\partial^2}{\partial x^2} + \frac{\partial^2}{\partial y^2} + \frac{\partial^2}{\partial z^2} \tag{3.5}$$

2. Spherical polar coordinates *(r, θ, φ)*

$$\nabla^2 = \frac{1}{r^2}\frac{\partial}{\partial r}\left(r^2\frac{\partial}{\partial r}\right) + \frac{1}{r^2\sin\theta}\frac{\partial}{\partial\theta}\left(\sin\theta\frac{\partial}{\partial\theta}\right) + \frac{1}{r^2\sin^2\theta}\frac{\partial^2}{\partial\phi^2} \tag{3.6}$$

3. Cylindrical polar coordinates *(r, θ, z)*

$$\nabla^2 = \frac{1}{r}\frac{\partial}{\partial r}\left(r\frac{\partial}{\partial r}\right) + \frac{1}{r^2}\frac{\partial^2}{\partial\phi^2} + \frac{\partial^2}{\partial z^2} \tag{3.7}$$

As shown in **Figure 3.1**, *r* has different meanings in equations (3.6) and (3.7). In spherical coordinates, *r* is the magnitude of the radius vector from the origin. In cylindrical coordinates, *r* is the perpendicular distance from the cylinder axes.

The Laplacian operator ∇^2 involves differentiation with respect to more than one variable, and so Poisson's equation is a partial differential equation that may be solved once we know the functional dependence of ρ and the boundary conditions appropriate to particular configuration of electrodes and applied potentials. Such problems are often called boundary value problems.

The vector potential $\vec{A}$ also satisfies Poisson's equation. Mathematically, therefore, the differential equations that determine the potential are structurally identical, and the differences occur mainly in the boundary conditions. The mathematics used to solve one problem may often be used to solve another.

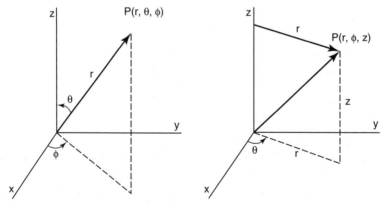

I FIGURE 3-1

The boundary value problems usually present formidable mathematical difficulties and sometimes can only be solved approximately. There are a few problems, however, that, because of the high degree of symmetry involved, can be solved exactly. These may be handled by the use of a series of known functions.

3.2 Solution of Laplace's Equation

Two solution methods for Laplace's equation are developed in some detail. The first is a method for compounding a general solution to equation (3.4) from particular solutions in a coordinate system dictated by the symmetry of the problem. The second is the method of images.

The solutions to Laplace's equation have two basic properties, given below.

a. Principle of Superposition

If V_1, V_2, . . . , V_n are all solution of Laplace's equation, then

$$V = C_1 V_1 + C_2 V_2 + \ldots + V_n \tag{3.8}$$

where the C is an arbitrary constant and is also a solution. This is because Laplace's equation is linear. The proof of this property follows immediately by direct substitution.

b. Uniqueness Theorem

Two solutions of Laplace's equation that satisfy the same boundary conditions differ at most by an additive constant. To prove it, we consider the closed region V_0 exterior to the surfaces S_1, S_2, . . . , S_N of the various conductors in the problem and bounded on the outside by a surface S that can be either a surface at infinity or a finite surface enclosing V_0. Suppose the solution is not unique and there are two solutions, Φ_1 and Φ_2, of Laplace's equation in V_0 with the same boundary conditions. These boundary conditions may be specified by assigning either Φ or the normal derivative $\partial \Phi / \partial n$ on the bounding surfaces. They are often called the Dirichlet condition and the Neumann condition, respectively.

Note that to avoid confusion with volume V_0, Φ has been used to represent the potential. Now let us define

$$\phi = \Phi_1 - \Phi_2 \tag{3.9}$$

Because Φ_1 and Φ_2 are both solutions of Laplace's equation,

$$\nabla^2 \Phi_1 = \nabla^2 \Phi_2 = 0 \tag{3.10}$$

and so

$$\nabla^2 \phi = \nabla^2 (\Phi_1 - \Phi_2) = 0 \tag{3.11}$$

Furthermore, either ϕ or $\partial\phi/\partial n$ vanishes on the boundaries:

$$\phi = \Phi_1 - \Phi_2 = 0, \quad \partial\phi/\partial n = \partial\Phi_1/\partial n - \partial\Phi_2/\partial n = 0 \tag{3.12}$$

We next apply the divergence theorem to the vector $\phi\nabla\phi$:

$$\int_{V_0} \nabla \cdot (\phi\nabla\phi)\, dV = \int_{S + S_1 + \ldots + S_N} \phi\partial\phi/\partial n\, dA = 0 \tag{3.13}$$

Using the vector identity

$$\nabla \cdot (\phi\nabla\phi) = \phi\nabla^2\phi + (\nabla\phi)^2 \tag{3.14}$$

we can rewrite the divergence integral as

$$\int_{V_0} \nabla \cdot (\phi\nabla\phi)\, dV = \int_{V_0} [\phi\nabla^2\phi + (\nabla\phi)^2]\, dV = 0 \tag{3.15}$$

But $\nabla^2\phi$ vanishes at all points in V_0, so the divergence integral reduces to

$$\int_{V_0} \nabla \cdot (\phi\nabla\phi)\, dV = \int_{V_0} (\nabla\phi)^2\, dV = 0 \tag{3.16}$$

Now $(\nabla\phi)^2$ must be either positive or zero at all points in V_0, and because its integral is zero, it is obvious that $\nabla\phi = 0$. Consequently, throughout the volume

$$\phi = \Phi_1 - \Phi_2 = C \tag{3.17}$$

where C is a constant.

If $\phi = 0$ at the boundary, $C = 0$ on the surface. Hence, it is zero throughout the region and

$$\Phi_1 = \Phi_2 \tag{3.18}$$

If, on the other hand, the Neumann condition ($\partial\phi/\partial n = 0$) is satisfied on the surface, then we have $\Phi_1 - \Phi_2 = C$. Because the constant is arbitrary, we can take it to be zero and the solution is again unique.

If the Dirichlet condition is satisfied over a part of the surface and the Neumann condition over the remaining part, the left-hand side of equation (3.15) still vanishes and the solution is unique.

3.3 Laplace's Equation in Rectangular Coordinates

a. Method of Variable Separation

In rectangular coordinates, Laplace's equation becomes

$$\nabla^2 V = \frac{\partial^2 V}{\partial x^2} + \frac{\partial^2 V}{\partial y^2} + \frac{\partial^2 V}{\partial z^2} = 0 \tag{3.19}$$

It is often possible to find the solution by the process of variable separation, that is, the solution can be expressed as the product of three functions $X(x)$, $Y(y)$, and $Z(z)$:

$$V(x, y, z) = X(x)Y(y)Z(z) \tag{3.20}$$

Substituting this in equation (3.19) and dividing through by $X(x)Y(y)Z(z)$, we obtain

$$\frac{1}{X(x)}\frac{d^2 X}{dx^2} + \frac{1}{Y(y)}\frac{d^2 Y}{dy^2} + \frac{1}{Z(z)}\frac{d^2 Z}{dz^2} = 0 \tag{3.21}$$

Notice that we replaced partial derivatives with total derivatives. This is permissible because each term now involves a function of one coordinate only. We now rewrite equation (3.21) in a slightly different form:

$$\frac{1}{X(x)}\frac{d^2 X}{dx^2} + \frac{1}{Y(y)}\frac{d^2 Y}{dy^2} = -\frac{1}{Z(z)}\frac{d^2 Z}{dz^2} \tag{3.22}$$

and we see that the left side is a function of x and y, whereas the right side is a function of z only. For equation (3.22) to be valid for arbitrary values of independent coordinates, both sides must be equal to a common constant, say C_3:

$$\frac{1}{X(x)}\frac{d^2 X}{dx^2} + \frac{1}{Y(y)}\frac{d^2 Y}{dy^2} = -C_3 \tag{3.23}$$

$$\frac{1}{Z(z)}\frac{d^2 Z}{dz^2} = C_3 \tag{3.24}$$

Similarly, we can write equation (3.23) as

$$\frac{1}{X(x)}\frac{d^2 X}{dx^2} = -\left(\frac{1}{Y(y)}\frac{d^2 Y}{dy^2} + C_3\right) = C_1$$

or

$$\frac{1}{X(x)}\frac{d^2X}{dx^2} = C_1 \quad \text{and} \quad \frac{1}{Y(y)}\frac{d^2Y}{dy^2} = -C_1 - C_3 = C_2 \tag{3.25}$$

where

$$C_1 + C_2 + C_3 = C_1 + (-C_1 - C_3) + C_3 = 0 \tag{3.26}$$

Equation (3.19) is now reduced to three one-dimensional ordinary differential equations. The problem then is to solve these ordinary differential equations, subject to condition (3.26) and to the boundary conditions.

Example 1

An infinitely long rectangular conductive cylinder has three sides that are grounded and the fourth side is held at potential V_0. Find the potential at all interior points.

Solution

Based on the geometry of the cylinder, we choose rectangular coordinates with the cylinder in the direction of the z-axis and origin at one of the corner, as shown in **Figure 3.2**. The problem is actually two-dimensional with all interior points lying within the region defined by $0 \le x \le a$, $0 \le y \le b$.

Laplace's equation then is

$$\nabla^2 V = \frac{\partial^2 V}{\partial x^2} + \frac{\partial^2 V}{\partial y^2} = 0$$

Assuming that

$$V(x,y) = X(x)Y(y)$$

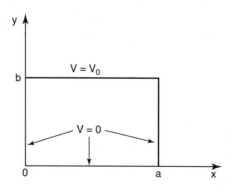

I FIGURE 3-2

we then have

$$YX'' + XY = 0$$

where primes denote derivatives.

Dividing both sides by XY, we obtain

$$\frac{X''}{X} = -\frac{Y''}{Y} = -k^2$$

or

$$X'' + k^2 X = 0, \quad \text{and} \quad Y'' - k^2 Y = 0$$

where k^2 is a separation constant.

We solve the X equation first; its solution has the form

$$X = A \sin kx + B \cos kx$$

Applying the boundary conditions, which are $X = 0$ at $x = 0$ and $x = a$, we get

$$0 = A \sin 0 + B \cos 0 \quad \text{or} \quad B = 0$$

and

$$0 = A \sin ka \quad \text{or} \quad ka = n\pi, \quad n = 1, 2, \ldots .$$

Then

$$X = A \sin\left(\frac{n\pi}{a} x\right)$$

The Y equation has a solution of the form

$$Y = C \sinh ky + D \cosh ky$$

Here k is no longer an arbitrary constant; it has been determined by the X equation: $k = n\pi/a$. Applying the boundary condition, $Y = 0$ at $y = 0$, we get

$$0 = C \sinh 0 + D \cosh 0 \quad \text{or} \quad D = 0$$

The solution, therefore, has the form

$$V(x, y) = X(x) Y(y) = \sum_n A_n \sin\frac{n\pi x}{a} \sinh\frac{n\pi y}{a}$$

The coefficients A_n are determined by the remaining boundary condition, $V = V_0$ at $y = b$. This gives

$$V_0 = \sum_n \left(A_n \sinh \frac{n\pi b}{a} \right) \sin \frac{n\pi x}{a}$$

The right side is a Fourier series and the coefficients (quantities in parentheses) are found by multiplying both sides by $\sin(m\pi x/a)$ and then by integrating over the range 0 to π:

$$\int_0^\pi V_0 \sin \frac{m\pi x}{a} \, dx = \sum_n \left(A_n \sinh \frac{n\pi x}{a} \right) \int_0^\pi \sin \frac{n\pi x}{a} \sin \frac{m\pi x}{a} \, dx$$

$$= \sum_n \left(A_n \sinh \frac{n\pi x}{a} \right) \frac{1}{2} \delta_{mn}$$

from which we get

$$\frac{1}{2} A_m \sinh \frac{m\pi b}{a} = V_0 \int_0^\pi \sin \frac{m\pi x}{a} \, dx$$

$$= \frac{a V_0}{m\pi} (1 - \cos m\pi) \quad a = \pi$$

$$= \begin{cases} \frac{2a V_0}{m\pi}, & \text{for } m \text{ odd} \\ 0, & \text{for } m \text{ even} \end{cases}$$

and

$$A_m = \frac{4 V_0}{m\pi \sinh (m\pi b/a)}, \quad (m \text{ odd})$$

The final solution is

$$V(x, y) = \frac{4 V_0}{\pi} \sum_{n \text{ odd}} \frac{1}{n} \left(\sin \frac{n\pi x}{a} \right) \frac{\sinh (n\pi y/a)}{\sinh (n\pi b/a)}$$

Unfortunately, the answer is in the form of an infinite series.

3.4 Laplace's Equation in Spherical Polar Coordinates

Although electrostatic fields can usually be calculated in Cartesian coordinates, many cases are best treated in other coordinates. When the problem has axial symmetry, it is usually more convenient to use spherical polar coordinates r, θ, ϕ and to take the axis of symmetry as the polar axis $\theta = 0$. Now Laplace's equation takes the form

$$\frac{1}{r^2} \frac{\partial}{\partial r} \left(r^2 \frac{\partial V}{\partial r} \right) + \frac{1}{r^2 \sin \theta} \frac{\partial}{\partial \theta} \left(\sin \theta \frac{\partial V}{\partial \theta} \right) + \frac{1}{r^2 \sin^2 \theta} \frac{\partial^2 V}{\partial \phi^2} = 0 \tag{3.27}$$

Its solutions are known as spherical harmonic functions.

We shall limit our discussion to those cases in which the potential V is a function of r and θ only; then the last term in (2.27) drops out. If we put $\mu = \cos\theta$, Laplace's equation becomes

$$\frac{\partial}{\partial r}\left(r^2\frac{\partial V}{\partial r}\right) + \frac{\partial}{\partial\mu}\left\{(1 - \mu^2)\frac{\partial V}{\partial\mu}\right\} = 0 \qquad (3.28)$$

a. Legendre's Equation and Legendre Polynomials

Equation (3.28) has a solution of the form

$$V = r^n P_n$$

where P_n is a function of μ alone. Substituting in equation (3.28) gives the differential equation for P_n:

$$\frac{d}{dt}\left\{(1 - \mu^2)\frac{dP_n}{d\mu}\right\} + n(n+1)P_n = 0 \qquad (3.29)$$

which is known as Legendre's equation. When n is an integer, its solutions are polynomials in $\cos\theta$, known as Legendre polynomials, and n is called the degree of the polynomials. The first six terms of $P_n(\cos\theta)$ are given in Appendix 1 and are reproduced in **Table 3.1** for the reader's convenience.

We now show that P_n/r^{n+1} is also a solution of equation (3.28). This can be seen in the following way. Let us replace n in equation (3.29) by $-(n+1)$, then the coefficient of the last term becomes

$$\{-(n+1)\}\{-(n+1)+1\} = n(n+1)$$

|Table 3-1   **First Six Legendre Polynomials**

n	$P_n(\cos\theta)$
0	1
1	$\cos\theta$
2	$\frac{3}{2}\cos^2\theta - \frac{1}{2}$
3	$\frac{5}{2}\cos^3\theta - \frac{3}{2}\cos\theta$
4	$\frac{35}{8}\cos^4\theta - \frac{15}{4}\cos^2\theta + \frac{3}{8}$
5	$\frac{63}{8}\cos^5\theta - \frac{35}{4}\cos^3\theta + \frac{15}{8}\cos\theta$

That is, Legendre's equation for $P_{-(n+1)}$ is the same as that for P_n; $P_{-(n+1)}$ and P_n are identical. Therefore, every P_n satisfying Legendre's equation (3.29) provides us with two solutions of Laplace's equation, namely

$$r^n P_n \quad \text{and} \quad P_n/r^{n+1}$$

For the reader's convenience in solving some problems later, the first five solutions of Laplace's equation in spherical coordinates for field with axial symmetry (i.e., for field where V is independent of the angle ϕ), equation (3.28), are listed in **Table 3.2**. These solutions suffice for the applications that we shall make later. Appendix give a full coverage of the solutions of equation (3.28).

The functions listed in Table 3.2 are known as zonal harmonics. We can check these easily. By substituting in equation (3.28), we find that $V_0 = 1/r$ and $P_0 = 1$ is the first solution of Laplace's equation. Now we take the z-axis in the direction of the polar axis of the spherical coordinates, so that $z = r\cos\theta$. We can get a second solution from the first solution by differentiating partially with respect to z. Remembering that $\partial r/\partial z = z/r = \cos\theta$, we find that

$$V_1 = z/r^3 = \cos\theta/r^2, \quad P_1 = \cos\theta$$

Differentiating again with respect to z for a third solution,

$$V_2 = \frac{1}{2}\left(\frac{3z^2}{r^5} - \frac{1}{r^3}\right) = \frac{3\cos^2\theta - 1}{2r^3}, \quad P_2 = \frac{1}{2}\left(3\cos^2\theta - 1\right)$$

and so on. The solutions we obtained are of the form P_n/r^{n+1}, but we can write down at once solutions of the form $r^n P_n$.

Table 3-2 **Solutions of Laplace's Equation in Spherical Polar Coordinates with Axial Symmetry**

n	$r^n P_n (\cos\theta)$	$r^{-(n+1)} P_n \cos\theta$
0	1	r^{-1}
1	$r\cos\theta$	$r^{-2}\cos\theta$
2	$\frac{1}{2}r^2(3\cos^2\theta - 1)$	$\frac{1}{2}r^{-3}(3\cos^2\theta - 1)$
3	$\frac{1}{2}r^3(5\cos^3\theta - 3\cos\theta)$	$\frac{1}{2}r^{-4}(5\cos^3\theta - 3\cos\theta)$
4	$\frac{1}{8}r^4(35\cos^4\theta - 30\cos^2\theta + 3)$	$\frac{1}{8}r^{-5}(35\cos^4\theta - 30\cos^2\theta + 3)$

Example 2

An uncharged conducting sphere of radius a is placed in a uniform field E_0. Find the potential at any point in space exterior to the conducting sphere.

Solution

Let the uniform field $\vec{E}_0$ point to the right. At any point the electric field intensity is $\vec{E}_0$ plus that due to the induced charges. The induced charges arrange themselves on the conducting sphere so that the total field is zero inside. It is more convenient to use spherical polar coordinates with the origin at the center of the sphere. If we choose the polar axis along $\vec{E}_0$, the problem then has axial symmetry. Because the potential is a relative quantity, we may take the potential of the sphere as zero. Although the sphere distorts the field in its neighborhood, the field at a great distance retains its original uniform character. Thus, we now look for a solution of Laplace's equation that satisfies the two boundary conditions

$$V = -E_0 r \cos\theta \qquad \text{for } r \to \infty$$
$$V = 0 \qquad\qquad \text{for } r = a$$

To satisfy the first of these conditions, the zonal harmonic $r \cos\theta$ is required, and to satisfy the second condition, another harmonic involving only the first power of $\cos\theta$ must be added (**Figure 3.3**). Thus, a glance at Table 3.2 shows that the potential at a point P is of the form

$$V = Ar\cos\theta + \frac{B\cos\theta}{r^2} = \left(Ar + \frac{B}{r^2}\right)\cos\theta$$

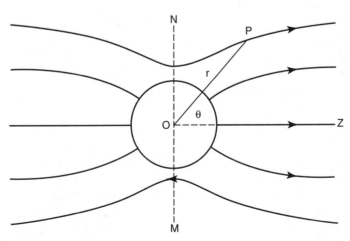

I FIGURE 3-3

where the arbitrary constants A and B are to be determined so as to satisfy the boundary conditions. Applying the first condition at $r = \infty$, we find that the second term in V vanishes and $A = -E_0$. To satisfy the second condition, we put $r = a$ and equal the coefficient of $\cos\theta$ to zero, and then we find $B = E_0 a^3$. Finally, we find the potential outside the sphere is

$$V(r,\theta) = -E_0 r\cos\theta + E_0\frac{a^3\cos\theta}{r^3} = -E_0\left(1 - \frac{a^3}{r^3}\right)r\cos\theta \tag{3.30}$$

It is obvious that the first term on the right side of equation (3.30) is the potential corresponding to the uniform field $\vec{E}_0$. Have you noticed that the second term has the form of potential due to a dipole? In fact, if we replace the sphere by a dipole of moment

$$p = 4\pi e_0 E_0 a^3$$

located at the center, the field outside the surface previously occupied by the sphere is unchanged.

The field components are

$$E_r = -\frac{\partial V}{\partial r} = E_0\left(1 + \frac{2a^3}{r^3}\right)\cos\theta \tag{3.31a}$$

$$E_\theta = -\frac{1}{r}\frac{\partial V}{\partial\theta} = -E_0\left(1 - \frac{a^3}{r^3}\right)\sin\theta \tag{3.31b}$$

where E_r is along the radius vector and E_θ is at right angle to the radius vector in the direction of increasing θ.

As the radius vector is normal to the surface of the sphere, the surface density of induced charge on the sphere is equal to

$$\sigma = \varepsilon_0 E_r|_{r-a} = 3\varepsilon_0 E_0\cos\theta \tag{3.32}$$

and the total induced charge on the sphere is zero

$$Q = \int_0^\pi \sigma(\theta)\,a^2\sin\theta\,d\theta = 0$$

because it is positive on the right side of the hemisphere and negative on the left.

Figure 3.4 is a reproduction of Figure 3.3 with equipotential lines added on; it reveals an interesting feature. The plane MN through the center of the sphere is an equipotential surface at the same potential as the sphere. Hence, we can replace the portion of this plane outside the sphere by a conducting

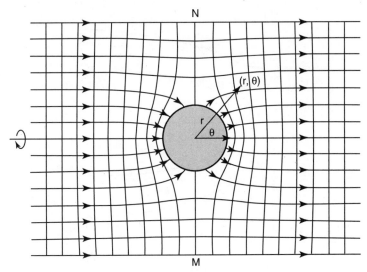

I FIGURE 3-4

surface, wiping out the field to the left and leaving that to the right unaltered. We have, then, the case of an infinite conducting plane with a hemispherical boss, as illustrated in **Figure 3.5**. The lines of forces originate on a positive charge on the surface of plane instead of starting from positive charges at an infinite distance to the left. The potential is still given by equation (3.30) and the charge per unit area on the hemisphere by equation (3.32). To find the charge per unit area on the plane portion of the surface, we rewrite equation (3.30) in the form

$$V = -E_0\left(1 - \frac{a^3}{r^3}\right)r\cos\theta = -E_0\left(1 - \frac{a^3}{r^3}\right)z$$

Because the z-axis is normal to the plane,

$$\sigma = -\varepsilon_0\left(\frac{\partial V}{\partial z}\right)_{z=0} = \left(1 - \frac{a^3}{r^3}\right)\varepsilon_0 E_0$$

Inspection of this equation and of equation (3.32) shows that the charge density decreases from $3\varepsilon_0 E_0$ at the tip of the boss to zero along the line of intersection of the hemisphere and the plane and then increases to $\varepsilon_0 E_0$ a great distance from the boss. This problem illustrates the general rule that the charge density and therefore the adjacent field are greatest where the surface of the conductor is most convex and least where the surface is most concave.

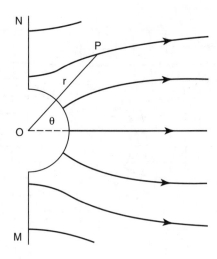

I FIGURE 3-5

If the sphere has a charge Q, we may add $Q/4\pi\varepsilon_0 r$ to equation (3.30), getting

$$V(r,\theta) = -E_0\left(1 - \frac{a^3}{r^3}\right)r\cos\theta + \frac{Q}{4\pi\varepsilon_0 r}$$

The radial component of the field is increased by $Q/4\pi\varepsilon_0 r^2$ and the charge per unit area by $Q/4\pi a^2$.

3.5 Laplace's Equation in Cylindrical Polar Coordinates

In some problems such as those having to do with a long straight wire, it is more convenient to use cylindrical polar coordinates. Laplace's equation takes the form

$$\frac{1}{r}\frac{\partial}{\partial r}\left(r\frac{\partial V}{\partial r}\right) + \frac{1}{r^2}\frac{\partial^2 V}{\partial \phi^2} + \frac{\partial^2 V}{\partial z^2} = 0 \tag{3.33}$$

where the meanings of r and ϕ are illustrated in Figure 3.1.

Once again we use the method of separation of variables and write

$$V(r,\phi,z) = R(r)\,\Phi(\phi)\,Z(z)$$

The final solution is given in terms of Bessel and Neumann functions. However, in some problems the potential is not a function of coordinate z, and then the problem reduces to a two-dimensional one and we can use polar coordinates in the plane perpendicular to the z-axis. In this case the last term in equation (3.33) drops out and Laplace's equation reduces to

$$\frac{1}{r}\frac{\partial}{\partial r}\left(r\frac{\partial V}{\partial r}\right)+\frac{1}{r^2}\frac{\partial^2 V}{\partial\phi^2}=0 \tag{3.34}$$

We can now work out solutions rather easily. Substituting $V = R(r)\Theta(\phi)$ gives

$$\frac{r}{R}\frac{d}{dr}\left(r\frac{dR}{dr}\right)=-\frac{1}{\Theta}\frac{d^2\Theta}{d\phi^2}=k^2 \tag{3.35}$$

where k is a separation constant. The ϕ-equation is very simple; it has the solutions $\cos k\phi$ and $\sin k\phi$ (we could use exponential form, but it has no advantage). To ensure that Θ is single valued, these solutions must be single valued:

$$\cos k\,(\phi+2\pi)=\cos k\phi, \quad \sin k\,(\phi+2\pi)=\sin k\phi$$

After ϕ has gone through its full range from 0 to 2π, the function joins smoothly to its value at $\phi = 0$. Such is the case only if k is an integer. We may further require k to be positive without losing any of these solutions.

We now consider the r-equation. It can be simplified by the substitution $r = e^t$ to

$$\frac{d^2 R}{dt^2}-k^2 R=0 \tag{3.36}$$

or

$$(D-k)(D+k)R=0, \quad \text{with} \quad D=d/dt \tag{3.37}$$

Let

$$(D+k)R=Y \tag{3.38}$$

Then equation (3.36) becomes

$$(D-k)Y=0$$

Integration gives $Y = e^{kt}$. Substituting this in equation (3.38) we get

$$(D+k)R=e^{kt} \tag{3.39}$$

This is not an exact equation, but we can make it exact by multiplying an integrating factor $\int k dt = kt$, and then its solution is

$$R = e^{-kt}\left[\int e^{2kt}\,dt + C\right] = \frac{1}{2k}\,e^{kt} + Ce^{-kt} = Ar^k + Br^{-k}$$

Thus, in general we can write

$$R = A_k r^k + B_k r^{-k}$$

remembering that k is a positive integer. When $k = 0$, we can go back to equation (3.35), which reduces to

$$R' + rR'' = 0$$

Its solution is of the form

$$R = C \ln r + D \tag{3.40}$$

Hence, the required solutions to Laplace's equation (34), known as cylindrical harmonics, are shown in **Table 3.3**.

Example 3: Conducting Cylinder in Uniform Electric Field

Consider an infinitely long uncharged conducting cylinder of circular cross-section placed in a uniform electric field E_0 with its axis at right angles to the lines of force. Denote the radius of the cylinder by a and take the x-axis in the direction of the field and the z-axis along the axis of the cylinder. It is clear from symmetry that the potential is not a function of z, and therefore the problem is one in two dimensions for the solution of which polar coordinates are indicated. The boundary conditions are evidently

1. $V = -E_0 r \cos\phi$ for $r \to \infty$;   **2.** $V = 0$ for $r = a$

Table 3-3   **Cylindrical Harmonics**

1;	$\ln r$
$r\cos\phi$, $r\sin\phi$;	$\cos\phi/r$, $\sin\phi/r$
$r^2\cos2\phi$, $r^2\sin2\phi$;	$\cos2\phi/r^2$, $\sin2\phi/r^2$
$r^3\cos3\phi$, $r^3\sin3\phi$;	$\cos3\phi/r^3$, $\sin3\phi/r^3$

where we take the potential of the conductor to be zero. Referring to Table 3.3, it is clear that the potential function must be of the form

$$V = Ar\cos\phi + \frac{B\cos\phi}{r}$$

It is clear that the condition at infinity is satisfied by taking $A = -E_0$ and the condition at the surface of the cylinder is also satisfied by making $B = E_0 a^2$. Hence, the potential outside the cylinder is

$$V = -\left(1 - \frac{a^2}{r^2}\right)E_0 r\cos\phi$$

and the components of electric field intensity in the direction of increasing r and ϕ are

$$E_r = -\frac{\partial V}{\partial r} = \left(1 + \frac{a^2}{r^2}\right)E_0\cos\phi, \quad E_\phi = -\frac{\partial V}{\phi\partial r} = -\left(1 - \frac{a^2}{r^2}\right)E_0\sin\phi$$

The charge per unit area of the conducting surface is

$$\sigma = -\varepsilon_0\left(\frac{\partial V}{\partial r}\right)_{r=a} = 2\varepsilon_0 E_0\cos\phi$$

3.6 Solution of Poisson's Equation

Having discussed the methods for solving Laplace's equation, we now turn our attention to Poisson's equation. In cases where the charge is not localized on conductors or distributed over discrete point charges, the potential satisfies Poisson's equation:

$$\nabla^2 V = -\rho/\varepsilon_0 \tag{3.41}$$

Given a charge distribution and some boundary conditions, the solution of equation (3.41) consists of two parts: the complimentary solution that is the solution of Laplace's equation (the homogeneous part of equation (3.41)) and the particular solution $1/4\pi\varepsilon_0 \int \rho d\vec{r'}/|\vec{r}-\vec{r'}|$, where the integration is carried over the given charge distribution. Hence, boundary value problems involving Poisson's equation are somewhat more difficult. In general, its solution can be obtained by what is known as Green's function. Before we discuss the Green function method, let us first illustrate the use of Poisson's equation by considering some simple examples.

Example 4: Uniformly Charged Sphere

Given a charge q distributed over a sphere of radius R with a constant volume charge density ρ, find the potential inside and outside the sphere.

Solution

Because the charge is distributed in a spherically symmetric way, it is convenient to spherical polar coordinates. Obviously, outside the sphere $(r < R)$ the potential satisfies Laplace's equation

$$\frac{1}{r^2}\frac{d}{dr}\left(r^2\frac{dV}{dr}\right) = 0$$

Inside the sphere $(r \leq R)$, the potential satisfies Poisson's equation

$$\frac{1}{r^2}\frac{d}{dr}\left(r^2\frac{dV}{dr}\right) = -\frac{\rho}{\varepsilon_0}$$

Laplace's equation has the solution

$$V(r) = \frac{A_1}{r} + B_1 \quad r \geq R$$

One can easily show that the particular solution of Poisson's equation is $-\rho r^2/6\varepsilon_0$ (to save time, you can check Example 10, Chapter 2). Thus, Poisson's equation has the solution

$$V(r) = \frac{-\rho}{6\varepsilon_0}r^2 + \frac{A_2}{r} + B_2 \quad r \leq R$$

The boundary conditions are

1. $V = 0$ as $r \to \infty$;
2. V is finite as $r \to 0$ because there is no point charge at the center of the sphere;
3. The two potentials match at $r = R$;
4. The total charge of this distribution is $(4\pi/3)R^3\rho$.

The first condition gives $B_1 = 0$, and the second condition requires that $A_2 = 0$. The third condition imposes a relation between B_2 and A_1:

$$A_1/R = -\rho R^2/6\varepsilon_0 + B_2$$

To solve for A_1 and B_2 we need a second relation between them; this is provided by condition 4. Taking a Gaussian surface centered at the center of the sphere with a radius $r > R$, Gauss' law gives

$$\oint \vec{E} \cdot \hat{n}dS = \frac{4\pi}{3}R^3\frac{\rho}{\varepsilon_0}$$

The electric field outside the sphere is given by

$$\vec{E} = -\nabla\left(\frac{A_1}{r}\right) = \frac{A_1}{r^2}\hat{r}$$

from which we find

$$\oint \vec{E} \cdot \hat{n}\,dS = 4\pi A_1$$

and thus

$$\frac{4\pi}{3} R^3 \frac{\rho}{\varepsilon_0} = 4\pi A_1$$

$$A_1 = \frac{R^3}{3}\frac{\rho}{\varepsilon_0}$$

Then

$$B_2 = A_1/R + PR^2/6\varepsilon_0 = PR^2/2\varepsilon_0$$

The potential is therefore

$$V(r) = \frac{\rho R^2}{2\varepsilon_0}\left(1 - \frac{r^2}{3R^2}\right) \quad r < R$$

or

$$V(r) = \frac{\rho R^3}{3\varepsilon_0}\frac{1}{r} \quad r > R$$

Example 5: The Potential Distribution in a Space Charge Limited Diode

Consider a parallel plate diode and assume that the surface area of the electrodes is large enough so the edge effect may be neglected. Under operating conditions, electrons leave the anode ($V = 0$ at $x = 0$) and are accelerated toward the anode ($V = V_0$ at $x = d$). During the process the space charge builds up in the region between the electrodes and limits the flow of the current. Because the diode is vacuum, Poisson's equation has the form

$$\frac{d^2V}{dx^2} = -\frac{\rho}{\varepsilon_0}$$

where ρ is the charge density. Now the current density $\vec{j}$ is related to electron velocity $\vec{u}$

$$\vec{j} = -\rho\vec{u}$$

If the emission velocity of electron is negligible, then energy conservation gives

$$mu^2 = eV$$

Poisson's equation can be rewritten as

$$\frac{d^2V}{dx^2} = -\frac{j}{u\varepsilon_0} = \frac{j}{u}\sqrt{\frac{m}{2eV}}$$

Integration gives

$$\frac{1}{2}\left(\frac{dV}{dx}\right)^2 = \frac{2J}{\varepsilon_0}\left(\frac{m}{2e}\right)^{1/2}V^{1/2}$$

or

$$\frac{dV}{dx} = \frac{2j^{1/2}}{\varepsilon_0^{1/2}}\left(\frac{mV}{2e}\right)^{1/4}$$

By integrating once and applying the boundary conditions, we get

$$V = (3/2)^{4/3}(j/\varepsilon_0)^{2/3}(m/2e)^{1/3}x^{4/3}$$

3.7 Formal (Green's Function) Solution to Poisson's Equation

The method of Green's function enables a differential equation with appropriate boundary conditions to be solved by quadrature. We first discuss the basic concept of the method of Green's function and then apply it to solve Poisson's equation.

a. Green's Function

For simplicity and clarity, we consider the one-dimensional case first, where $\mathscr{L}$ is a differential operator and $f(x)$ is a continuous function. Let us see how a function $g(x)$ that satisfies the inhomogeneous differential equation

$$g(x) = f(x) \tag{3.42}$$

and certain specified boundary conditions are found. If there exists a unique solution for each $f(x)$, there must exist an inverse operator $\mathscr{L}^{-1}$ such that for all $f(x)$ we have

$$g(x) = \mathscr{L}^{-1}f(x) \tag{3.43}$$

which is the formal solution of equation (3.42). The operator $\mathscr{L}^{-1}$ does not only represent the operation that is inverse to that represented by $\mathscr{L}$ but also the application of the associated boundary conditions. As a simple example, consider the equation

$$dy/dx = 2x$$

subject to the condition that $y = -1$ when $x = 0$. The operator inverse to $\mathscr{L} = d/dx$ is $\int dx = \mathscr{L}^{-1}$. Thus

$$y = \mathscr{L}^{-1}(2x) = \int 2x\,dx = x^2 + c$$

where the constant c is determined by the boundary condition $c = 1$.

Let us now rewrite equation (3.43) in terms of Dirac's δ-function:

$$g(x) = \int \mathscr{L}^{-1} \delta(x - x') f(x')\,dx' \tag{3.44}$$

Now, let us call the solution of equation (3.42), when $f(x) = \delta(x - x')$, Green's function $G(x, x')$ for the operator $\mathscr{L}$ and the given boundary conditions. That is, Green's function $G(x, x')$ satisfies the equation

$$G(x, x') = \delta(x - x') \tag{3.45}$$

with the same boundary conditions as the function $g(x)$. Obviously, the application of $\mathscr{L}^{-1}$ on equation (3.45) gives

$$G(x, x') = \mathscr{L}^{-1} \delta(x - x') \tag{3.46}$$

Hence, the solution $g(x)$ of equation (3.42) is

$$g(x) = \int \mathscr{L}^{-1} \delta(x, x') f(x')\,dx' = \int G(x, x') f(x')\,dx' \tag{3.47}$$

Once the proper Green's function is found, the solution $g(x)$ can be obtained by integration. Thus, the problem now is reduced to obtaining proper Green's function for a given operator. The solution is then given by equation (3.47).

For three dimensions, equations (3.46) and (3.47) can be easily extended:

$$G(\vec{r}, \vec{r}') = \mathscr{L}^{-1} \delta(\vec{r}, \vec{r}') \tag{3.48}$$

and

$$g(r) = \int G(\vec{r},\vec{r}') \, g(\vec{r}') \, d\vec{r}' \tag{3.49}$$

We now show how the solution of Poisson's equation with appropriate boundary conditions could be found with Green's function method. Let us start with the expression $\nabla^2(1/r)$. We can show that it is equal to zero when $r \neq 0$:

$$\nabla^2\left(\frac{1}{r}\right) = \frac{1}{r^2}\frac{d}{dr}\left(r^2\frac{d}{dr}\frac{1}{r}\right) = \frac{1}{r^2}\frac{d}{dr}\left(r^2 \cdot (-1)r^{-2}\right) = 0. \tag{3.50}$$

Obviously,

$$\int \nabla^2\left(\frac{1}{r}\right) dr = 0 \quad \text{when} \quad r \neq 0$$

At $r = 0$, we can find the value of $\int \nabla^2(1/r)dr$ by a limiting procedure:

$$\int \nabla^2\left(\frac{1}{r}\right) d\vec{r} = \lim_{\alpha \to 0} \int \nabla^2\left(\frac{1}{\sqrt{r^2+\alpha^2}}\right) d\vec{r}$$

Now

$$\nabla^2\left(\frac{1}{\sqrt{r^2+\alpha^2}}\right) = \frac{1}{r^2}\frac{d}{dr}\left(r^2\frac{d}{dr}(r^2+\alpha^2)^{-1/2}\right) = -\frac{3\alpha^2}{(r^2+\alpha^2)^{5/2}}$$

So

$$\int \nabla^2\left(\frac{1}{r}\right) d\vec{r} = \lim_{\alpha \to 0} \int \nabla^2\left(\frac{1}{\sqrt{r^2+\alpha^2}}\right) d\vec{r}$$

$$= \lim_{\alpha \to 0} \iiint \left[-\frac{3\alpha^2 r^2}{(r^2+\alpha^2)^{5/2}}\right] \sin\theta \, d\theta d\phi dr$$

$$= \lim_{\alpha \to 0}\left[-12\pi \int_0^\infty \frac{\alpha^2 r^2}{(r^2+\alpha^2)^{5/2}} \, dr\right]$$

Setting $r = \alpha p$, we get

$$\int \nabla^2\left(\frac{1}{r}\right) d\vec{r} = \lim_{\alpha \to 0} \int \nabla^2\left(\frac{1}{\sqrt{r^2+\alpha^2}}\right) d\vec{r} = -12\pi \int_0^\infty \frac{p^2 \, dp}{(p^2+1)^{5/2}} = -4\pi$$

$$\int \nabla^2\left(\frac{1}{r}\right) d\vec{r} = -4\pi \tag{3.51}$$

We can express equations (3.50) and (3.51) in a single relation:

$$\nabla^2\left(\frac{1}{r}\right) = -4\pi\delta(r)$$

or more generally

$$\nabla^2\left(\frac{1}{4\pi|\vec{r}-\vec{r}'|}\right) = -\delta(\vec{r}-\vec{r}') \tag{3.52}$$

which is Poisson's equation, with $V = 1/\{4\pi\varepsilon_0|\vec{r}-\vec{r}'|\}$ produced by a unit charge at $\vec{r}'$. Moreover, a comparison of equation (3.52) with equation (3.45) shows that $1/\{4\pi\varepsilon_0|\vec{r}-\vec{r}'|\}$ is Green's function of the Laplacian operator ∇^2

$$\nabla^2 G(\vec{r},\vec{r}') = -\delta(\vec{r}-\vec{r}')/\varepsilon_0 \tag{3.53}$$

where

$$G(\vec{r},\vec{r}') = \frac{1}{4\pi\varepsilon_0|\vec{r}-\vec{r}'|}$$

This is one of Green's functions of ∇^2; the general form is

$$G(\vec{r},\vec{r}') = \frac{1}{4\pi\varepsilon_0|\vec{r}-\vec{r}'|} + H(\vec{r},\vec{r}') \tag{3.54}$$

where H is a harmonic function, that is, it satisfies Laplace's equation.

The solution of Poisson's equation has to satisfy certain boundary conditions. To see how the boundary conditions can determine a solution, we convert Poisson's equation into integral equation with the help of Green's theorem:

$$\int_V \left(V\nabla^2\psi - \psi\nabla^2 V\right) d\tau = \int_S \left(V\frac{\partial\psi}{\partial n} - \psi\frac{\partial V}{\partial n}\right) dS$$

Now let

$$\psi = \frac{1}{4\pi\varepsilon_0|\vec{r}-\vec{r}'|} \quad \text{and} \quad \nabla^2 V = -\rho/\varepsilon_0$$

Green's theorem gives

$$\int_V \left(V\nabla^2\left[\frac{1}{4\pi\varepsilon_0|\vec{r}-\vec{r}'|}\right] + \frac{\rho}{4\pi\varepsilon_0^2|\vec{r}-\vec{r}'|}\right) d\tau = \int_S \left(V\frac{\partial}{\partial n'}\left[\frac{1}{4\pi\varepsilon_0|\vec{r}-\vec{r}'|}\right] - \frac{1}{4\pi\varepsilon_0|\vec{r}-\vec{r}'|}\frac{\partial V}{\partial n'}\right) dS'$$

which becomes, with the aid of equation (3.52)

$$\int_V \left(V\left[-\frac{\rho}{\varepsilon_0}\delta(\vec{r}-\vec{r}')\right] + \frac{\rho}{4\pi\varepsilon_0^2|\vec{r}-\vec{r}'|}\right) d\tau = \int_S \left(V\frac{\partial}{\partial n'}\left[\frac{1}{4\pi\varepsilon_0|\vec{r}-\vec{r}'|}\right] - \frac{1}{4\pi\varepsilon_0|\vec{r}-\vec{r}'|}\frac{\partial V}{\partial n'}\right) dS'$$

or

$$V(\vec{r}) = \int_V \frac{\rho}{4\pi\varepsilon \, |\vec{r} - \vec{r}'|} \, d\tau + \frac{1}{4\pi} \int_S \left(\frac{1}{|\vec{r} - \vec{r}'|} \frac{\partial V}{\partial n'} - V \frac{\partial}{\partial n'} \left[\frac{1}{|\vec{r} - \vec{r}'|} \right] \right) dS' \tag{3.55}$$

We can rewrite equation (3.55) in terms of Green's function $G(\vec{r} - \vec{r}')$,

$$V(\vec{r})/\varepsilon_0 = \int_V \frac{\rho}{\varepsilon_0} G(\vec{r} - \vec{r}') \, d\tau + \int_S \left(G(\vec{r} - \vec{r}') \frac{\partial V}{\partial n'} - V \frac{\partial G(\vec{r} - \vec{r}')}{\partial n'} \right) dS' \tag{3.56}$$

where Green's function $G(\vec{r} - \vec{r}')$ is given by equation (3.54). Poisson's equation is an elliptic equation and as such is overspecified if both V and $\partial V/\partial n$ are prescribed on the boundary of the region V. Appropriate boundary conditions for elliptic equations are either Dirichlet or Neumann. The solution to Poisson's equation over a finite volume V with either Dirichlet or Neumann boundary conditions on S may be found from equation (3.56) by choosing the harmonic function H to eliminate one or the other of the surface integrals in equation (3.56). For Dirichlet boundary conditions, for example, $G(\vec{r} - \vec{r}') = 0$ for r' on S' and

$$V(\vec{r}) = \int_V \rho G(\vec{r} - \vec{r}') \, d\tau - \varepsilon_0 \int_S V(\vec{r}') \frac{\partial G(\vec{r} - \vec{r}')}{\partial n'} \, dS' \tag{3.57}$$

When S is earthed so that $V(\vec{r}') = 0$, this gives

$$V(\vec{r}) = \int_V \rho G(\vec{r} - \vec{r}') \, d\tau \tag{3.58}$$

which is an expression of the superposition principle applied to an ensemble of point sources within the region V.

3.8 Method of Image Charges

Method of image is very useful for solving some electrostatic problems, without specifically solving a differential equation. It involves the conversion of an electric field into another equivalent field that is simpler to calculate. It is very useful for point charges near conductors. It is possible in some cases to replace the conductors by one or more charges in such a way that the conductor surfaces are replaced by equivalent surfaces at the same potentials. Because the boundary conditions are conserved, the electric field thus found is correct for the region outside the conductor. Let us illustrate this image charge method with some simple examples.

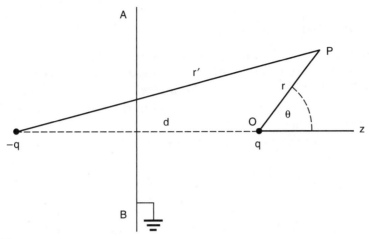

I FIGURE 3-6

Example 6: Point Charge Near an Infinite Grounded Conducting Plane

As shown in **Figure 3.6**, a point charge q is placed at a distance $d/2$ in front of an infinite grounded conducting plane AB, which may be taken to be at zero potential. We want to find the potential in front of the plane. Instead of considering the point charge and the conducting plane, we remove the conducting plane and replace it by a point charge $-q$ at a distance $d/2$ behind the plane. Evidently, the median plane AB is an equipotential plane of zero potential. In the region to the right of AB, the two charges must therefore give the proper solution for the point and the conducting plane. The charge $-q$ is said to be the image of the charge q in the plane.

The potential at P due to the two point charges is

$$V = \frac{1}{4\pi\varepsilon_0}\left[\frac{q}{r} + \frac{-q}{r'}\right]$$

$$= \frac{1}{4\pi\varepsilon_0}\left[\frac{q}{r} - \frac{q}{\sqrt{r^2 + 2rd\cos\theta + d^2}}\right]$$

Notice that $V = 0$ when $r = r'$. That is, the median plane AB is an equipotential surface of zero potential, as expected.

The components of the electric field intensity at P are given by the components of $-\nabla V$:

$$E_r = -\frac{\partial V}{\partial r} = \frac{1}{4\pi\varepsilon_0}\left[\frac{q}{r^2} - \frac{q\,(r + d\cos\theta)}{(r^2 + 2rd\cos\theta + d^2)^{3/2}}\right]$$

and

$$E_\theta = -\frac{\partial V}{r\partial\theta} = \frac{1}{4\pi\varepsilon_0}\left[\frac{qd\sin\theta}{(r^2 + 2rd\cos\theta + d^2)^{3/2}}\right]$$

The lines of force and equipotentials are shown in **Figure 3.7**.

Because the median plane AB is an equipotential surface of zero potential, if we transfer the charge on the image $-q$ to a grounded conducting plane placed to coincide with AB, the field to the right of AB remains unaltered and that to the left is wiped out.

The density of the charge induced on the grounded conducting plane by q is **(Figure 3.8)**

$$\sigma = -\varepsilon_0\left(\frac{\partial V}{\partial z}\right)_{z=-d/2} = \varepsilon_0(E_z)_{z=-d/2} = \varepsilon_0(E_r\cos\theta - E_\theta\sin\theta)_{r=r'} = -\frac{qd}{4\pi r^3}$$

The total induced charge on the plane is, as expected

$$Q = \int_0^\infty \sigma 2\pi\, r dr = \int_0^\infty \frac{-qd}{2}\frac{2r dr}{(d^2+s^2)^{3/2}} = -qd\left.\left(\frac{-1}{\sqrt{r^2+d^2}}\right)\right|_0^\infty = -q$$

where $2\pi r dr$ is the area of an annulus of width dr on the plane at distance s from the foot of the perpendicular.

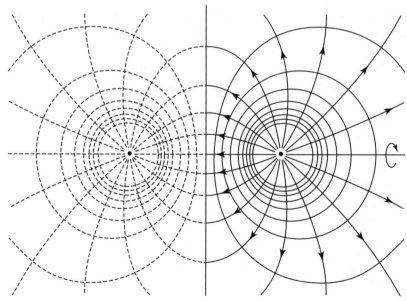

I FIGURE 3-7

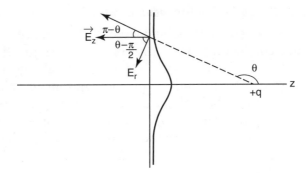

I FIGURE 3-8

The charge q is attracted toward the plane because of the induced negative charge. What is the force of attraction? Because the field in the vicinity of q is not changed by the substitution of the conducting plane for the image $-q$, the force of attraction is therefore simply given by

$$\vec{F} = -\frac{q^2}{4\pi\varepsilon_0 d^2}\,\hat{z}$$

Then from the law of action and reaction, the force on the conducting plane due to the attraction of q is $-\vec{F}$.

Beware that not everything is the same in the two problems; the energy is not the same. With the point charges and no conducting plane, the energy is

$$W = \frac{1}{4\pi\varepsilon_0}\frac{q(-q)}{2d} = -\frac{1}{4\pi\varepsilon_0}\frac{q^2}{2d}$$

On the contrary, for a single charge and conducting plane the energy is just half of the above value:

$$W = \frac{1}{2}\left(-\frac{1}{4\pi\varepsilon_0}\frac{q^2}{2d}\right) = -\frac{1}{4\pi\varepsilon_0}\frac{q^2}{4d}$$

To explain this difference, let us recall equation (2.24) of Chapter 2:

$$W = \frac{1}{2}\varepsilon_0 \int E^2\, d\tau.$$

For the two point charges both the left region ($z < 0$) and the right region ($z > 0$) contribute to the energy integral, and by symmetry they contribute equally. On the contrary, for the point charge and the conducting plane only the right region contributes, and so the energy is just half as much. You can also compute the energy by integrating the force to find the work required to bring q in from $z = \infty$ to $z = d$. We shall leave this as homework.

Example 7: Point Charge Near a Grounded Conducting Sphere

This simple system consists of a point charge q at a distance d from the center of a grounded conducting sphere (**Figure 3.9a**). We want to find the potential at any point P exterior to the sphere by the image charge method. So we remove the sphere and try to find the position and magnitude of an image charge q' such that the spherical surface is still potential zero in the field produced by q and q'. By symmetry q' must lie on the line connecting q and the center of the sphere, say at a distance b from the center, then $V = 0$ at P_1, P_2, and so we have

$$0 = \frac{1}{4\pi\varepsilon_0}\left(\frac{q}{d+a} + \frac{q'}{a+b}\right), \quad 0 = \frac{1}{4\pi\varepsilon_0}\left(\frac{q}{d-a} + \frac{q'}{a-b}\right)$$

Solving these two equations we find

$$q' = -\frac{a}{d}\,q, \quad b = \frac{a^2}{d}$$

As a double check, try to find potential at any general point on the spherical surface, such as point P_3. It should be zero.

We can now put back the grounded conducting sphere and transfer the image charge q' to it. This annuls the field inside the sphere but leaves the field outside unaltered. So the potential at any point $P(r,\theta)$ exterior to the sphere due to the point charge q and the grounded conducting sphere is

$$V(r,\theta) = \frac{1}{4\pi\varepsilon_0}\left[\frac{q}{R} - \frac{(a/d)q}{R'}\right]$$

$$= \frac{1}{4\pi\varepsilon_0}\left[\frac{q}{\sqrt{d^2 + r^2 + 2dr\cos\theta}} - \frac{(a/d)q}{\sqrt{(a^2/d)^2 + r^2 + 2(a^2/d)r\cos\theta}}\right]$$

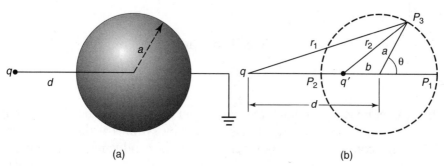

(a)   (b)

I FIGURE 3-9

where R and R' are the distances of q and q' from P (**Figure 3.10**).

The components of the electric field intensity at P are given by the components of $-\nabla V$. The lines of force and equipotentials are shown in **Figure 3.11**.

The induced charge per unit area on the surface of the sphere is

$$\sigma = -\varepsilon_0 \left(\frac{\partial V}{\partial r} \right)_{r=a} = -\frac{q}{4\pi a} \frac{d^2 - a^2}{(d^2 + 2da\cos\theta + a^2)^{3/2}}$$

If we integrate σ over the surface of the sphere, we get the total induced charge

$$q' = \int_0^\pi \sigma 2\pi a^2 \sin\theta d\theta = -\frac{a}{d} q$$

as expected. As in the previous example, the force between the charge q and the conducting sphere is equal to that between q and its image q' in the sphere.

Let us see what happens if we send the point charge to infinity while increasing its charge at the same time in such a way that field it produces, viz

$$E_0 = \frac{q}{4\pi\varepsilon_0 d^2}$$

retains a finite value. In this process the image charge q' $(= -aq/d$, at $b = a^2/d)$ goes to the center of the sphere, whereas $q'b$ $(= qa^3/d^2 = 4\pi\varepsilon_0 E_0 a^3)$ retains a finite value $4\pi\varepsilon_0 E_0 a^3$. We thus obtain an electric dipole at the center of the sphere, which is defined by the relation $\vec{p} = 4\pi\varepsilon_0 a^3 \vec{E}_0$. The field of E_0 of

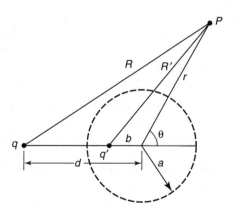

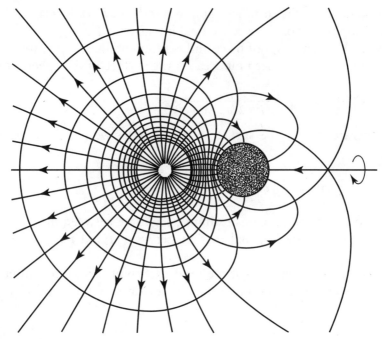

I FIGURE 3-11

the infinitely distant and infinitely large point charge is, of course, in the vicinity of the sphere and it is homogeneous. What do we see? An uncharged conducting sphere in a homogeneous electric field becomes polarized in such a way that its surface charge acts in the exterior space like a dipole of moment $4\pi\varepsilon_0 E_0 a^3$ located at the center of the sphere (see Example 2).

So far we have considered the potential of the sphere to be zero. If the sphere has a potential other than zero, we can add a uniformly distributed charge Q to the surface of the sphere; the effect of such a charge is merely to increase the potential of every point on the surface by $Q/4\pi\varepsilon_0 a$ and therefore the surface remains equipotential. The potential at an exterior point P is then also increased by the same term $Q/4\pi\varepsilon_0 a$.

Example 8: Charged Sphere Near a Grounded Conducting Plane

This is an example that shows some electrostatic fields may be calculated by the method of image through successive approximation. **Figure 3.12** shows a charged spherical conductor of radius a at a distance $d/2$ from a grounded

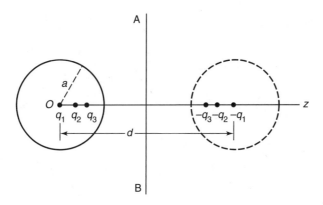

I FIGURE 3-12

conducting plane AB. We replace both the sphere and the plane by a set of charges that maintains these surfaces as equipotentials. First, put a charge q_1 at the center of the sphere. This charge makes the sphere an equipotential. Next, we add the image $-q_1$ of q_1 to the right of the plane AB. This makes the plane an equipotential, but the sphere is no longer so. We now add the image q_2 of $-q_1$ in the sphere. This makes the sphere again an equipotential but disturbs that of the plane. Adding image $-q_2$ of q_2 to the right of the plane AB restores the plane as an equipotential surface again but disturbs the potential over the surface of the sphere. Each one of the added fields is smaller than the preceding one, so continuation of the process brings us ever closer to the desired state in which both the plane and the spherical surface are equipotential surfaces. Finally, we transfer the charges inside the sphere to the surface and the others to the plane. The total charge on the sphere is then the sum of the series $q_1 + q_2 + q_3 + \ldots$, and the charge on the plane is equal but opposite in sign.

The distance of $-q_1$ from the center of the sphere is d, so it follows from the previous example that q_2 is a charge $(a/d)q_1$ at a distance a^2/d from O. Hence, $-q_2$ is at a distance $d(1 - a^2/d^2)$ from O and q_3 is a charge

$$\frac{a/d}{1 - a^2/d^2}\, q_2 = \frac{a^2/d^2}{1 - a^2/d^2}\, q_1$$

at a distance $a^2/d/1 - a^2/d^2$ from O. If we set $a/d = m$, then we have the following point charges at the distances specified from the center of the sphere O:

Charge	Distance from O
q_1	0
mq_1	ma
$\dfrac{m^2}{1-m^2}\,q_1$	$\dfrac{m}{1-m^2}\,a$
$\dfrac{m^3}{(1-m^2)\left(1-\dfrac{m^2}{1-m^2}\right)}\,q_1$	$\dfrac{m}{1-\dfrac{m^2}{1-m^2}}\,a$
$\dfrac{m^4}{(1-m^2)\left(1-\dfrac{m^2}{1-m^2}\right)\left(1-\dfrac{m^2}{1-\dfrac{m^2}{1-m^2}}\right)}\,q_1$	$\dfrac{m}{1-\dfrac{m^2}{1-\dfrac{m^2}{1-m^2}}}\,a$

The total charge on the sphere is

$$Q = q_1\left\{1 + m + \frac{m^2}{1-m^2} + \frac{m^2}{(1-m^2)\left(1-\dfrac{m^2}{1-m^2}\right)} + \ldots\right\}$$

$$= q_1(1 + m + m^2 + m^3 + 2m^4 + 3m^5 + \ldots)$$

Because q_2 is the image of $-q_1$ in the sphere, the potential of the surface of the sphere due to this pair of charges is zero, as shown in the previous example. The same is true of the pairs q_3 and $-q_2$, q_4 and $-q_3$, and so on. Consequently, the potential of the sphere is

$$V = \frac{q_1}{4\pi\varepsilon_0 a}$$

due to the charge q_1 at the center. The potential of the conducting plane AB is also zero, because equal charges of opposite signs are placed symmetrically on the two sides of AB. If we want to calculate the capacitance of the sphere relative to the plane, it is equal to

$$C = Q/V = 4\pi\varepsilon_0 a\,(1 + m + m^2 + m^3 + 2m^4 + 3m^5 + \ldots)$$

Figure 3.13 shows the ratio C/a as function of the ratio m (= a/d). For $m = 0$, the sphere is infinitely distant from the grounded plate, and C is simply the capacitance of an isolated sphere. For $m \to 1/2$, the sphere comes infinitely close to the grounded plane, and the capacitance tends to infinity.

The analysis presented above provides us with the solution of the problem of two conducting spheres of the same radius a at a distance d apart, the charge on the second sphere being equal but opposite in sign to that on the first. The negatively charged sphere is indicated by a broken circle in Figure 3.11. Its potential is

$$V' = -\frac{q_1}{4\pi\varepsilon_0 a}$$

The field on the right of AB is the reflection in the plane of the field on the left. The capacitance is half that for the sphere and plane.

Example 9: Point Charge Near a Semi-Infinite Dielectric

Consider a point charge q at a distance d from a semi-infinite block of homogeneous linear dielectric, as shown in **Figure 3.14**. Let us first calculate the potential at P' due to the point charge q. It is not the one that q would produce at P' in free space. We assume for the moment that this potential is the same as that produced by a charge Q at A in free space and the dielectric was not there. Thus,

$$V_{P'} = \frac{Q}{4\pi\varepsilon_0\sqrt{x'^2 + y'^2 + (z' - d)^2}}$$

This reduces to, in the z-plane,

$$V_{P'}|_{z=0} = \frac{Q}{4\pi\varepsilon_0\sqrt{x'^2 + y'^2 + d^2}}$$

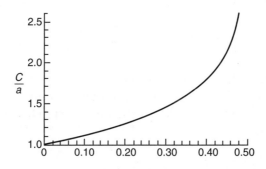

I FIGURE 3-13

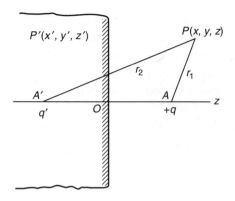

I FIGURE 3-14

The value of Q is to be determined later.

We now turn our attention to the electric field outside the dielectric. The field of the charge q polarizes the dielectric that in turn produces its own field and affect field outside the dielectric. To find the resultant field outside the dielectric, we assume that the field produced by the polarization charges at any point outside the dielectric is the same as that produced by a charge q' located at A', the image charge of q. That is, we replace the dielectric by the image charge q'. Now we have to find the value of q'. To this end, let us first write down the potential at P due to q and its image q':

$$V_P = \frac{q}{4\pi\varepsilon_0 r_1} + \frac{q'}{4\pi\varepsilon_0 r_2}$$

or, in rectangular coordinates,

$$V_P = \frac{1}{4\pi\varepsilon_0}\left[\frac{q}{\left\{x^2 + y^2 + (z-d)^2\right\}^{1/2}} - \frac{q'}{\left\{x^2 + y^2 + (z+d)^2\right\}^{1/2}}\right]$$

This reduces in the plane $z = 0$ to

$$V_P = \frac{1}{4\pi\varepsilon_0}\left[\frac{q + q'}{\left\{x^2 + y^2 + d^2\right\}^{1/2}}\right]$$

On the boundary (i.e., at the surface of the dielectric), $V_P = V_{P'}$ that automatically satisfies the boundary condition that the tangential components of E must be the same on both sides of the boundary. This condition gives

$$Q = q + q'$$

The second boundary condition is that the normal components of the electric displacement must be continuous at the boundary. That is,

$$\frac{\partial V_P}{\partial z} = \kappa \frac{\partial V_{P'}}{\partial z}, \quad at \; z = 0$$

where κ is the dielectric constant of the homogeneous linear dielectric block. Now

$$\left. \frac{\partial V_P}{\partial z} \right|_{z=0} = \frac{1}{4\pi\varepsilon_0} \frac{d(q - q')}{(x^2 + y^2 + d^2)^{3/2}}$$

and

$$\left. \frac{\partial V_{P'}}{\partial z} \right|_{z=0} = \frac{1}{4\pi\varepsilon_0} \frac{Qd}{(x^2 + y^2 + d^2)^{3/2}}$$

Thus, the second boundary condition gives

$$\frac{1}{4\pi\varepsilon_0} \frac{d(q - q')}{(x^2 + y^2 + d^2)^{3/2}} = \frac{1}{4\pi\varepsilon_0} \frac{\kappa Qd}{(x^2 + y^2 + d^2)^{3/2}}$$

or $q - q' = \kappa Q$.

We also have from the first boundary condition

$$Q = q + q'$$

Solving these two equations we get

$$Q = \frac{2q}{\kappa + 1}, \quad \text{and} \quad q' = -\frac{q(\kappa - 1)}{\kappa + 1}$$

3.9 Conjugate Functions and Two-Dimensional Electrostatic Problems

A two-dimensional electrostatic problem whose Laplace's equation is of the form

$$\frac{\partial^2 V}{\partial x^2} + \frac{\partial^2 V}{\partial y^2} = 0 \tag{3.59}$$

can be solved either in x, y or in polar coordinates r and θ. Now we describe a method involving what are known as conjugate functions.

Let us first introduce the complex quantity z:

$$z = x + iy, \quad i = (-1)^{1/2}$$

Note that we use bold italic letters to denote complex quantities.

Consider any function of z such as $F(z)$:

$$F(z) = F(x + iy) = g(x, y) + ih(x, y)$$

where g and h are, respectively, the real and the imaginary part of $F(z)$, and they are called conjugate functions of x and y.

Now

$$\partial F/\partial x = (dF/dz)(\partial z/\partial x) = dF/dz, \quad \partial^2 F/\partial x^2 = d^2 F/dz^2 \tag{3.60a}$$

$$\partial F/\partial y = (dF/dz)(\partial z/\partial y) = idF/dz, \quad \partial^2 F/\partial y^2 = -d^2 F/dz^2 \tag{3.60b}$$

Therefore

$$\frac{\partial^2 F}{\partial x^2} + \frac{\partial^2 F}{\partial y^2} = 0 \tag{3.61}$$

This shows that any function of the complex variable $x + iy$ satisfies Laplace's equation (3.59) for two dimensions. Consequently, both the real part and the imaginary part of $F(z)$ must satisfy Laplace's equation. Hence, either the real part of $F(z)$ or the imaginary part may be taken as the potential function in a possible electrostatic field.

From equation (3.60) we have

$$\frac{\partial F}{\partial y} = i \frac{\partial F}{\partial x}$$

or

$$\frac{\partial g}{\partial y} + i \frac{\partial h}{\partial y} = i \left(\frac{\partial g}{\partial x} + i \frac{\partial h}{\partial x} \right)$$

So, as g and h are real functions,

$$\frac{\partial g}{\partial y} = -\frac{\partial h}{\partial x}, \quad \frac{\partial h}{\partial y} = \frac{\partial g}{\partial x}$$

and

$$\frac{\partial g/\partial x}{\partial g/\partial y} = -\frac{\partial h/\partial y}{\partial h/\partial x} \tag{3.62}$$

Now consider the two families of curves: $g(x, y) = $ constant and $h(x, y) = $ constant.

$$\left(\frac{dy}{dx} \right)_g = -\frac{\partial g/\partial x}{\partial g/\partial y}$$

and that to a curve of the second family is

$$\left(\frac{dy}{dx}\right)_b = -\frac{\partial h/\partial x}{\partial h/\partial y}$$

But equation (3.62) states that the one slope is the negative reciprocal of the other, that is, the two families of curves intersect orthogonally. So if one family represents the traces of equipotential surfaces on the XY plane, the other represents lines of force.

Example 10: Field Due to Two Conducting Planes Intersecting at Right Angles

The function of the complex variable z needed for the solution of this problem is

$$F(z) = A(x + iy)^2 = A(x^2 - y^2) + 2iAxy$$

where A is an arbitrary constant. We choose the imaginary part for the solution of the problem

$$V(x, y) = 2Axy$$

The equipotential surfaces are the equilateral hyperbolic cylinders (**Figure 3.15**, solid lines). The conducting planes (the XZ and YZ coordinate planes) are at zero potential. The equation of the lines of force is the hyperbolas (the broken lines)

$$x^2 - y^2 = \text{constant}$$

The components of the electric field intensity are

$$E_x = -\partial V/\partial x = -2Ay, \quad E_y = -\partial V/\partial y = -2Ax$$

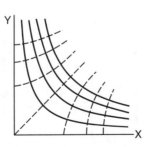

I FIGURE 3-15

If a dielectric of dielectric constant κ occupies the region that contains the field, the charge per unit area on the XZ plane is

$$\sigma = -\varepsilon_0 \kappa \left(\frac{\partial V}{\partial y} \right)_{y=0} = -2\varepsilon_0 \kappa A x$$

with a similar expression for that on the YZ plane.

Problems

1. Show that if $V_1, V_2, \ldots, V_n$ are all solution of Laplace's equation, then

$$V = C_1 V_1 + C_2 V_2 + \ldots + V_n$$

where C is an arbitrary constant and is also a solution.

2. Consider a set of wires parallel to the y-axis in the plane $z = 0$. The wires are equally spaced at $x = \pm n(a/2)$, where $n = 0, 1, 2, 3, \ldots$. Wires at odd positions (i.e., when n is odd) are at potential $-V_0$, whereas those at even positions are at V_0. Find the potential at points in space (**Figure 3.16**).

3. **Figure 3.17** shows two grounded, semi-infinite, parallel electrodes separated by a distance b. At $x = 0$ an electrode is maintained at a potential V_0. Find the potential V at any point between the plates.

4. If R is the distance of a point P from the origin O, a the distance of a point Q from O, and r the distance of P from Q,

$$R = \sqrt{r^2 + 2ar\cos\theta + a^2}$$

where θ is the angle between OQ and QP. Expand the reciprocal of R by the binomial theorem both for $r < a$ and for $r > a$. Show that the successive terms in the expansions are zonal harmonics.

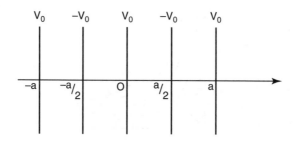

I FIGURE 3-16

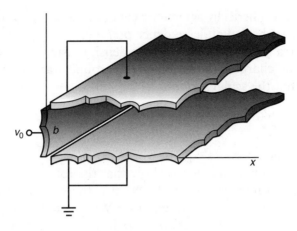

I FIGURE 3-17

5. In Example 2 we solved the problem of an uncharged conducting sphere in a uniform electric field. Now the sphere has a charge Q on its surface; find the potential and the electric field at any point (r, θ).

6. Find the stress on the surface of the conducting sphere treated in Example 2, when there is a charge Q on the sphere.

7. Two point charges q and q' are placed in very small spherical cavities in a rigid infinite dielectric of dielectric constant K, at a distance r apart. The charges do not touch the walls of the cavities. Find the force between them.

8. In Example 3, if the cylinder instead of being uncharged has a charge per unit length equal to λ, find the potential outside the cylinder and the components of electric field intensity in the direction of increasing r and ϕ.

9. In a coaxial cable the potential of the outer cylinder of radius b is held at zero and that of the inner cylinder of radius a is V_1. Find the potential at a point in the region between the two cylinders.

10. Discuss the field of a charged conducting plane of infinite extent having a cylindrical ridge of semicircular cross-section; find the charge per unit area on the surface of the plane.

11. A voltage V is applied to a thin parallel-plate capacitor of plate separation d filled with a cloud of charge of constant density ρ. Find the potential inside the capacitor with respect to the positive plate, the electric field vector inside the capacitor, and the surface charge density σ on the inner surface of the plates.

12. Consider a spherically symmetric charge distribution of total charge q, which has the radial dependence $\rho(r) = \rho_0 e^{-\alpha r}$. Find the potential at an arbitrary point r. (This charge density distribution describes the electronic charge distribution in the ground state of hydrogen.)

13. Show that the scalar potential

$$V(r) = \int \frac{\rho(r)\, d\vec{r}}{4\pi\varepsilon_0 r}$$

satisfies the Poisson's equation $\nabla^2 V = -\rho/\varepsilon_0$.

14. Referring to Example 6, (a) show that the total induced charge on the conducting plane is $-q$. (b) What is the energy for the single charge q and the conducting plane? What is the energy for the point charge q and the image charge $-q$? If there is difference, explain why there is such difference.

15. Given two infinitely long straight parallel wires of radius a, one has a charge per unit length equal to λ and the other an equal negative charge distribution. Find the electric field intensity and the capacitance per unit length between the two wires.

16. Using conjugation functions, find the electric field at the edge of a conducting plane.

17. In Example 6, we found that for a single charge and conducting plane the energy is just half of that for the two point charges and no conducting plane. Compute the energy for a single charge and conducting plane by integrating the force to find the work required to bring q in from $z = \infty$ to $z = d$.

4 Magnetostatics

The science of electricity and magnetism evolved from their beginning until the early 19th century quite independently of one another. The use of a natural magnet or lodestone to indicate direction was known to the Chinese more than 4000 years ago. The interaction of magnets can be explained on the basis of an inverse square law similar to Coulomb's law for static charges. Until 1820 the connection between magnetism and electricity was unknown, but in that year an event occurred to change that and sparked a period of rapid development. What happened in that year was the discovery by the Danish physicist Christian Oersted that an electric current could exert forces precisely similar to that exerted by permanent magnets. Since then, the point of view taken by physicists toward magnetism has changed. We now attribute all magnetic effects to charges in motion. This viewpoint eliminates the need for such a concept as the magnetic pole, and as a result it makes Coulomb's law for the force between magnetic poles unnecessary. Thus, we study in this chapter the motion of the electric charges (i.e., electric current) and the forces associated with them.

4.1 Electric Current

A free charge placed in an electric field experiences a force and moves under the influence of this force in accordance with the laws of mechanics. In many processes of physics and chemistry, the outer weakly bounded atomic electrons can migrate from one atom to another. A neutral atom or molecule that loses one or more electrons then becomes a positive ion. A migrating electron can be captured by another neutral atom or molecule to form a negative ion, or it may move more or less freely as a conduction electron, capable of transporting its charge through the system. In a solid conductor it is usually only electrons that contribute to charge transport; in a liquid or plasma, positive and negative ions may contribute to and may dominate in charge transport.

At the atomic level, we can define an atomic current density

$$\vec{j}_{at}(\vec{r}, t) = \sum_i Q_i \vec{v}_i (\vec{r} - r_i(t)) \tag{4.1}$$

where $v_i = dr_i/dt$ is the velocity of the ith particle, and Q_i is the charge on the particle at r_i. Obviously, the current density is a vector field.

From the definition, we can show that the current density is related to charge density ρ_{at} by the continuity equation

$$\frac{\partial \rho_{at}}{\partial t} + \nabla \cdot \vec{j}_{at} = 0 \tag{4.2}$$

To see this, let us take a partial derivative of $\delta(r - r_i(t))$:

$$\frac{\partial}{\partial t} \delta(\vec{r} - \vec{r}_i(t)) = \frac{\partial \delta}{\partial x_i} \frac{dx_i}{dt} + \frac{\partial \delta}{\partial y_i} \frac{dy_i}{dt} + \frac{\partial \delta}{\partial z_i} \frac{dz_i}{dt}$$

$$= -\frac{\partial \delta}{\partial x} \frac{dx_i}{dt} - \frac{\partial \delta}{\partial y} \frac{dy_i}{dt} - \frac{\partial \delta}{\partial z} \frac{dz_i}{dt}$$

$$= -\vec{v}_i \cdot \nabla \delta = -\nabla \cdot (\vec{v}_i \delta).$$

Multiplying the last equation by Q, and then summing over i for all particles, we obtain equation (4.2), where

$$\rho_{at}(\vec{r}) = \sum_i Q_i \delta(\vec{r} - \vec{r}_i) \tag{4.3}$$

is the charge density at the point r_i.

Although ρ_{at} and $\vec{j}_{at}$ may vary rapidly with both time and position, macroscopic charge densities and currents can be quite slowly varying or even constant in both space and time because of the mechanical retardation the charges experience from frequent collisions with the atoms of which the conductor is composed. We consider, for simplicity, only one kind of charged particle responsible for the current; then the macroscopic current density $\vec{j}$ is the macroscopic charge density ρ of those particles, multiplied by their mean velocity v within the average volume.

The continuity equation is still valid on the macroscopic scale. This is most easily seen by integrating equation (4.2) over a volume V' that contains the charges

$$\int \left[\frac{\partial \rho_{at}(\vec{r} - \vec{r}')}{\partial t} + \nabla \cdot \vec{j}_{at}(\vec{r} - \vec{r}') \right] dV' = \frac{\partial}{\partial t} \int \rho_{at}(\vec{r} - \vec{r}') dV' + \nabla \cdot \int \vec{j}_{at}(\vec{r} - \vec{r}') dV' = \frac{\partial \rho}{\partial t} + \nabla \cdot \vec{j}$$

both the $\partial/\partial t$ and the ∇ operator can be taken outside the integral that is over V'. Now the quantity in the square brackets vanishes at each point in space. Hence,

$$\frac{\partial \rho}{\partial t} + \nabla \cdot \vec{j} = 0 \tag{4.4}$$

This important equation is based on the conservation of charge.

In the steady state, which is our concern in this chapter, $\partial \rho/\partial t = 0$ and the continuity equation reduces to

$$\nabla \cdot \vec{j} = 0 \tag{4.5}$$

This means that in every volume element as much current flows out as flows in; in other words, equation (4.5) is valid in the region that does not contain a source or a sink of current. Thus, a static current distribution contained in a finite volume must flow in closed loops.

We are all familiar with the idea of electric currents flowing in conductors, say a conducting wire. The electric current I through a cross-section of the wire is the charge per unit time passing through the cross-section and is given by

$$I = \int_S \vec{j} \cdot \vec{dS} = \int_S \vec{j} \cdot \hat{n} dS \tag{4.6}$$

where $\hat{n}$ is a unit vector normal to dS. In a steady state the current I is constant along the wire.

The SI unit of current is the ampere (A), after the French physicist André M. Ampere, and corresponds to a charge flow of one coulomb per second. The SI unit of current density is A/m^2.

4.2 The Lorentz Force and Magnetic Fields

Having established what we mean by an electric current, we investigate the effects of currents on one another and on electric charges. Experiments such as those of Ampere in 1820 showed (**Figure 4.1**) that parallel currents in adjacent wires attract one another and antiparallel currents repel. The wires themselves are electrically neutral to a very high degree (about 1 in 10^{25}), so the force acting is not an electrostatic or Coulomb force. Further experiments have shown no force acting between a stationary charge and a current (**Figure 4.2a**), but charges moving parallel to a current do attract it (Figure 4.2b), whereas charges circulating round in a plane normal to the current produce no effect (Figure 4.2c).

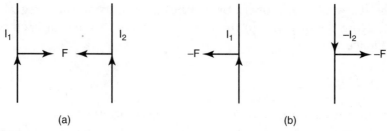

I FIGURE 4.1

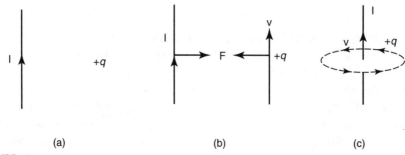

I FIGURE 4.2

Exactly similar effects can be produced when currents are placed between the poles of a permanent magnet (**Figure 4.3**): The currents are repelled out of the magnet in a direction depending on the direction of the current. Today we describe all these effects by attributing a magnetic field $\vec{B}$ to both the current in the wire and the (atomic) currents in the magnet. Then, how do we define a magnetic field?

The definition of the magnetic field is slightly more difficult to formulate. The electric field at a given point in space is defined in terms of the force the electric field exerts on a small stationary test charge placed there. We may define the magnetic field $\vec{B}$ in a similar manner. Magnetic forces only acting on moving charges, so it is natural to use the magnetic force $\vec{F}$ exerted on a positive charge q whose velocity is $\vec{v}$ to define $\vec{B}$. Experiments such as those shown in Figures 4.1 to 4.3 all demonstrate that the magnetic force acting on a moving charge in a magnetic field is proportional to

1. The charge q;
2. The speed v; and
3. The sine of the angle between $\vec{v}$ and $\vec{B}$.

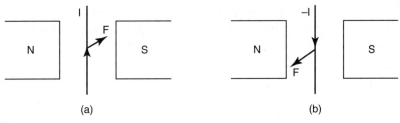

I FIGURE 4.3

Furthermore, it is always perpendicular to the plane containing $\vec{v}$ and $\vec{B}$. This can be summarized in vector form

$$\vec{F} \propto q\vec{v} \times \vec{B}$$

In the SI system the constant of proportionality is unity. Thus,

$$\vec{F} = q\vec{v} \times \vec{B} \tag{4.7}$$

In principle, $\vec{B}(\vec{r})$ at a point $\vec{r}$ can be determined from observations of the force on moving charged particles. Hence, we use equation (4.7) to give an "operational definition" of the field $\vec{B}(\vec{r})$. We begin our definition of $\vec{B}$ by choosing its direction at any point as that in which a moving charge would experience no magnetic force, and its magnitude is specified by

$$B = \frac{F}{qv \sin\theta}$$

where θ is the angle between $\vec{B}$ and $\vec{v}$.

The unit of magnetic field in the SI unit of B may be expressed as newtons/ampere-meter = newton-seconds/coulomb-meter = volt-seconds/meter². The name tesla (T) or weber/meter² has been given to this unit. To give an idea of the magnitude of this unit, the earth's magnetic field is about 5×10^{-5} T, a good iron-core electromagnet can produce a few teslas, and the best superconducting solenoids produce somewhat over 10 T. It is also very common to measure magnetic fields in gauss, a cgs unit, and 1 gauss = 10^{-4} tesla.

If both electric and magnetic fields are present, the magnetic force experienced by a moving charge q is

$$\vec{F} = q(\vec{E} + \vec{v} \times \vec{B}) \tag{4.8}$$

This force is known as the Lorentz force and is found to be true even for particles moving at speeds close to the speed of light. Thus, a charged particle q of momentum $\vec{p}$ has the classical equation of motion

$$\frac{d\vec{p}}{dt} = q\,(\vec{E} + \vec{v} \times \vec{B}) \tag{4.9}$$

It is interesting to note that the magnetic field does no mechanical work on a moving charged particle. Why? The vector product $\vec{v} \times \vec{B}$ in the Lorentz force and in equation (4.7) or (4.8) incorporates the fact that the force due to the $\vec{B}$ field is perpendicular both to $\vec{B}$ and $\vec{v}$, and so $\vec{B} \cdot (\vec{v} \times \vec{B}) = 0$.

We take the Lorentz force on a charged particle to be the defining measure of a magnetic field; there are many situations in which such a measurement is inconvenient or impossible. We explain later other more practical methods of measuring $\vec{B}$.

We have introduced $\vec{B}$ as the magnetic field, whereas in some books it is called the magnetic induction and another vector $\vec{H}$ is called the magnetic field. We introduce the vector $\vec{H}$ later when we consider the magnetic properties of materials. The relation between $\vec{B}$ and $\vec{H}$ is discussed there in detail.

4.3 Gauss' Law for the Magnetic Field

It was an early observation that magnetic fields, whether produced by electric currents or by magnets, are characterized by a dipole (or higher multipole) form at large distances; despite intensive search, no magnetic monopole has been found. So far all the experimental evidence is consistent with the total magnetic flux through any closed surface being exactly zero:

$$\int_S \vec{B} \cdot d\vec{S} = 0 \tag{4.10}$$

This is the magnetic equivalent of Gauss' law in integral form for the electric field. Application of the divergence theorem to equation (4.10) gives

$$\int_V \nabla \cdot \vec{B}\, d\vec{r} = 0$$

This holds that for any arbitrary volume V, the integrand must vanish identically and we have

$$\nabla \cdot \vec{B} = 0 \tag{4.11}$$

This is Gauss' law in differential form for the magnetic field.

The principle of superposition holds for magnetic fields as it holds for electric fields: Magnetic fields produced by different sources simply add vectorially.

4.4 Forces on Current-Carrying Conductors

We now apply the Lorentz force law to investigate the motion of a current element in magnetic fields. A current element is a short length dl of a conductor carrying a current I. We take the sense of dl in the direction of the current, and then $\vec{dl}$ is parallel to the velocity $\vec{v}$ of the charge carries in the conductor. If the charge density is N, the force on the element dl is

$$d\vec{F} = (NAdlq)\,\vec{v} \times \vec{B} = I\vec{dl} \times \vec{B} \tag{4.12}$$

where we assume that $\vec{E} = 0$, A is the cross-sectional area of the conductor, and $\vec{v}$ is the velocity of the charge. And

$$I = Nadlq \tag{4.13}$$

If the current-carrying conductor forms a closed loop, then the force on this closed loop is

$$\vec{F} = \oint I\vec{dl} \times \vec{B} = I \oint \vec{dl} \times \vec{B} \tag{4.14}$$

where $\vec{B}$ depends on position. If $\vec{B}$ is uniform, independent of position, then equation (4.17) becomes

$$\vec{F} = I \left\{ \oint \vec{dl} \right\} \times \vec{B} = 0 \tag{4.15}$$

because the vector sum $\oint \vec{dl}$ vanishes for a closed loop.

4.5 The Magnetic Field of Steady Current: Ampere's Circuital Law

The magnetic fields produced by steady currents flowing in thin wires were an early subject of investigation, and the laws governing them are well established. After he performed and analyzed various experiments, Ampere discovered the $(1/\rho)$ dependence of the magnetic field due to a steady current I flowing in a long straight wire of circular cross-section (**Figure 4.4**). In cylindrical coordinates (ρ, ϕ, z), with the wire as z-axis, at point $\vec{r}$ outside the wire it is found experimentally

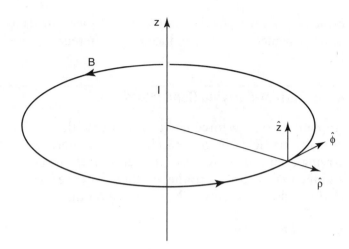

$$\vec{B} = \frac{\mu_0}{2\pi} \frac{I}{\rho} \hat{\phi} \tag{4.16}$$

B is at right angle both to I and ρ. We can write it in a different form

$$\vec{B} = \frac{\mu_0}{2\pi} \frac{\vec{I} \times \hat{\rho}}{\rho} \tag{4.17}$$

where μ_0 is a constant called the permeability of free space.

Let us find the line integral $\oint \vec{B} \cdot d\vec{l}$ along a path Γ_1 that encloses the wire:

$$\oint \vec{B} \cdot d\vec{l} = \oint \frac{\mu_0}{2\pi} \frac{I}{\rho} \hat{\phi} \cdot (d\rho\hat{\rho} + \rho d\phi\hat{\phi} + dz\hat{z}) = \frac{\mu_0}{2\pi} I \oint d\phi$$

For the path Γ_1, $\oint_{\Gamma_1} d\phi = 2\pi$. Hence,

$$\oint_{\Gamma_1} \vec{B} \cdot d\vec{l} = \mu_0 I \tag{4.18}$$

For a path Γ_2 that does not enclose the wire, $\oint_{\Gamma_2} d\phi = 0$, and so

$$\oint_{\Gamma_2} \vec{B} \cdot d\vec{l} = 0 \tag{4.19}$$

Equation (4.18) is often called the circuital form of Ampere's law.

It is remarkable that equations (4.18) and (4.19) are found to be true not only in the special case we have just considered, but for *any* closed path Γ threading through *any* circuits in which steady currents are flowing. In the general case, I is the total current that flows through a surface $S(\Gamma)$ bounded by Γ. The "right-hand rule" linking the sense of the line integral and the direction of I is explained in Chapter 1.

Ampere's law now recognizes only one source of a magnetic field: moving electric charges. We see in a later chapter that there is a second origin of a magnetic field: a changing electric field, also called a displacement current.

Example 1: Toroidal Solenoid

Consider a doughnut-shaped core wound uniformly with a wire carrying a current I (**Figure 4.5**). Find the magnetic field.

Solution

Let there be N turns and N/l turns per unit length, where l is the mean circumference around the interior. In applying Ampere's law we take for our path Γ this circular path of length l. From symmetry considerations we know that the magnetic field is constant and in the direction of the path, so that

$$\oint_{\Gamma} \vec{B} \cdot d\vec{l} = Bl$$

Because the path Γ threads the current N times, we have, from equation (4.18),

$$B = \mu_0 NI/l$$

As l increases, a toroidal solenoid becomes more and more similar to a very long straight solenoid. Hence, our result also gives B for a very long straight solenoid with NI/l ampere-turns per meter.

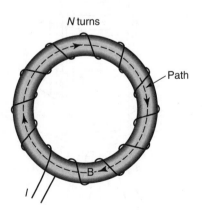

I FIGURE 4.5

Example 2: Solenoids

A solenoid is a long piece of wire wound tightly on a cylindrical surface in the form of a helix. We now apply Ampere's circuital law to such a solenoid, carrying current I. Take cylindrical polar coordinates with the axis of the coil as the z-axis and the length of the solenoid is much greater than its diameter. Because the coil is tightly wound, the field has axial symmetry and so does not depend on ϕ, to a good approximation.

Outside the coil the field is small except near the ends. This can be seen by replacing each turn of the coil by its equivalent dipole sheet: The field outside the coil can be thought of as due to the superposition of these sheets, which cancel out except at the ends of the solenoid, where equivalent magnetic "poles" appear. For a long solenoid these apparent poles are well separated, and the field they produce is small, except near the ends.

Applying Ampere's circuital law to a rectangular path C (**Figure 4.6**) away from the ends of the solenoid,

$$\int_C \vec{B} \cdot d\vec{l} = BL + B_{outside}L = \mu_0 nIL$$

where n is the number of turns of the coil per unit length, L the length of the vertical path, and we have assumed that the B field in this region has no

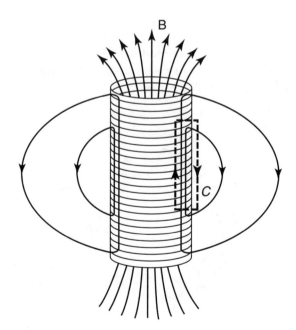

I FIGURE 4.6

radial component, so $\vec{B} \cdot d\vec{l} = 0$ along the two horizontal parts of the path C. Because $B \gg B_{\text{outside}}$, we are left with

$$B_z \cdot L = \mu_0 nIL$$

Hence, inside the coil B_z is constant and is given by

$$B_z = \mu_0 nI$$

Taking a circular path around the axis, Ampere's law gives $B_\phi \approx 0$ inside the coil and $B_\phi (r) \approx \mu_0 I/2\pi r$ outside the coil. Thus, the field is essentially constant inside the solenoid. At the ends of the solenoid the $\boldsymbol{B}$ lines diverge and because $\nabla \cdot \boldsymbol{B} = 0$, the field lines must return to the opposite end of the solenoid, joining up to give closed curves.

As shown in Example 9, a more exact expression for the field without making the approximation at any point on the axis can be found fairly easily using the Biot-Savart law.

We now return to equation (4.18) and write

$$I = \int_S \vec{J} \cdot d\vec{S}$$

where $\vec{J}$ is the current density in the wire so that

$$\oint_\Gamma \vec{B} \cdot d\vec{l} = \mu_0 \int_S \vec{J} \cdot d\vec{S} \tag{4.20}$$

We shall assume that equation (4.20) also holds for paths passing *through* a current distribution.

We can transform the line integral to an integral over the surface S with the help of Stokes' theorem. Then equation (4.20) becomes

$$\int_S \nabla \times \vec{B} \cdot dS = \mu_0 \int_S \vec{J} \cdot dS \tag{4.21}$$

Because this is true for *all* surfaces S, it follows that the magnetic field $\vec{B}(\vec{r})$ produced by a distribution of steady current $\vec{J}$ satisfies the field equation

$$\nabla \times \vec{B}(\vec{r}) = \mu_0 \vec{J} \tag{4.22}$$

This is the differential form of Ampere's circuital law. Thus, given a field $\vec{B}(\vec{r})$ the current density $\vec{J}$ that gives rise to it is determined. Conversely, the field equations (4.11) and (4.22) determine the magnetic field produced by a given distribution of steady currents $\vec{j}$; this is shown in Chapter 5, where the field equations are solved. Equation (4.22) is the magnetic equivalent of $\nabla \times \vec{E} = 0$ (see equation 2.9). Because the right-hand side of equation (4.22) does not vanish, we cannot *in general* express $\vec{B}(\vec{r})$ as the gradient of a potential function.

Example 3: The Magnetic Field Outside and Inside a Linear Current Circuit

Solution

To get some experience with equation (4.22), let us show that the curl of the magnetic field outside the wire is zero and inside the wire it is rotational. Let the z-axis be along the wire in the direction of the current or out from the page in **Figure 4.7**. From equation (4.16) we have the field outside and at a distance r from the axis of the wire

$$B = \frac{\mu_0}{2\pi} \frac{I}{r}$$

where I is the current flowing in the wire. Because the field is directed circularly about the wire, at the point $P(x,y)$ we have

$$\frac{B_x}{B} = -\frac{y}{r}, \quad B_x = -\frac{\mu_0 I}{2\pi} \frac{y}{r^2}$$

$$\frac{B_y}{B} = \frac{x}{r}, \quad B_y = \frac{\mu_0 I}{2\pi} \frac{x}{r^2}; \quad B_z = 0$$

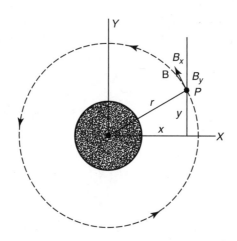

I FIGURE 4.7

Because B_x and B_y are independent of z and B is zero, $\nabla \times \vec{B}$ reduces to

$$\nabla \times \vec{B} = \hat{k} \left(\frac{\partial B_y}{\partial x} - \frac{\partial B_x}{\partial y} \right)$$

Putting $b = \mu_0 I/2\pi$ and $r^2 = x^2 + y^2$, we get

$$\frac{\partial B_y}{\partial x} = \frac{\partial}{\partial x} \left(\frac{bx}{x^2 + y^2} \right) = \frac{b}{x^2 + y^2} - \frac{2bx^2}{(x^2 + y^2)^2}$$

$$-\frac{\partial B_x}{\partial y} = \frac{\partial}{\partial y} \left(\frac{by}{x^2 + y^2} \right) = \frac{b}{x^2 + y^2} - \frac{2by^2}{(x^2 + y^2)^2}$$

Thus,

$$\nabla \times \vec{B} = \hat{k} \left(\frac{2b}{r^2} - \frac{2br^2}{r^4} \right) = 0.$$

The same conclusion follows from equation (4.22), because outside the wire j is everywhere zero. In this region the field does not rotate about itself, but the field does circulate about the wire. Inside the conductor, the situation is different. The current density j inside the wire is

$$j = I/\pi a^2$$

and from equation (4.22) we have now

$$\nabla \times \vec{B} = \frac{\mu_0}{\pi a^2} \vec{I}$$

Thus, within the conductor the field is rotational, that is, it turns about itself.

4.6 The Vector Potential

The basic equations for the magnetic field are (4.11) and (4.22):

$$\nabla \cdot \vec{B} = 0 \tag{4.21}$$

$$\nabla \times \vec{B} = \mu_0 \vec{j} \tag{4.22}$$

One indicates the absence of magnetic monopole and the other relates $\vec{B}$ to $\vec{j}$.

Because the divergence of a curl is always zero, we can therefore express $\vec{B}$ as a curl of a vector field we call $\vec{A}$:

$$\vec{B} = \nabla \times \vec{A} \tag{4.23}$$

The field $\vec{A}$ is called the vector potential. It is easy to see that $\vec{A}$ is not uniquely defined by equation (4.23): Adding any field that is the gradient of some scalar field to $\vec{A}$ will yield the same $\vec{B}$. To prove this, let $\vec{A}$ and $\vec{A}'$ give the same magnetic field $\vec{B}$:

$$\vec{B} = \nabla \times \vec{A} = \nabla \times \vec{A}'$$

Therefore,

$$\nabla \times \vec{A} - \nabla \times \vec{A}' = \nabla \times (\vec{A} - \vec{A}') = 0$$

Thus, $(\vec{A} - \vec{A}')$ must be the gradient of some scalar field, say φ,

$$\vec{A}' - \vec{A} = \nabla \varphi \quad \text{or} \quad \vec{A}' + \vec{A} = \nabla \varphi$$

We can take some of the "latitude" out of $\vec{A}$ by arbitrarily placing some other condition on it. Recall that in electrostatics the scalar potential V was not completely specified by its definition $\vec{E} = -\nabla V$. If V is the potential for some problem, a different potential $V' = V + C$, where C is a constant, gives the same field

$$-\nabla V' = -\nabla (V + C) = -\nabla V = \vec{E}$$

Now, to make $\vec{A}$ more specific by imposing an additional restriction on it, we have to do it without affecting $\vec{B}$. In magnetostatics a convenient condition that $\vec{A}$ has to satisfy is

$$\nabla \cdot \vec{A} = 0 \tag{4.24}$$

This is known as Coulomb gauge. The reason for this choice is that it makes calculations easier than with any other choice.

Thus, $\vec{B} = \nabla \times \vec{A}$ and $\nabla \cdot \vec{A} = 0$ together define the vector potential $\vec{A}$ that satisfies the fundamental equation $\nabla \cdot \vec{B} = 0$.

Because $\vec{B}$ is determined by currents, so also is the vector potential $\vec{A}$. To see this, let us substitute equation (4.23) into the basic equation (4.22) to obtain

$$\nabla \times (\nabla \times \vec{A}) = \nabla (\nabla \cdot \vec{A}) - \nabla^2 \vec{A} = \mu_0 \vec{j}$$

Thanks to our choice of the condition (4.24), the last equation now reduces to

$$\nabla^2 \vec{A} = -\mu_0 \vec{j} \tag{4.25}$$

Writing in components, we have three equations:

$$\nabla^2 A_x = -\mu_0 j_x, \quad \nabla^2 A_y = -\mu_0 j_y, \quad \nabla^2 A_z = -\mu_0 j_z \tag{4.26}$$

Each of these equations is mathematically identical to Poisson's equation in electrostatics:

$$\nabla^2 V = -\rho/\varepsilon_0$$

And all we have learned about solving for potential when ρ is known can be used here for solving for each component of $\vec{A}$ when $\vec{j}$ is known. So we can write down the general solution for A_x immediately.

$$A_x(1) = \frac{\mu_0}{4\pi} \int \frac{j_x(2)}{r_{12}} dV_2$$

and similarly for A_y and A_z. The general solution in the vector form is

$$\vec{A}(1) = \frac{\mu_0}{4\pi} \int \frac{\vec{j}(2)}{r_{12}} dV_2 \tag{4.27}$$

Figure 4.8 shows our conventions for r_{12} and dV_2.

We have now a general method for finding the magnetic field of steady currents. The field can be computed by first determining the vector potential $\vec{A}$ from equation (4.27) and then substituting it in equation (4.23). To get some experience with the vector potential, let us look at two simple examples.

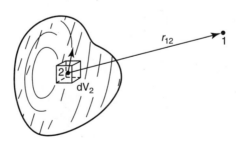

I FIGURE 4.8

Example 4: The Magnetic Field Due To a Current Flowing in a Thin Long Straight Wire

Solution

The magnetic field $\vec{B}$ due to a current flowing in a thin long straight wire is given by equation (4.16). We now use the vector potential $\vec{A}$ to calculate $\vec{B}$. That is, we first find $\vec{A}$ and then calculate $\vec{B}$ from equation (4.23), $\vec{B} = \nabla \times \vec{A}$. As shown in **Figure 4.9**, elements $I\,d\vec{l}$ of the current contribute to the vector potential elements

$$d\vec{A} = \frac{\mu_0 I}{4\pi} \frac{d\vec{l}}{r}$$

We first calculate $\vec{A}$ for a current wire of finite length $2L$ and then take the limit $L \gg \rho$.

$$A_z = 2 \frac{\mu_0 I}{4\pi} \int_0^L \frac{dl}{\sqrt{\rho^2 + l^2}} = \frac{\mu_0 I}{2\pi} \ln\left[l + (\rho^2 + l^2)^{1/2} \right]$$

$$\approx \frac{\mu_0 I}{2\pi} \ln \frac{2L}{\rho}, \quad \text{for} \quad \rho^2 \ll L^2$$

Next,

$$\vec{B} = \nabla \times \vec{A} = \begin{vmatrix} \hat{e}_\rho & \rho\hat{e}_\phi & \hat{e}_z \\ \frac{\partial}{\partial \rho} & 0 & \frac{\partial}{\partial z} \\ 0 & 0 & A_z \end{vmatrix}$$

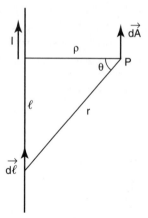

I FIGURE 4.9

from which we find

$$B_\rho = 0, \quad B_Z = 0, \quad \text{and}$$

$$B_\phi = -\partial A_Z/\partial\rho \rightarrow \mu_0 I/(2\pi\rho) \quad \rho^2 \ll L^2$$

Example 5: Vector Potential of a Uniform Magnetic Field

Solution

Suppose we have a uniform magnetic field B_0 in the direction of the z-axis. Then we have, from $\vec{B} = \nabla \times \vec{A}$,

$$\begin{pmatrix} 0 \\ 0 \\ B_0 \end{pmatrix} = \begin{pmatrix} \hat{i} & \hat{j} & \hat{k} \\ \dfrac{\partial}{\partial x} & \dfrac{\partial}{\partial y} & \dfrac{\partial}{\partial z} \\ A_x & A_y & A_z \end{pmatrix}$$

from which we obtain

$$B_x = \frac{\partial A_z}{\partial y} - \frac{\partial A_y}{\partial z} = 0, \quad B_y = \frac{\partial A_x}{\partial z} - \frac{\partial A_z}{\partial x} = 0, \quad B_z = \frac{\partial A_y}{\partial x} - \frac{\partial A_x}{\partial y} = B_0$$

It is clear for any particular magnetic field $\vec{B}$, the vector potential $\vec{A}$ given by equation (4.23) is not unique. This shows up by this simple example. By inspection, we see three possible solutions:

1. $A_y = xB_0, \quad A_x = 0, \quad A_z = 0$
2. $A_x = -yB_0, \quad A_y = 0, \quad A_z = 0$
3. $A_x = -\frac{1}{2}yB_0, \quad A_y = \frac{1}{2}xB_0, \quad A_z = 0$

The third solution is more interesting. Because the x-component is proportional to $-y$ and the y component is proportional to x, $\vec{A}$ must be at right angle to the vector from the z-axis, which we call $\vec{r}'$. Note that $\vec{r}'$ is not the vector displacement from the origin. Furthermore, the magnitude of $\vec{A}$ is proportional to $(x^2 + y^2)^{1/2}$ and, so to r'. Thus, $\vec{A}$ can be written as

$$\vec{A} = \frac{1}{2}\vec{B} \times \vec{r}'$$

and it has the magnitude $Br'/2$ and rotates about the z-axis as shown in **Figure 4.10**.

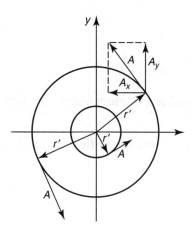

I FIGURE 4.10

Example 6: Vector Potential and Magnetic Field for a Slowly Moving Charge

For a charge q moving with a constant velocity $\vec{v}$, we may put $q\vec{v}$ for $\vec{j}\,dV$, and then we have

$$\vec{A} = \frac{\mu_0}{4\pi}\frac{q\vec{v}}{r}$$

The magnetic field is given by equation (4.23)

$$\vec{B} = \nabla \times \vec{A}$$

and remember that ∇ operates differentially only on the $1/r$ factor. We then get

$$\vec{B} = \frac{\mu_0 q}{4\pi}\nabla\left(\frac{\vec{v}}{r}\right) = \frac{\mu_0 q}{4\pi}\frac{\vec{v}\times\hat{r}}{r^2}$$

where $\hat{r}$ is a unit vector from the charge to the point P where the field is computed.

4.7 Field of a Small Current Loop: The Magnetic Dipole

The vector potential $\vec{A}$ may also be used to advantage in determining the magnetic field $\vec{B}$ due to a small current loop. What do we mean here by small? We mean that we are only interested in knowing $\vec{B}$ at distances from the loop that are large compared with the dimensions of the loop. We find that any small loop is equivalent to a magnetic dipole; that is, it gives rise to a $\vec{B}$ field just like the $\vec{E}$ field found for an electric dipole.

Consider a small loop with a current I flowing in it; we may then replace $\vec{j}\,dV_2$ in equation (4.27) by $I\,d\vec{r}'$, where $d\vec{r}'$ is an element of the loop (**Figure 4.11**). The result is

$$\vec{A}\,(\vec{r}) = \frac{\mu_0 I}{4\pi} \oint \frac{d\vec{r}'}{|\vec{r} - \vec{r}'|}$$

where $\vec{r}'$ is the position vector of the element $d\vec{r}'$. Because we are interested in the far field (at large distances from the loop), i.e., $|\vec{r}| \gg |\vec{r}'|$, we may expand $|\vec{r} - \vec{r}'|^{-1}$ in powers of r'/r. To first order, we have

$$|\vec{r} - \vec{r}'|^{-1} = \frac{1}{r}\left[1 + \frac{\vec{r}' \cdot \vec{r}}{r^2} + \ldots\right]$$

then

$$\vec{A}\,(\vec{r}) = \frac{\mu_0 I}{4\pi}\left[\frac{1}{r^3}\oint (\vec{r}' \cdot \vec{r})\,d\vec{r}' + \ldots\right]$$

because the first term in the expansion gives no contribution to $\vec{A}$. We may rewrite this expression in the form

$$\vec{A}\,(\vec{r}) = \frac{\mu_0 I}{8\pi r^3}\oint (\vec{r}' \times d\vec{r}') \times \vec{r} \tag{4.28}$$

To show this, we first write the triple vector product as

$$(\vec{r}' \times d\vec{r}') \times \vec{r} = d\vec{r}'(\vec{r} \cdot \vec{r}') - \vec{r}'(\vec{r}' \cdot d\vec{r}')$$

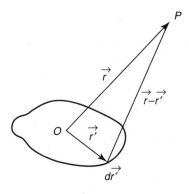

I FIGURE 4.11

Next we rewrite the second term on the right in the form of the first term. To do this we note that

$$d\left[(\hat{r} \cdot \vec{r}')\,\vec{r}'\right] = (\hat{r} \cdot d\vec{r}')\,\vec{r}' + (\hat{r} \cdot \vec{r}')\,d\vec{r}'$$

From this we find

$$\oint (\hat{r} \cdot d\vec{r}')\,\vec{r}' + (\hat{r} \cdot \vec{r}')\,d\vec{r}' = \oint d\left[(\hat{r} \cdot \vec{r}')\,\vec{r}'\right] = 0$$

because the total change in $[(\hat{r} \cdot \vec{r}')\,\vec{r}']$ around a closed loop is zero. Thus,

$$\oint (\vec{r} \cdot d\vec{r}')\,\vec{r}' = -\oint (\vec{r} \cdot \vec{r}')\,d\vec{r}' \quad \text{and} \quad \oint (\vec{r}' \cdot d\vec{r}') \times \vec{r}' = 2\oint (\vec{r} \cdot \vec{r}')\,d\vec{r}'$$

We now define

$$\vec{m} = \tfrac{1}{2}\oint \vec{r}' \times d\vec{r}' \tag{4.29}$$

as the magnetic moment of the loop, which is equal to current I times the area of the loop, $\tfrac{1}{2}\oint \vec{r}' \times d\vec{r}'$. The expression for $\vec{A}$, equation (4.28), then becomes

$$\vec{A}(\vec{r}) = \frac{\mu_0}{4\pi}\frac{\vec{m} \times \vec{r}}{r^3} \tag{4.30}$$

The magnetic field $\vec{B}$ is determined by calculation $\nabla \times \vec{A}$; from equation (4.30) we have

$$\nabla \times \vec{A} = \nabla \times \left(\frac{\mu_0}{4\pi}\frac{\vec{m} \times \vec{r}}{r^3}\right) = \frac{\mu_0}{4\pi}\left[\vec{m}\nabla \cdot \left(\frac{\vec{r}}{r^3}\right) - (\vec{m} \cdot \nabla)\frac{\vec{r}}{r^3}\right]$$

$$\tag{4.31}$$

$$= -\frac{\mu_0}{4\pi}(\vec{m} \cdot \nabla)\frac{\vec{r}}{r^3}$$

because $\nabla \cdot (\vec{r}/r^3) = 0$. Now

$$(\vec{m} \cdot \nabla)\frac{\vec{r}}{r^3} = \left(m_x\frac{\partial}{\partial x} + m_y\frac{\partial}{\partial y} + m_z\frac{\partial}{\partial z}\right)\left(\frac{\hat{i}x + \hat{j}y + \hat{k}z}{(x^2 + y^2 + z^2)^{3/2}}\right)$$

$$= \frac{\vec{m}}{r^3} - \frac{3(\vec{m} \cdot \vec{r})\vec{r}}{r^5}$$

and equation (4.31) reduces to

$$\vec{B} = \nabla \times \vec{A} = -\frac{\mu_0}{4\pi}\left[\frac{\vec{m}}{r^3} - \frac{3(\vec{m}\cdot\vec{r})\vec{r}}{r^5}\right] \tag{4.32}$$

Comparing equation (4.32) with equation (2.39) of Chapter 2 for the electric dipole field, we see that the formulae are structurally identical. As a result of this, $\vec{m}$ is often called the *magnetic dipole moment* even though there are no individual magnetic poles to correspond to the charges in the electric dipole. In particular, because we have shown that the forces on small coils are of the same type as the forces on electric dipoles discussed in Chapter 2, it follows that in an external field $\vec{B}$, a small coil of magnetic dipole moment $\vec{m}$ has potential energy U

$$U = -\vec{m}\cdot\vec{B} \tag{4.33}$$

and experiences a torque τ

$$\tau = -\vec{m}\times\vec{B} \tag{4.34}$$

Likewise, the mutual potential energy W between two small coils of magnetic dipole moment $\vec{m}_1$ and $\vec{m}_2$ separated by a distance $\vec{r}$ is simply the analogue of equation (2.43) of Chapter 2, namely

$$W = \frac{\mu_0}{4\pi r^3}\left[\vec{m}_1\cdot\vec{m}_2 - 3\frac{(\vec{m}_1\cdot\vec{r})(\vec{m}_2\cdot\vec{r}_2)}{r^2}\right] \tag{4.35}$$

We make two remarks. The first is that perhaps you might wonder why (4.32) is identical in form to its electric counterpart when there is such a difference between the basic differential equations: $\nabla\cdot\vec{B} = 0$, $\nabla\times\vec{B} = \mu_0\vec{j}$; $\nabla\cdot\vec{E} = \rho/\varepsilon_0$, $\nabla\times\vec{E} = 0$. The answer to this apparent anomaly lies in the far-field approximation used in both cases so that we actually have $\nabla\cdot\vec{B} = 0, \nabla\times\vec{B} = 0$ and $\nabla\cdot\vec{E} = 0, \nabla\times\vec{E} = 0$. The second remark is regarding equation (4.29), derived for a simple magnetic dipole. It also applies to more complicated current distributions. Without repeating the previous calculations, it is reasonably clear that the circulation of any localized charge distribution can be analyzed into a sum of idealized current loops. The far-field limit of such a more general current distribution is therefore a vector sum of contributions like (4.29). As a result, the **B** field has the same far-field form, characterized by a total magnetic dipole moment that can be given in a very general form as a volume integral over the localized current density:

$$\vec{m} = \frac{1}{2} \int \vec{r}\,' \times \vec{j}(\vec{r}\,')\, dV'$$

(4.29a)

where $\vec{j}(\vec{r}\,')$ is the current density at $\vec{r}\,'$.

4.8 Magnetic Dipole in an External Magnetic Field

We now consider a magnetic dipole (or a small current loop) placed in a magnetic field B (**Figure 4.12**), produced by some fixed external current distribution. This external field does not include the field of the dipole itself.

The external magnetic field B exerts a force dF on each differential line element dl:

$$\vec{dF} = dq\vec{v} \times \vec{B}$$

where dq is the charge within the line element that moves at some average velocity v. Now

$$I = dq/dt, \quad \vec{v} = dl/dt$$

From these we find

$$dq\,\vec{v} = I\,\vec{dl}$$

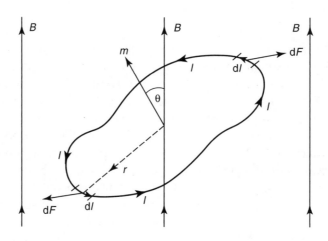

I FIGURE 4.12

and we can rewrite dF as

$$\vec{dF} = I d\vec{l} \times \vec{B} = I d\vec{r} \times \vec{B} \tag{4.36}$$

If **B** is uniform, then the total force acting on the dipole (i.e., the whole current loop) is zero:

$$\vec{F} = \oint \vec{dF} = -I\vec{B} \times \oint \vec{dl} = -I\vec{B} \times \oint \vec{dr} = 0$$

However, the total torque, Γ, about some point is not zero:

$$\vec{\Gamma} = \oint \vec{d\Gamma} = \oint \vec{r} \times \vec{dF} = I \oint \vec{r} \times (d\vec{r} \times \vec{B}) \tag{4.37}$$

Now the sum (or integral) around a closed path of the changes of any vector field is zero. Thus,

$$\oint d(\vec{r} \times (\vec{r} \times \vec{B})) = \oint d\vec{r} \times (\vec{r} \times \vec{B}) + \oint \vec{r} \times (d\vec{r} \times \vec{B}) = 0$$

It follows that

$$\oint \vec{r} \times (d\vec{r} \times \vec{B}) = -\oint d\vec{r} \times (\vec{r} \times \vec{B})$$

Using the vector identity

$$\vec{a} \times (\vec{b} \times \vec{c}) = \vec{b} \times (\vec{a} \times \vec{c}) + (\vec{a} \times \vec{b}) \times \vec{c}$$

we find

$$\oint \vec{r} \times (d\vec{r} \times \vec{B}) = \oint d\vec{r} \times (\vec{r} \times \vec{B}) + \oint (\vec{r} \times d\vec{r}) \times \vec{B}$$

$$= -\oint \vec{r} \times (d\vec{r} \times \vec{B}) + \oint (\vec{r} \times d\vec{r}) \times \vec{B}$$

or

$$2 \oint \vec{r} \times (d\vec{r} \times \vec{B}) = \oint (\vec{r} \times d\vec{r}) \times \vec{B}$$

Substituting this result into equation (4.37), we obtain

$$\vec{\Gamma} = I \oint \vec{r} \times (d\vec{r} \times \vec{B}) = \frac{1}{2} \oint (\vec{r} \times d\vec{r}) \times \vec{B} = I\vec{A} \times B = \vec{m} \times \vec{B} \tag{4.38}$$

where $\vec{A} = \frac{1}{2} \oint (\vec{r} \times d\vec{r})$.

Just as in the corresponding electric case, this torque acts to try to turn the magnetic dipole into alignment with the external B field, reducing the angle θ in Figure 4.11. The work done in changing the orientation of the dipole can be calculated directly from the torque. In turning the dipole moment of Figure 4.10 to increase the angle θ by $d\theta$, an external agency must do work dW against the torque, thus increasing the energy of orientation, U, of the dipole by

$$dU = dW = \Gamma \quad d\theta f = mB\sin\theta\, d\theta$$

Integration of this gives

$$U = -mB\cos\theta = -\vec{m} \cdot \vec{B} \tag{4.39}$$

The constant of integration has been chosen here to fix the zero of the orientational energy (4.39) according to the convention set previously for the electric dipole; U is zero when m is perpendicular to B, and it achieves its maximum value when m and B are in opposite alignment.

Some dipole moments may arise from rather large currents associated with small loops. If the physical size is small enough, it may be that any variation of B in (4.39) over the area of the current loop is for most purposes negligible. Whenever such an approximation applies, the dipole would be called a point dipole. The dipoles associated with currents in individual atoms or molecules are examples where a point-like idealization could be valid. For such cases it is possible to take the B field in (4.39) acting on some dipole of moment m_1 to be due to the proximity of a second magnetic dipole of moment m_2 a vector distance r away. A mutual energy of interaction of these two magnetic dipoles, U_{12}, can then be defined by combining (4.39) with (4.32). Thus,

$$\begin{aligned}
U_{12} &= -\vec{m}_1 \cdot \vec{B} \\
&= \frac{\mu_0}{4\pi r^3}\left[-3\,(\vec{m}_1 \cdot \hat{r})(\vec{m}_2 \cdot \hat{r}) + \vec{m}_1 \cdot \vec{m}_2 \right]
\end{aligned} \tag{4.40}$$

which is analogous to that for the corresponding electric dipole.

The total force acting on a magnetic dipole is only non-zero when it experiences a nonuniform external magnetic field. Integration of (4.36) is clearly the direct way of calculating such a force. However, for point-like dipoles, a simpler method is available, because, in this case, only the first-order changes in B across the dipole are important (changes proportional to the linear dimensions of the dipole). For point dipoles, the total force is related to the gradient of their potential energy in (4.39) in the standard way.

$$\vec{F} = -\nabla U = -\nabla(\vec{m} \cdot \vec{B})$$

$$= -(\vec{B} \cdot \nabla)\vec{m} + (\vec{m} \cdot \nabla)\vec{B} + \vec{B} \times (\nabla \times \vec{m}) + \vec{m} \times (\nabla \times \vec{B})$$

Because m is fixed and $\nabla \times B = 0$, the above equation reduces to

$$\vec{F} = (\vec{m} \cdot \nabla)\vec{B} \tag{4.41}$$

which again has an exact electric analogue.

4.9 The Law of Biot-Savart

In studying magnetostatics, many books start with Ampere's law of force between current elements, from which the Biot-Savart law is established. Then the laws of magnetostatics, $\nabla \cdot \vec{B} = 0$ and $\nabla \times \vec{B} = \mu_0 \vec{j}$, are derived. But, to save you from the burden of involved vector calculus, we have chosen a different approach. Ampere's law of force and the other laws are presented briefly below.

After performing and analyzing many experiments on the forces exerted by current-carrying circuits on one another, Ampere found the following law of force:

$$\vec{F}_2 = \frac{\mu_0}{4\pi} I_1 I_2 \oint_1 \oint_2 \frac{d\vec{l}_2 \times (d\vec{l}_1 \times \vec{r}_{12})}{r_{12}^3} \tag{4.42}$$

This is the force on circuit 2 (**Figure 4.13**) due to circuit 1; hence, the subscript 2 on $\vec{F}$; I_1 and I_2 are the currents flowing in the circuits, and dl_1 and dl_2 are respective elements of the circuits at a distance $\vec{r}_{12} = \vec{r}_2 - \vec{r}_1$ apart.

At first sight equation (4.42) appears not to conform to Newton's third law because of the asymmetry of the integrand. To demonstrate that $\vec{F}_2 = -\vec{F}_1$, as it must, we expand the integrand in equation (4.42) to give

$$\vec{F}_2 = \frac{\mu_0}{4\pi} I_1 I_2 \oint_1 \oint_2 \left[\frac{(d\vec{l}_2 \cdot \vec{r}_{12})\,d\vec{l}_1}{r_{12}^3} - \frac{(d\vec{l}_1 \cdot d\vec{l}_2)\,\vec{r}_{12}}{r_{12}^3} \right]$$

This expression can be further reduced. The first term in the integrand is a perfect differential with respect to $d\vec{l}_2$:

$$\oint_2 \frac{d\vec{l}_2 \cdot \vec{r}_{12}}{r_{12}^3} = \oint_2 \nabla_2\left(\frac{1}{r_{12}}\right) \cdot d\vec{l}_2 = 0$$

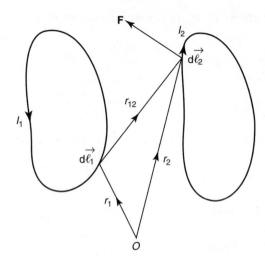

I FIGURE 4.13

and the expression for $\vec{F}_2$ reduces to

$$\vec{F}_2 = \frac{\mu_0}{4\pi} I_1 I_2 \oint_1 \oint_2 \frac{(\vec{dl}_1 \cdot \vec{dl}_2)\vec{r}_{12}}{r_{12}^3} = -\vec{F}_1 \tag{4.43}$$

Now let us return to equation (4.42) and cast it into a different form:

$$\vec{F}_2 = I_2 \oint_2 \vec{dl}_2 \times \vec{B}_2 \tag{4.44}$$

where

$$\vec{B}_2 = \frac{\mu_0}{4\pi} I_1 \oint_1 \frac{\vec{dl}_1 \times \vec{r}_{12}}{r_{12}^3} \tag{4.45}$$

is the magnetic field produced by I_1, the current flowing in circuit 1, at circuit 2. Equation (4.45) is known as the Biot-Savart law. The Biot-Savart law was originally expressed in differential form, i.e.,

$$\vec{dB}_2 = \frac{\mu_0}{4\pi} \frac{I_1 \vec{dl}_1 \times \vec{r}_{12}}{r_{12}^3}$$

Equation (4.45) should be properly called the generalization of the Biot-Savart law.

There is a similar integral in electrostatics that relates the electric field to a known charge distribution (see equation 2.6). It is usually more work to evaluate integrals. The following examples are exceptional simple cases.

Example 7: Find the Magnetic Field due to a Long Straight Current Wire.

Solution

After performing and analyzing various experiments, Ampere found the magnetic field of a steady current flowing in a long straight wire, and this is often called Ampere's law [equation (4.16)]. We now derive it from the Biot-Savart law. Let us compute $\vec{B}$ at a point P whose perpendicular distance from the axis of the current wire is R. Referring to **Figure 4.14**, in which θ is the angle between direction of the current and that of $\hat{r}$, we see that

$$\frac{\vec{dl} \times \hat{r}}{r^2} = \frac{dl\sin\theta}{r^2}$$

into the paper. Let l be the distance of the element $\vec{dl}$ from the point O, where OP is perpendicular to the wire. Then $R/l = \tan\theta$, or

$$l = R\cot\theta, \quad dl = -R\csc^2\theta d\theta, \quad r = R\csc\theta$$

We now take θ as the independent variable, its limit being 0 and π for a very long wire. Equation (4.45) becomes

$$B = -\frac{\mu_0}{4\pi}\frac{I}{R}\int_0^\pi \sin\theta d\theta$$

and

$$\vec{B} = \frac{\mu_0}{2\pi}\frac{I}{R}$$

into the paper at P, which is identical to equation (4.16).

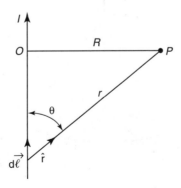

I FIGURE 4.14

Example 8: Axial Magnetic Field of a Circular Loop of Current Wire
Solution

We now consider a circular loop of wire carrying a current I. For simplicity we only compute field at a point (P) on the axis of symmetry; it is very difficult to compute the field at an arbitrary point. Equation (4.45) now gives

$$\vec{B} = \frac{\mu_0}{4\pi} \oint \frac{\vec{dl} \times \vec{r}}{r^3}$$

From **Figure 4.15** we see that

$$\vec{dl} = ad\theta \left(-\hat{i}\sin\theta + \hat{j}\cos\theta\right)$$

$$\vec{r} = -\vec{i}a\cos\theta - \vec{j}a\sin\theta + \hat{k}z, \quad r = |\vec{r}| = (a^2 + z^2)^{1/2}$$

then we have

$$\vec{B}(z) = \frac{\mu_0 I}{4\pi} \int_0^{2\pi} \frac{\left(\hat{i}za\cos\theta + \hat{j}za\sin\theta + \hat{k}a^2\right)}{(z^2 + a^2)^{3/2}}\, d\theta$$

The first two terms integrate to zero, and we finally obtain

$$\vec{B}(z) = \frac{\mu_0 I}{2} \frac{a^2}{(z^2 + a^2)^{3/2}}\, \hat{k}$$

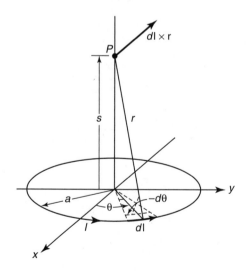

I FIGURE 4.15

Example 9: Axial Field of a Helmholtz Coil

Helmholtz coils are often used to produce a relatively uniform magnetic field over a small region of space. The configuration of a Helmholtz coil is depicted in **Figure 4.16.** Using the result of the previous example, we find the magnetic field at the midpoint P on the axis

$$B(z) = \frac{N\mu_0 I a^2}{2}\left[\frac{1}{(z^2+a^2)^{3/2}} + \frac{1}{\{(2b-z)^2+a^2\}^{3/2}}\right]$$

where N is number of turns of each coil.

Now let us take the first derivative of B_z with respect to z and we find

$$\frac{dB_z}{dz} = \frac{\mu_0 N I a^2}{2}\left[-\frac{3}{2}\frac{2z}{(z^2+a^2)^{5/2}} - \frac{3}{2}\frac{2(z-2b)}{\{(2b-z)^2+a^2\}^{5/2}}\right]$$

which vanishes at $z = b$. The second derivative with respect to z is

$$\frac{d^2 B_z}{dz^2} = -\frac{3\mu_0 N I a^2}{2}\left[\frac{1}{(z^2+a^2)^{5/2}} - \frac{5}{2}\frac{2z^2}{(z^2+a^2)^{7/2}}\right.$$

$$\left. + \frac{1}{\{(2b-z)^2+a^2\}^{5/2}} - \frac{5}{2}\frac{2(z-2b)^2}{\{(2b-z)^2+a^2\}^{7/2}}\right]$$

which reduces to, at $z = b$,

$$\frac{d^2 B_z}{dz^2} = -\frac{3\mu_0 N I a^2}{2}\left[\frac{b^2+a^2-5b^2+b^2+a^2-5b^2}{(b^2+a^2)^{7/2}}\right]$$

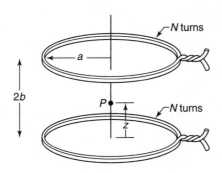

I FIGURE 4.16

This second derivative vanishes if $a^2 - 4b^2 = 0$. With $b = a/2$, the magnetic field at the midpoint P is

$$B_z = \frac{8}{\sqrt[3]{5}} \frac{\mu_o NI}{a}$$

What is the magnetic field at a point on the axis near the midpoint P? To this end, let us expand $B_z(z)$ in a Taylor series about the point $z = 1/2a$:

$$B_z(Z) = B_z(a/2) + \frac{1}{24}(z - a/2)^4 \frac{\partial^4 B_z}{\partial z^4}\bigg|_{z = a/2} + \dots$$

The first three derivatives vanish. After evaluating the fourth term explicitly, we obtain

$$B_z(z) = B_z(a/2)\left[1 - \frac{144}{125}\left(\frac{z - a/2}{a}\right)^4\right]$$

From this result we see that for the region where $(z - a/2) < a/10$, B_z deviates from $B_z(a/2)$ by less than one and a half parts in 10,000.

If we measure I in ampere, a in centimeters, and B in gauss, the magnetic field B at the midpoint of the Helmholtz coil is

$$B_z = \frac{32\pi N}{\sqrt[3]{5}} \frac{I}{10}$$

Example 10: Axial Magnetic Field of a Solenoid
Solution

A solenoid of N turns uniformly wound on a cylindrical form of radius a and length L, as shown in **Figure 4.17a**. We can find the magnetic field at P that is at a distance z_0 from the left end of the solenoid by dividing the length L into elements dz as shown in Figure 4.17b and applying the result of Example 7 to each element and then summing the result. As the element dz contains Ndz/L turns, we find that

$$B_z(z_0) = \frac{\mu_0 NIa^2}{2L} \int_0^L \frac{dz}{\left[(z_0 - z)^2 + a^2\right]^{3/2}}$$

Now let $z - z_0 = a\cos\alpha$, and we obtain

$$B_z(z_0) = \frac{\mu_0 NI}{2L} \int_{\alpha_2}^{\pi - \alpha_1} \sin\alpha \, d\alpha = \frac{\mu_0 NI}{2L}\left[-\cos(\pi - \alpha_1) + \cos(\alpha_2)\right]$$

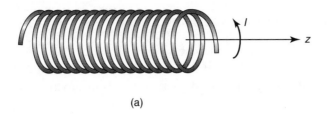

(a)

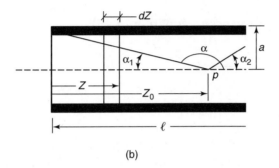

(b)

I FIGURE 4.17

where the angles α_1 and α_2 are less than $\pi/2$ (Figure 4.17b). $B_z(z_0)$ finally becomes

$$B_z(z_0) = \frac{\mu_0 NI}{L}\left[\frac{\cos\alpha_1 + \cos\alpha_2}{2}\right]$$

For a long solenoid (compared with its radius) and if z_0 is not too close to the ends of the solenoid, then both α_1 and α_2 are small angles and may be approximated by

$$\alpha_1 = a/z_0, \quad \alpha_2 = a/(L - z_0)$$

and $B_z(z_0)$ reduces to

$$B_z(z_0) = \frac{\mu_0 NI}{L}\left[1 - \frac{a^2}{4z_0^2} - \frac{a^2}{4(L - z_0)^2}\right]$$

where we have only kept quadratic terms in the expansion of $\cos\alpha_1$ and $\cos\alpha_2$.

4.10 The Laws of Magnetostatics from the Biot-Savart Law

We now revisit laws of magnetostatics, $\nabla \cdot \vec{B} = 0$ and $\nabla \times \vec{B} = \mu_0 \vec{j}$, from the Biot-Savart law. For the general case of a volume current, it is more convenient to introduce the current density $\vec{j}$; the Biot-Savart law now takes the form

$$\vec{B}(\vec{r}) = \frac{\mu_0}{4\pi} \int \frac{\vec{j}(\vec{r}') \times (\vec{r} - \vec{r}')}{|\vec{r} - \vec{r}'|^3} d\vec{r}' \tag{4.46}$$

We now apply the divergence to equation (4.46):

$$\nabla \cdot \vec{B}(\vec{r}) = \frac{\mu_0}{4\pi} \nabla \cdot \int \frac{\vec{j}(\vec{r}') \times (\vec{r} - \vec{r}')}{|\vec{r} - \vec{r}'|^3} d\vec{r}'$$

$$= -\frac{\mu_0}{4\pi} \nabla \cdot \int \vec{j}(\vec{r}') \times \nabla \left(\frac{1}{|\vec{r} - \vec{r}'|} \right) d\vec{r}' \tag{4.47}$$

Interchanging the order of the divergence operator and the integration and remembering that the former acts only on terms involving $\vec{r}$, we find that

$$\nabla \cdot \vec{B} = \frac{\mu_0}{4\pi} \int \vec{j}(\vec{r}') \cdot \nabla \times \nabla \left(\frac{1}{|\vec{r} - \vec{r}'|} \right) d\vec{r}'$$

where we have used the vector identity $\nabla \cdot (\vec{A} \times \vec{C}) = \vec{C} \cdot \nabla \times \vec{A} - \vec{A} \cdot \nabla \times \vec{C}$. But

$$\nabla \times \nabla \left(\frac{1}{|\vec{r} - \vec{r}'|} \right) = 0$$

so we finally obtain $\nabla \cdot \vec{B} = 0$. Similarly we can show that, starting with equation (4.46)

$$\nabla \times \vec{B} = \mu_0 \vec{j}$$

4.11 The Magnetic Scalar Potential

Because $\nabla \times \vec{B} = \mu_0 \vec{j}$, $\vec{B}$ is not a conservative field, and so it makes no sense to introduce a magnetic scalar potential in the same sense as we introduced an electrostatic potential. But $\nabla \times \vec{B} = 0$ is zero whenever the current density $\vec{j}$ is zero. When this is the case, the magnetic field $\vec{B}$ in such region, say τ, can be written as the gradient of a scalar potential V_m:

$$\vec{B} = -\mu_0 \nabla V_m \tag{4.48}$$

Substituting this into $\nabla \cdot \vec{B} = 0$, we obtain Laplace equation for V_m:

$$\nabla^2 V_m = 0 \tag{4.49}$$

Thus, if we have a current-carrying conductor surrounded by empty space, then outside of the conductor $\nabla \times \vec{B} = 0$ and we have $\vec{B} = -\mu_0 \nabla V_m$.

There is an explicit expression for V_m in terms of the current distribution outside τ. To find this expression, let us consider a single current loop shown in **Figure 4.18**, which produces a magnetic field $\vec{B}$ and subtended a solid angle Ω at the point P. If we displace P by a small distance $\vec{d\lambda}$, the solid angle changes by an amount $d\Omega$. The same change $d\Omega$ results if, instead of moving P, we displace the current loop by $-\vec{d\lambda}$. In the latter event the change in the solid angle is the sum of all the elementary changes brought about as each element $\vec{ds}$ of the current circuit is displaced by $-\vec{d\lambda}$ and thus sweeps out a small parallelogram of surface $-\vec{d\lambda} \times \vec{ds}$. If this surface lies with its plane normal to the vector $\vec{r}$ (from $\vec{ds}$ to P), then it subtends at P a solid angle equal to its area, $|-\vec{d\lambda} \times \vec{ds}|$, divided by r^2. In general, the solid angle is $(-\vec{d\lambda} \times \vec{ds}) \cdot \hat{r}_1 r^2$, where $\hat{r}_1$ is a unit vector in the direction of $\vec{r}$, and this may be written as $-\vec{d\lambda} \cdot (\vec{ds} \times \hat{r}_1)/r^2$, because the dot and cross are interchangeable in a triple

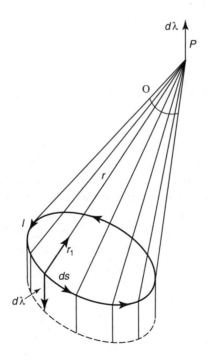

I FIGURE 4.18

vector product. Integrating this expression around the current loop, that is, with respect to $\vec{s}$, we have for the change in Ω

$$d\Omega = -\vec{d\lambda} \cdot \oint_{\vec{s}} \frac{\vec{ds} \times \hat{r}_1}{r^2} \tag{4.50}$$

Comparing equation (4.50) with equation (4.45) and Figure 4.18 with Figure 4.13, we see that $\vec{ds}$ here is $\vec{dl_1}$ there and r is r_{12}, and the integral in equation (4.50) is just $\vec{B}(4\pi/\mu_0 I)$. Equation (4.50) now becomes

$$d\Omega = -\vec{d\lambda} \cdot \vec{B}\,(4\pi/\mu_0 I)$$

or

$$\vec{B} \cdot \vec{d\lambda} = -\frac{\mu_0 I}{4\pi}\,d\Omega \tag{4.51}$$

The minus sign means that we take Ω to be positive on that side of the current loop for which the magnetic field $\vec{B}$ is directed away from the loop.

Now let $\vec{d\lambda}$ have the components dx, dy, and dz. Then

$$d\Omega = \frac{\partial\Omega}{\partial x}\,dx + \frac{\partial\Omega}{\partial y}\,dy + \frac{\partial\Omega}{\partial z}\,dz = \nabla\Omega \cdot \vec{d\lambda}$$

Substituting this in equation (4.51), we obtain

$$\vec{B} \cdot \vec{d\lambda} = -\frac{\mu_0 I}{4\pi}\,\nabla\Omega \cdot \vec{d\lambda}$$

or $\vec{B} = -\dfrac{\mu_0 I}{4\pi}\,\nabla\Omega$. Comparing this with $\vec{B} = -\mu_0 \nabla V_m$, we get

$$V_m = \frac{I}{4\pi}\,\Omega \tag{4.52}$$

In charge-free space outside a current-carrying conductor we can compute V_m from equation (4.52) and then $\vec{B}$ from equation (4.48). We emphasize to the reader again that the magnetic field $\vec{B}$ is not a conservative one. Consequently, the line integral of $\vec{B}$ between any two points depends on the path followed between the two points. Thus, it makes no sense to try to assign absolute values of V_m to various points in the field or to associate V_m with any concept of potential energy.

As an application of the magnetic scalar potential, let us consider the following example.

Example 11: Find the Magnetic Field at a Point P at a Distance from a Small Current Loop.

Solution

As shown in **Figure 4.19**, a current I flows in a small plane circuit of area A. The radius vector $\vec{r}$ of point $P(x, y, z)$ makes an angle θ with the normal to the loop, and r is assumed to be large compared with the dimension of the loop. V_m at P is

$$V_m = \frac{1}{4\pi} I\Omega = \frac{1}{4\pi}\left(\frac{IA\cos\theta}{r^2}\right)$$

Then from $\vec{B} = -\mu_0 \nabla V_m$, we obtain

$$B_r = -\frac{\partial V_m}{\partial r} = \frac{2\mu_0 IA}{4\pi}\frac{\cos\theta}{r^3}$$

$$B_\theta = -\frac{1}{r}\frac{\partial V_m}{\partial \theta} = \frac{\mu_0 IA}{4\pi}\frac{\sin\theta}{r^3}$$

Comparing these expressions with those for the electric field of an electric dipole of Chapter 2, we see that, as regards dependence on r and θ, the magnetic field due to a small current loop is similar to the electric field of an electric dipole.

This example also demonstrates that a small current loop is equivalent to a magnetic dipole. We know magnetic monopole does not exist. Historically, the concept of magnetic poles was used for a long time in the approach to the magnetostatic problems. Let us review this old concept

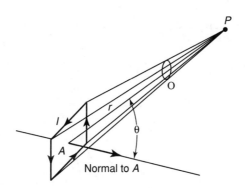

I FIGURE 4.19

briefly, as it helps us to explain the equivalence of a small current loop and a magnetic dipole. There were two magnetic poles, positive and negative; the force between the two poles, m and m', obeyed a Coulomb's law

$$F = \frac{m'm}{4\pi\mu_0 r^2}$$

The force F on a pole m placed in a magnetic field B is given by $F = mB$. Combining this with Coulomb's law (setting $m' = 1$), the magnetic field B at a distance r from m is

$$B = \frac{m}{4\pi r^2}$$

The potential V at a distance r from a point pole m is

$$V = \frac{m}{4\pi r}$$

which is obtained by computing the work done by the field of m on a unit positive pole as it moved from the given point to infinity.

Now let us consider a magnetic dipole, which consists of two equal and opposite point poles a short distance apart, as shown in **Figure 4.20**. The magnetic potential at point P at a distance r from the center of the dipole large compared with l is

$$V = \frac{-m}{4\pi r_2} + \frac{m}{4\pi r_1}$$
$$= \frac{m}{4\pi}\left(\frac{1}{r-(l/2)\cos\theta} - \frac{1}{r+(l/2)\cos\theta}\right)$$
$$= \frac{1}{4\pi}\left(\frac{ml\cos\theta}{r^2-(l^2/4)\cos^2\theta}\right) \cong \frac{p\cos\theta}{4\pi r^2}.$$

where $p = ml$, the magnetic moment.

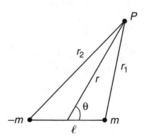

I FIGURE 4.20

Comparing V with V_m of Example 11 for a small current loop, we see they are identical if we define IA as the magnetic dipole moment of the loop. The magnetic phenomena of small current loops at points distant from the small loops are indistinguishable from the magnetic phenomena of magnetic dipoles. A large current may be pictured as being made up of a great number of small elementary loops. If the same current circulates in each small loop in the same direction, the currents cancel each other on all interior boundaries and so are equivalent to the current in the large circuit. Thus, the magnetic field of the large circuit must equal that of all the small loops. Because the field of each small loop is the same as that of a magnetic dipole at its center, with its axis normal to the surface, the field of the large circuit is the same as that of uniform surface layer of magnetic dipoles, the surface being bounded by the circuit. Such a layer is actually a double layer, a layer of positive poles adjacent to a parallel layer of negative poles.

4.12 A Comparison Between Electrostatics and Magnetostatics

The reader may have already observed some similarities and differences between electrostatics and magnetostatics. It would be instructive to carry the comparison between these two static fields further.

1. Both $\vec{E}$ and $\vec{B}$ are defined in terms of forces: $\vec{E}$ is defined via $\vec{F} = q\vec{E}$ on a stationary charge and is consequently a polar vector. The cross-product of two polar vectors is an axial vector or pseudovector. The cross-product of an axial vector and a polar vector is a polar vector. Now $\vec{B}$ is defined via Lorentz force $\vec{F} = q\vec{v} \times \vec{B}$ on a moving charge and is an axial vector; therefore, it cannot possess the same mathematical properties as $\vec{E}$.

    The difference between an axial vector and a polar vector is contained in their transformation properties under a reflection and may be demonstrated as follows. Consider a polar vector whose Cartesian components are (x, y, z) in a specified coordinate system. If the coordinate axes are then reflected in the "mirror" formed by the x-y plane, the Cartesian components of the vector in the new primed coordinate system are given by the transformation relations: $x' = x$, $y' = y$, $z' = -z$ (**Figure 4.21**). If the polar vector was parallel to the z-axis in the original unprimed system, the polar vector will be pointing in the negative z-direction in the reflected system. On the other hand, an axial vector defined by the cross-product between a polar vector parallel to the x-axis and a polar vector parallel to the y-axis still point in the positive z-direction in the reflected system.

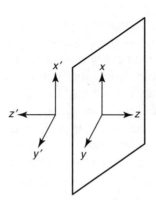

I FIGURE 4.21

2. Gauss's laws appear different for these two static fields. Their differential forms are

$$\nabla \cdot \vec{E} = \rho/\varepsilon_0, \quad \nabla \cdot \vec{B} = 0$$

The $\vec{B}$ field is divergence-less at all points and it is solenoid. Its field lines form closed loops. This is completely different from the structure of the electrostatic field, which by $\nabla \times \vec{E} = 0$ is cur-free or irrotational. Solenoid $\vec{E}$ fields are easy to generate. (We shall see in Chapter 5 the induced $\vec{E}$ field is divergence-less.) For an irrotational $\vec{B}$ field, we would need to have a magnetic charge, known as magnetic monopole (magnetic equivalent to electric charge). So far, no magnetic charge has been found in nature or in laboratory. If a magnetic charge did exist, it would have an exceedingly peculiar property. Notice that the left-hand side of Gauss' magnetic law changes sign when a transformation is carried out from a right-handed to a left-handed set of coordinate axes. This means that the magnetic charge would also have to change sign under such a transformation. A quantity that has all the properties of a scalar except that it has this peculiarity under reflections is called a pseudoscalar. If magnetic charge did exist, it would be the only quantity in classical physics that possessed this property.

Modern particle theory seems to demand the existence of magnetic monopoles, and efforts have been made to detect them, but the searches to date have been fruitless and so the equation $\nabla \cdot \vec{B} = 0$ must be regarded as the fundamental property of all magnetic fields.

3. The electric and magnetic dipole moments are related in a similar way as $\vec{E}$ and $\vec{B}$. The dipole moment

$$\vec{p} = \int \rho \vec{r} d\tau$$

is a polar vector, whereas the magnetic dipole moment

$$\vec{m} = \frac{1}{2} \int \vec{r} \times \vec{j} d\tau$$

is an axial vector.

4. In addition to the behavior of physical quantities under rotation of the coordinate axes and the reflection of the coordinate axes in a plane is their behavior under time reversal or time inversion. This time reversal operation consists of replacing t in all the equations by $-t$. Under time reversal operation, position vector $(\vec{r})$, acceleration $(\vec{a} = d^2\vec{r}/dt^2)$, and force $(\vec{F} = m\vec{a})$ all remain unchanged. But velocity $(\vec{v} = d\vec{r}/dt)$ is reversed. Because $\vec{E}$ is defined as the force per unit charge, it is invariant with respect to the time reversal operation. But $\vec{B}$ changes sign under time reversal. In the above discussion, it is assumed that the initial mass m and the charge q are invariant under time reversal operation, or at least their ratio is invariant.

Although invariance under space rotation, reflection, and time reversal are of little more than academic interest in classical physics, transformation properties of such physical quantities play a very important role in the construction of modern physical theories.

4.13 Magnetic Dipole Moments at the Atomic Level

An electron of mass m and charge $-e$ moving with a velocity v in a circular orbit of radius r_n around the nucleus traces out a current loop. Associated with such a loop is a magnetic field that, at large distances from the loop, is effectively the same as that due to a magnetic dipole $\vec{\mu}$ at the center of the loop and at right angles to its plane.

The magnitude of the orbital magnetic dipole moment for a current i in a loop of area A is

$$\mu = |\vec{\mu}| = iA$$

Now the current in the loop is $(-e)$ times the number of revolutions per unit time made by the electron

$$i = (-e)v/(2\pi r_n)$$

Therefore,

$$\vec{\mu} = \frac{-ev}{2\pi r_n} \pi r_n^2 = -\frac{e}{2m}(mvr_n) = -\gamma\vec{L}$$

(in vector notation $\vec{\mu} = -\gamma\vec{L}$)   (4.53)

where $L = mvr_n$ is the orbital angular moment of the electron and $\gamma = -e/2m$ is known as the orbital gyromagnetic ratio. The negative sign indicates that μ and L are in opposite directions for a negative circulating charge, the electron.

Equation (4.53) is commonly written as

$$\vec{\mu} = -\frac{e\hbar}{2m}\frac{\vec{L}}{\hbar}$$

(4.54)

Note that $(\vec{L}/\hbar)$ is dimensionless. And the quantity

$$\mu_B = e\hbar/2m_e \approx 9.27 \times 10^{-24} \, J/T$$

is called the Bohr magneton. We learn from quantum mechanics that any component of $(\vec{L}/\hbar)$, $L_x/\hbar$ say, takes on only integral value $0, \pm 1, \pm 2, \ldots$ so that the magnetic moment of an atom is conveniently measured in units of the Bohr magneton.

It is found experimentally that, in addition to any magnetic moment due to its orbital motion, an electron has a permanent *intrinsic magnetic moment*. A spinning electron should also generate a magnetic field and possess a magnetic moment. But the electron appears to behave like a point charge, and it is not possible to calculate a spin magnetic dipole moment in the same way as for the orbital motion. In analogy with equation (4.54) for orbital angular momentum, an electron's intrinsic magnetic dipole moment is given by

$$\mu_s = -g_s\frac{e\hbar}{2m}\frac{S}{\hbar}$$

(4.55)

where g_s is known as the spin g-factor and S is electron's intrinsic spin. To obtain agreement with the experiment, it is necessary to assign to g_s a value of 2. (Careful measurements of the Zeeman effect show that $g_s = 2.0023$.) The intrinsic magnetic moment of the electron is largely responsible for the phenomenon of ferromagnetism, to be discussed in Chapter 9. Protons and neutrons also possess permanent magnetic moments, but these are smaller in magnitude by a factor m_e/m_P.

Problems

1. A particle of charge q moves with velocity $\vec{v}$ relative to an observer. Find the electric and magnetic fields produced by the charge q at the point of the observer.

2. A positive charge q of mass m is accelerated into a magnetic field $\vec{B}$ with a velocity $\vec{v}$. If q enters the field of $\vec{B}$ normally, show that its path is a circle with radius R given by mv/qB.

3. Show that the total energy, kinetic plus potential, of a charge q moving in a magnetic field $\vec{B}$ is conserved.

4. Using the Biot-Savart law, find the magnetic field $\vec{B}$ due to a circular coil carrying a current I at a point on the axis of the coil.

5. A circuit carrying a current I in the form of a square of side l. Find the magnetic field at its center.

6. A circuit has the form of a regular polygon of n sides inscribed in a circle of radius a. Find the magnetic field at the center, and consider the case when n is indefinitely increased.

7. Two long parallel wires a distance $2d$ apart carry equal but oppositely directed current. Find the field at a point in the plane of the wires distant R from a line halfway between the two wires.

8. A current flowing along a surface is known as a current sheet. Find the magnetic field due to such a current sheet of uniform density k.

9. Two parallel plane current sheets of infinite extent, oppositely directed, have currents per unit width equal to k. Find the magnetic field (a) between the sheets and (b) outside.

10. Find the vector potential and the magnetic field due to an infinite wire carrying a current, at points (a) outside and (b) inside the wire.

11. Find the vector potential for a straight section of a current circuit.

12. Find a suitable vector for the field of an infinite straight wire carrying current I.

13. Show that $B_0 x \hat{j}$, $-B_0 y \hat{i}$, and $1/2 B_0 \times \vec{r}$ are all suitable vector potentials for a uniform field $B_0 \hat{k}$.

14. In the region $0 < r < 0.5$ m, in cylindrical coordinates, the current density is

$$\vec{j} = 4.5e^{-2r} \hat{z} \text{A/m}^2$$

and $\vec{j} = 0$ elsewhere. Use Ampere's law to find $\vec{B}$.

15. A long straight wire carries current I. An electron is shot parallel to the wire in the direction of the current. In what direction will it be deflected?

16. Show that, starting with equation (4.45), $\nabla \times \vec{B} = \mu_0 \hat{\jmath}$.

17. Find the force between two parallel wires, carrying currents I_a and I_b and separated by a distance ρ.

5 Time-Dependent Magnetic Fields and Faraday's Law of Induction

In Chapter 4 we discussed the magnetostatic fields associated with the flow of steady currents. We now consider magnetic fields that change with time. We learned in Chapter 4 that by causing a current to flow, an electric field is able to produce a magnetic field. Is there a converse effect? That is, is there any way in which a magnetic field can produce an electric field? This problem was tackled by a number of physicists without success. Finally, in 1831, the great British self-educated experimental physicist Michael Faraday and the American high school teacher Joseph Henry independently discovered that a changing magnetic field could induce a current in a conductor. This phenomenon is known as *electromagnetic induction*, a name given by Faraday. The discovery of Faraday and Henry was a technological turning point, and the world was never again the same. Two familiar applications of electromagnetic induction are the electric generator, which is the source of most electric power, and the transformer, which enables the electromotive force (*emf*) of an alternating current to be increased or decreased easily.

What Faraday and Henry discovered was that a current was induced to flow around a closed conducting circuit when a nearby magnet was moved, when the current in a nearby circuit was changed, or when the circuit was moved in a fixed magnetic field. In all these cases they established that the induced current was proportional to the rate of change of magnetic flux through the circuit. Let us explain the magnetic flux notion first.

5.1 Magnetic Flux

The concept of magnetic flux is mainly used for surfaces that are closed. Consider a wire loop moving in a magnetic field. If the area of the wire loop is A and it is perpendicular to $\vec{B}$, the total magnetic flux Φ through the loop is defined as $\Phi = BA$. The loop can have any shape; its area must be taken from its projection on a plane perpendicular to $\vec{B}$. Thus, in general the magnetic flux Φ is defined as

$$\Phi = \vec{B} \cdot \vec{A} \tag{5.1}$$

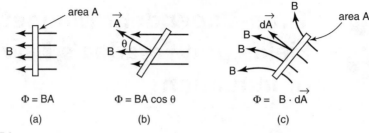

I FIGURE 5.1

Figure 5.1 illustrates magnetic flux for (a) coil normal to **B**; (b) coil at angle θ to **B**; and (c) nonuniform **B**.

5.2 Electromotive Force

The electromotive force, *emf*, in short, is not a force; it is the work per unit charge done by force and has the dimensions energy per charge. For example, for a steady current to flow in a conductor, a potential difference V must be maintained at its two ends by a source. If a battery is used to maintain a constant potential difference V between its terminals, its *emf* is equal to V. Although the term *emf* is not quite appropriate, the term is firmly entrenched into common usage, and changing it may cause more confusion. Thus, we stick with this term *emf* and denote it by the symbol $\mathscr{E}$.

The *emf* of a battery can also be expressed in terms of the electric field $\vec{E}$ established in the conductor:

$$\mathscr{E} = V = \int_a^b \vec{E} \cdot d\vec{l} \tag{5.2}$$

In terms of the work done on the charge in moving it from one terminal to the other, we can rewrite equation (5.2)

$$\mathscr{E} = V = \frac{1}{q}\int_a^b q\vec{E} \cdot d\vec{l} \tag{5.3}$$

5.3 Motional Electromotive Force and Lenz's Law

Now consider a conductor of length L moving with velocity $\vec{v}$ perpendicular to a uniform magnetic field $\vec{B}$ (**Figure 5.2**). The magnetic field exerts a force on charge (electron) q in the conductor that is given by

$$\vec{F} = q\vec{v} \times \vec{B}$$

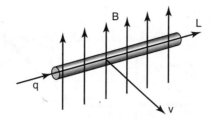

I FIGURE 5.2

The free charges move toward the end of the conductor and establish an electric field $\vec{E}$ given by

$$\vec{E} = \vec{v} \times \vec{B}$$

The potential difference between the ends of the conductor is

$$V_{ba} = \int_a^b \vec{E} \cdot d\vec{l} = \int_a^b (\vec{v} \times \vec{B}) \cdot d\vec{l}$$

and in analogue to equation (5.3) we can define the induced *emf* between the ends of the conductor to be

$$\mathscr{E} = \frac{1}{q} \int_a^b q\vec{E} \cdot d\vec{l} = \frac{1}{q} \int_a^b q\,(\vec{v} \times \vec{B}) \cdot d\vec{l} \tag{5.4}$$

For the mutually perpendicular configuration shown in Figure 5.2, we obtain

$$\mathscr{E} = vBL \tag{5.5}$$

where L is the length of the conductor. This *emf* is often called a motional *emf*, because it depends on the velocity of the conductor moving through the magnetic field. This potential difference (i.e., the motional *emf*) produces no current in the conductor. However, if the conductor forms a part of a circuit, as shown in **Figure 5.3**, a current will flow. Now we can also discuss this process in terms of the magnetic flux. In the time Δt for the charge to traverse the conductor, the conductor will move forward a distance $v\Delta t$. Thus, it sweeps an area $vL\Delta t$ in time Δt. The change in magnetic flux associated with this motion is

$$\Delta\Phi = BvL\Delta t$$

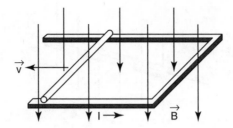

I FIGURE 5.3

and the rate of change of magnetic flux through the circuit is BvL. We can therefore write

$$|\mathscr{E}| = vBL = d\Phi/dt \qquad (5.6)$$

We have not specified the direction of the *emf*. This information is provided by Lenz's law, which states that

> The induced emf acts in such a way as to oppose the change in flux. That is, it acts to generate currents in the circuit whose associated magnetic effect would counteract the external change.

We may combine equation (5.6) and Lenz's law and state that the *emf* associated with a change in magnetic flux through a circuit is

$$|\mathscr{E}| = -\frac{d\Phi}{dt} = -\frac{d}{dt}\int_A \vec{B} \cdot d\vec{A} \qquad (5.7)$$

This is Faraday's law of electromagnetic induction.

Lenz's law is merely a particular case of a very general physical principle known as Le Chatelier's principle that states that a physical system always reacts to oppose any change that is imposed from outside. Now the change, in our case, is an alteration in flux Φ, and the reaction is an induced *emf* that opposes the change of Φ. The induced *emf* is often called the back *emf*.

Figure 5.4 illustrates how Lenz's law works. The flux through circuit C in Figure 5.4a is decreasing. By Lenz's law we know that the induced *emf* acts in this case to produce in the circuit a current whose associated magnetic field is so directed to increase the flux through C. Thus, the induced current in Figure 5.4a flows in an anticlockwise sense and the *emf* > 0. In Figure 5.4b, on the other hand, the flux through C is increasing so that the induced current now flows in a clockwise direction and *emf* < 0.

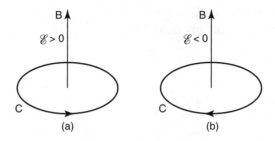

I FIGURE 5.4

Faraday's law of induction is an experimental law; it includes all possible changes of magnetic flux, for example, due to motion of the circuit. This is demonstrated in the following section.

5.4 Faraday's Law of Induction for a Moving Circuit

In the preceding section we considered a moving circuit with a particular configuration. Now consider a circuit of any arbitrary shape moving in a time-independent magnetic field $\vec{B}$ and an element PQ represented by $\vec{dl}$ moving with a velocity $\vec{v}$ that does not need to be uniform (**Figure 5.5**). In time dt it moves a distance $\vec{v}\,dt$ to a position $P'Q'$. If the velocity of the conducting electron relative to the wire is $\vec{u}$, its velocity relative to the field $\vec{B}$ is $\vec{u} + \vec{v}$. Thus, the magnetic field exerts on each electron a force $q(\vec{v} + \vec{u}) \times \vec{B}$. The component of this force along PQ is $[q(\vec{v} + \vec{u}) \times \vec{B}] \cdot \vec{dl}$. As $\vec{u}$ is parallel to $\vec{dl}$, $\vec{u} \times \vec{B} \cdot \vec{dl} = 0$, and so the component force along PQ is $[q\vec{v} + \vec{B}] \cdot \vec{dl}$. This shows that there is an electric field, $\vec{E} = \vec{v} \times \vec{B}$ induced in the wire, and

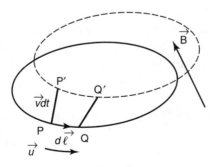

I FIGURE 5.5

its component along the wire is $\vec{v} \times \vec{B} \cdot \vec{dl}$. The corresponding induced *emf* is simply the line integral of this field round the circuit:

$$\mathcal{E} = \oint (\vec{v} \times \vec{B}) \cdot \vec{dl} \tag{5.8}$$

We now compute the change in the magnetic flux through the circuit as it moves in the magnetic field during the time interval dt. As shown in Figure 5.5, as the circuit moves, its surface sweeps a band in time interval dt of which $PP'QQ'$ is an element. The element $PP'QQ'$ has a vector area $\vec{v}\, dt \times \vec{dl}$, and the flux across this element is $\vec{v}\, dt \times \vec{dl} \cdot \vec{B}$. Thus, flux over the entire band is

$$d\Phi = \oint \vec{B} \cdot (\vec{v}\, dt \times \vec{dl})$$

from which we obtain

$$\frac{d\Phi}{dt} = \oint \vec{B} \cdot (\vec{v} \times \vec{dl}) = -\oint (\vec{v} \times \vec{B}) \cdot \vec{dl} \tag{5.9}$$

Comparing equation (5.8) and equation (5.9), we obtain Faraday's law of induction

$$\mathcal{E} = -\frac{d\Phi}{dt}$$

just as in the special case considered in the preceding section. The present result is very general: The circuit can have any shape and moves arbitrarily. Thus, velocity does not need to be uniform and the circuit might deform as it moves. Furthermore, the effects on a circuit are due only to *relative motion* between the circuit and the sources of the field. Changes of field at the circuit due to motion of the sources cannot be distinguished from changes due to any other cause such as changes in the strength of the sources.

Example 1

A circular loop of radius a is rotated at angular speed ω about one of its diameters in a uniform magnetic field perpendicular to the axis. Calculate the *emf* at the instant when the plane of the coil is at angle α relative to the field.

Solution

For the segment $ad\theta$ at angle θ (measured from the rotation axis, **Figure 5.6**), we have

$$\vec{v} = \omega\rho\hat{e}_\phi = \omega a \sin\theta\, \hat{e}_\phi, \qquad \vec{v} \times \vec{B} = \omega a B \sin\theta \cos\alpha\, (-\hat{e}_k)$$

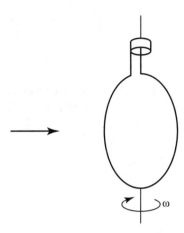

I FIGURE 5.6

Then by equation (5.8)

$$dE = -\omega aB \sin\theta \cos\alpha \hat{e}_k \cdot ad\theta \hat{e}_\theta = \omega a^2 B \cos\alpha \sin^2\theta d\theta$$

and

$$\mathcal{E} = \oint d\varepsilon = \omega a^2 B \cos\alpha \int_0^{2\pi} \sin^2\theta d\theta = \pi\omega a^2 B \cos\alpha$$

5.5 Integral and Differential Forms of Faraday's Law of Induction

Faraday and Henry discovered that the same effect resulted from varying the magnetic field while keeping the circuit still. But in a circuit at rest there is no Lorentz force, and we must conclude that when the magnetic flux through a circuit is changing, an electric field $\vec{E}$ is induced, giving a force $\vec{F} = q\vec{E}$ on a free charge. Because the induced current is the same, the *emf* must be the same, so that

$$\mathcal{E} = \oint (\vec{F}/q) \cdot d\vec{l} = \oint \vec{E} \cdot d\vec{l} = -\frac{d\Phi}{dt} \tag{5.10}$$

But

$$\frac{d\Phi}{dt} = \frac{d}{dt} \int_A \vec{B} \cdot d\vec{A}$$

Substituting this back into equation (5.10) we obtain

$$\oint_C \vec{E} \cdot d\vec{l} = -\frac{d}{dt} \int_A \vec{B} \cdot d\vec{A} \tag{5.11}$$

This is the integral form of Faraday's law of induction. The circuit "C" can be thought of as any closed geometric path in space, not necessarily coincident with an electric circuit. If a circuit is actually present, the induced current will flow.

Now our circuit is held fixed, and so

$$\frac{d}{dt} \int_A \vec{B} \cdot d\vec{A} = \int_A \frac{\partial \vec{B}}{\partial t} \cdot d\vec{A}$$

and equation (5.11) becomes

$$\oint_C \vec{E} \cdot d\vec{l} = -\int_A \frac{\partial \vec{B}}{\partial t} \cdot d\vec{A} \tag{5.12}$$

We can convert equation (5.12) into differential form. Using Stokes' theorem, we can transform the line integral of the electric field into a surface integral

$$\int_A \nabla \times \vec{E} \cdot d\vec{A} = -\int_A \frac{\partial \vec{B}}{\partial t} \cdot d\vec{A} \tag{5.12a}$$

or

$$\int_A \left(\nabla \times \vec{E} + \frac{\partial \vec{B}}{\partial t} \right) \cdot d\vec{A} = 0$$

Because this equation holds for any arbitrary surface, it follows that

$$\nabla \times \vec{E} + \frac{\partial \vec{B}}{\partial t} = 0 \tag{5.13}$$

or

$$\nabla \times \vec{E} = -\frac{\partial \vec{B}}{\partial t} \tag{5.13a}$$

This is the differential form of Faraday's law of electromagnetic induction. It replaces equation (2.9) of electrostatics. We see that $\vec{E}$ cannot in general be expressed as the gradient of a scalar potential.

Equation (5.13a) shows that the time derivative of B acts as the source of E in exactly the same way that a static current density j acts as the source of a magnetic field [$\nabla \times \vec{B} = \mu_0 j$]. Similarly, the integrated form of the law (equation 5.12a)] may be written as the analogue of Ampere's circuital law. Furthermore, both the types of sources (right sides of the equations) have the same basic structure: Both steady-state current and magnetic flux form closed loops. In other words, their densities are divergence-less. Thus, all the mathematical machinery for the calculation of magnetic field from current may be taken over for the calculation of induced electric field from the rate of change of flux. The only difference is that the constant of proportionality is -1 instead of μ_0.

5.6 Applications of Faraday's Law of Induction

a. The Betatron

A very illustrating application of Faraday's law is found in the betatron. This device, developed by D. W. Kerst in 1941, is used to accelerate electrons to high energies (so great that the mass m of each electron is many times its rest mass m_0). The instrument consists of a large electromagnet (**Figure 5.7a**) between the pole pieces of which is placed an evacuated toroidal chamber DD known as a *doughnut*, a top view of which is shown in Figure 5.7b. Electrons are injected at A into the evacuated region within the toroid. They are constrained to move in a circular path by the application of a magnetic field normal to the plan of the diagram. The magnetic field is time dependent and is produced by an alternating current in the windings of the electromagnet. The electrons are essentially at rest when the magnetic field is zero. As the B-field increases in magnitude with time, the changing magnetic *flux* linking with the path of the electron gives rise to an electric field E that is tangential to the electron orbit. Because of the symmetry, the lines of E are circular, with their common center on the axis of the electromagnet, which passes through O.

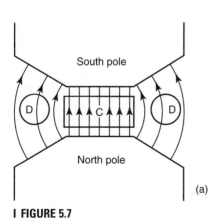

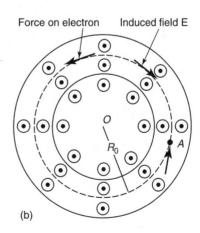

I FIGURE 5.7

During the cycle when B is directed out of the plane and $\partial B/\partial t > 0$, the sense of the induced E field is clockwise, and the tangential force that accelerates the negatively charged electron has the opposite sense. Let us see under what condition the electron can move in an orbit of constant radius R_0 inside the doughnut.

Let B be the varying magnetic field at any point in the median plane, and it is a function of the distance r from the axis. Then the time-varying flux through the area bounded by a circular orbit of radius R_0 is

$$\Phi_0 = \int_0^{R_0} B\,dA = \int_0^{R_0} B \cdot 2\pi r\,dr \tag{5.14}$$

Faraday's law gives the induced electric field E_0 tangential to the orbit of radius R_0:

$$\int_0^{2\pi R_0} \vec{E} \cdot d\vec{l} = 2\pi R_0 E_0 = -d\Phi_0/dt \tag{5.15}$$

Because the tangential force exerted on an electron by this field is qE_0, the equation governing the motion tangent to the circular path is

$$\frac{d}{dt}(mv) = -\frac{q}{2\pi R_0}\frac{d\Phi_0}{dt} \tag{5.16}$$

where v is the speed and m the variable mass of the electron. If $d\Phi_0/dt > 0$, the minus sign indicates that the tangential force is clockwise on a positive charge. Because $q = -e$, the sense of the force on an electron is counterclockwise, as shown in Figure 5.7b.

The magnetic field provides the deflecting force. Hence, the equation of motion that governs the transverse motion is

$$m\frac{v^2}{R_0} = qvB_0$$

or

$$mv = qR_0B_0 \tag{5.17}$$

Now the integral of equation (5.16), subject to the initial condition that $v = 0$ when $\Phi_0 = 0$, is

$$mv = -\frac{q\Phi_0}{2\pi R_0}$$

Combining this with equation (5.17), we see that the electron can move in a stable circular orbit of radius R_0, provided B_0, the absolute value of B at the orbit, is given by

$$B_0 = \frac{1}{2}\frac{\Phi_0}{\pi R_0^2} \tag{5.18}$$

The equilibrium orbit is that circle whose radius R_0 satisfies this equation. Evidently, the magnetic field must be stronger at the axis of the magnet than at the distance, as indicated by the lines of force in Figure 5.7. This is accomplished in part by a proper shaping of the pole pieces and in part by the insertion of the soft iron cylinder C in the center of the gap. The nonuniform character of the field is actually an advantage in holding the beam to the proper path. First, the deflecting force is in such a direction as to bring back to the median plane any electron that may have strayed from this plane. Second, the required weakening of the magnetic field near the periphery can be used to produce focusing in the median plane. For this purpose the pole pieces are designed so that B falls off less rapidly than $1/r$ in the neighborhood of the equilibrium path. Then, as the centrifugal reaction is proportional to $1/r$, the electron is brought back to the equilibrium path if it deviates to either side of it. Actually, a more detailed analysis shows that in such circumstances the electron oscillates about the equilibrium path with rapidly decreasing amplitude.

The upper limit of a betatron's performance is determined by the radiant energy emitted by the electron. If this loss per turn becomes comparable with the energy gained from the field per turn, the electron orbit shrinks. Assuming that the total energy E of the electron is much greater than its rest energy m_0c^2 than ΔE, the radiation loss per turn has been shown to be

$$\Delta E = \frac{8.85 \times 10^{-8}E^2}{R_0}\,\text{eV}$$

where E is expressed in MeV and R_0 in meters. A betatron can accelerate electrons up to energies of many MeV, when they begin to radiate significantly.

b. Electric Generators

We now use the electric generator as a second illustrative example of the application of Faraday's law. Consider the plane coil abc (**Figure 5.8**) rotating about an axis PQ lying its plane, and there is a uniform magnetic field at right angles to the axis PQ. If θ is the angle that the normal to the plane of the circuit makes with B, then the magnetic flux is

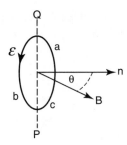

I **FIGURE 5.8**

$$\Phi = \vec{B} \cdot \vec{A} = BA \cos\theta$$

where A is the area of the circuit. The induced *emf* is

$$\mathscr{E} = -\frac{d\Phi}{dt} = BA \sin\theta \frac{d\theta}{dt}$$

If the coil consists of N turns, each of which embraces the same flux, the *emf* is N times as great. Both the flux and the induced *emf* reverse their sense relative to the circuit twice each revolution. If the rotation is uniform, the one is maximum when the other is zero and vice versa (**Figure 5.9**). In this case $\theta = \omega t$, where ω is the constant angular velocity of the coil and $\varepsilon = NBA\omega \sin\omega t$.

The ends of the coil may be connected to collective rings, C, C (**Figure 5.10**), located on the axis of rotation PQ, from which the current is carried by means of brushes through an external circuit R. These are the essentials of an alternating-current generator. The magnetic field in which the armature coil rotates is supplied by the field magnet whose poles are indicated.

If we replace the collector rings in Figure 5.10 by a commutator, it reverses the sense in which the external circuit is connected to the rotating coil every half-revolution. With a single coil, as shown in Figure 5.10, this

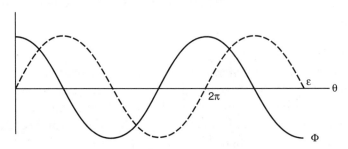

I **FIGURE 5.9**

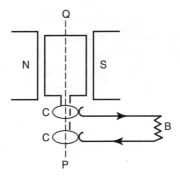

I FIGURE 5.10

arrangement gives a unidirectional but fluctuating current. By using a number of armature coils uniformly oriented about the axis PQ and connecting each to the external circuit only during that part of its revolution when the *emf* is near its maximum, an effectively constant current may be produced, and we have a direct-current generator.

c. Faraday's Disc

Faraday's disc is a first simple direct generator. The disc may be a circular copper disc rotating at a constant angular velocity in a steady magnetic field. **Figure 5.11** shows a different construction: an insulating disc with a conducting ring around its circumference, a conducting axle OR, and a conducting wire OP embedded in it. The brushes on the moving parts at Q and R complete the circuit OPQR. The magnetic field, not shown in Figure 5.11, is uniform and perpendicular to the plane of the disc, directed out of the paper.

The motional *emf* is given by

$$\mathcal{E} = \oint \vec{E} \cdot d\vec{l} = \oint (\vec{v} \times \vec{B}) \cdot d\vec{l}$$

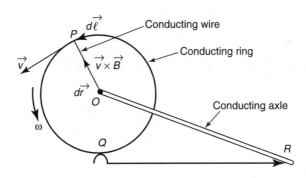

I FIGURE 5.11

The ORQ part of the circuit is stationary with respect to $\vec{B}$ and so does not contribute to the *emf*. But OP and PQ are not stationary with respect to $\vec{B}$. Let us examine their contribution to the *emf*. For PQ, $\vec{v} \times \vec{B}$ is normal to $d\vec{l}$ at all points and so the contribution to the above integral is zero. The only contribution is from OP; now $\vec{v} \times \vec{B}$ is along $\vec{r}$ and $d\vec{l} = d\vec{r}$; moreover, we can write $v = r\omega$. Thus,

$$\mathscr{E} = \oint \vec{E} \cdot d\vec{l} = \oint (\vec{v} \times \vec{B}) \cdot d\vec{l} = \int_0^a r\omega B \, dr = \frac{1}{2} a^2 \omega B$$

where a is radius of the disc.

We can also analyze this problem with Faraday's law of induction:

$$\mathscr{E} = -d\Phi/dt$$

Now what is the flux Φ through the circuit? It is not the one through the whole disc. The flux is cut as the conducting wire OP sweeps around through an angle POQ. If this angle is θ radians, then $\Phi = B\pi r^2(\theta/2\pi)$, and

$$\mathscr{E} = -\frac{d\Phi}{dt} = -\frac{Ba^2}{2} \frac{d\theta}{dt} = -\frac{1}{2} a^2 \omega B$$

A machine constructed on this principle is known as a homopolar generator.

5.7 Electromagnetic Potentials

Potentials for static fields were introduced in Chapter 4. The electrostatic field was expressed in terms of scalar potential, $\vec{E} = -\nabla\phi$, and the magnetic field in terms of a vector potential, $\vec{B} = \nabla \times \vec{A}$. We now consider potentials of electromagnetic fields that are time varying. From $\nabla \cdot \vec{B} = 0$ we can express $\vec{B}$ in terms of vector potential $\vec{A}$, just as the static case

$$\vec{B} = \nabla \times \vec{A} \tag{5.19}$$

Substituting this in Faraday's law of induction, we obtain

$$\nabla \times \vec{E} = -\frac{\partial \vec{B}}{\partial t} = -\frac{\partial}{\partial t}\left(\nabla \times \vec{A}\right) = -\nabla \times \frac{\partial \vec{A}}{\partial t}$$

or

$$\nabla \times \left(\vec{E} + \frac{\partial \vec{A}}{\partial t}\right) = 0$$

Because the curl of the gradient of a scalar function vanishes, we can express the quantity within the brackets as a gradient of a scalar function ϕ:

$$\vec{E} + \frac{\partial \vec{A}}{\partial t} = -\nabla \phi$$

or

$$\vec{E} = -\nabla \phi - \frac{\partial \vec{A}}{\partial t} \qquad (5.20)$$

Thus, once $\vec{A}$ and ϕ are determined, $\vec{B}$ and $\vec{E}$ can be found.

Just as in the static case, equation (5.19) does not completely define the vector potential $\vec{A}$. If we add the gradient of an arbitrary scalar function χ to $\vec{A}$

$$\vec{A}' = \vec{A} + \nabla \chi \qquad (5.21)$$

The magnetic field $\vec{B}$ remains unchanged:

$$\nabla \times \vec{A}' = \nabla \times (\vec{A} + \nabla \chi) = \nabla \times \vec{A} = \vec{B}$$

However, $\vec{E}$ has been changed:

$$\vec{E} = -\nabla \phi - \frac{\partial}{\partial t}\left(\vec{A}' - \nabla \chi\right) = -\nabla\left(\phi - \frac{\partial \chi}{\partial t}\right) - \frac{\partial \vec{A}'}{\partial t}$$

To keep $\vec{E}$ unchanged, the scalar potential ϕ must be simultaneously transformed to ϕ' where

$$\phi' = \phi - \frac{\partial \chi}{\partial t} \qquad (5.22)$$

Equations (5.21) and (5.22) are called gauge transformations. Equations involving electromagnetic potentials $\vec{A}$ and ϕ must be gauge invariant.

5.8 Self-Inductance

We saw in Chapter 4 that for a given electrical circuit, the **B** field produced at any point is proportional to the current I flowing in the circuit. Accordingly, the magnetic flux Φ liking any closed path is also proportional to I. The constant of proportionality between Φ and I is a very important property of the circuit, called the inductance. If the closed path is taken to be the circuit of current I, then for the flux Φ linked with the circuit itself, we may write

$$\Phi = LI \tag{5.23}$$

where the constant of proportionality L is the self-inductance of the circuit. When no confusion can arise, self-inductance L is simply referred to as inductance. The self-inductance of a circuit depends on the geometry of the circuit and the permeability of the medium. A wire made in the form of a solenoid has a much larger self-inductance than when it is unwound. The circuit symbol for inductance is in fact a solenoid, ⌇⌇⌇⌇⌇, crossed by an arrow if variable and with parallel lines adjacent to it if the inductor has a ferromagnetic core.

If the circuit is stationary but the current in it varies with time, then the magnetic field and, hence, the flux varies with time. Then we have

$$\frac{d\Phi}{dt} = \frac{d\Phi}{dI}\frac{dI}{dt} = L\frac{dI}{dt} \tag{5.24}$$

An *emf* $\mathscr{E}$ is also induced in the circuit as current I varies with time:

$$\mathscr{E} = -\frac{d\Phi}{dt} = -L\frac{dI}{dt} \tag{5.25}$$

Again, the negative sign reminds us that the induced *emf* tends to oppose the change of current. From equation (5.25) we see that a circuit has unit self-inductance when the *emf* of 1 volt is induced in it by a current vary at the rate of 1 ampere per second. The unit of L is called the henry, in honor of Joseph Henry. From equation (5.19) we may identify the henry as

1 henry = 1 weber/ampere

or from equation (5.25) we may have written an equivalent expression:

1 henry = 1 volt · second/ampere

In a network with a varying current, an inductor acts as an *emf* (LdI/dt) with an internal resistance that cannot be avoided. To see the growth and decay of current in an inductor, let us examine a simple circuit as shown in **Figure 5.12**, where R includes the internal resistance of the inductor and of the cell. If I is the current flowing at any time t after the switch K_1 is closed, then we have

$$\mathscr{E} - LdI/dt - RI = 0 \quad \text{or} \quad LdI/dt + RI = \mathscr{E}$$

Solving this differential equation with the initial condition $I = 0$ when $t = 0$, we get the current growth equation

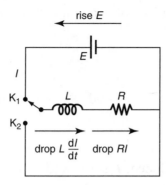

rise E

E

I

K_1

L R

K_2

drop $L\dfrac{dI}{dt}$ drop RI

I FIGURE 5.12

$$I = I_0\left[1 - \exp(-Rt/L)\right]$$

where $I_0 = \mathscr{E}/R$ and is the current in the circuit when $t \to \infty$. The time L/R is the time constant or relaxation time of the circuit.

When the switch is connected to K_2, we have

$$LdI/dt + RI = 0$$

Integration gives the current decay equation

$$I = I_0\exp(-Rt/L)$$

where the initial condition is $I = I_0$ when $t = 0$.

Because any circuit has self-inductance, a broken contact when a current is flowing is equivalent to the introduction of a very high resistance. The time constant L/R thus suddenly decreases to a very small value and the *rate of collapse* of current is very high even if the current itself is small. The resultant induced *emf* is often sufficient to cause a spark to jump across the gap. A quantitative analysis is more complex than appears at first sight, because of the capacitance also introduced.

Example 2

Calculate the inductance L of a long solenoid.

Solution

Magnetic field inside a long solenoid, neglecting end effects, is

$$B = \mu_0 IN/S$$

where N is the total number of turns and S is the length of the solenoid. Then

$$\Phi = (\mu_0 I N / S)(\pi R^2)$$

where R is the radius of the solenoid. The inductance L is

$$L = N\Phi/I = \mu_0 \pi N^2 R^2 / S$$

When end effects are included, L is reduced by a factor K that is a function of R/S.

5.9 Mutual Inductance and Neumann's Formula

If we consider more than one circuit, then we may generalize equation (5.24) to

$$\frac{d\Phi_{kl}}{dt} = \frac{d\Phi_{kl}}{dI_l}\frac{dI_l}{dt} = M_{kl}\frac{dI_l}{dt} \tag{5.26}$$

where $M_{kl} = d\Phi_{kl}/dI_l$ is the mutual inductance between circuits k and l. The unit of M is again the henry. Note that $M_{kk} = d\Phi_{kk}/dI_k = L_k$. Now let us look at a simple system in **Figure 5.13**. Circuit 1 carries a current I_1 that changes with time. Circuit 2 has no current of its own, but there is a magnetic flux Φ_{21} through it due to the current I_1 in the first, where

$$\Phi_{21} = M_{21}I_1 \quad \text{or} \quad M_{21} = \Phi_{21}/I_1 \tag{5.27}$$

and the induced *emf* in circuit 2 is given by

$$\mathscr{E} = -M_{21}\frac{dI_1}{dt} \tag{5.28}$$

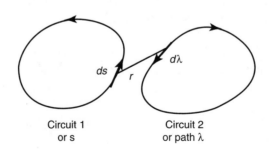

Circuit 1
or s

Circuit 2
or path λ

I FIGURE 5.13

Although M_{21} is usually calculated from equation (5.27) (in which Φ is found from Ampere's law), it can also be found from a general expression called Neumann's formula, which shows that M depends only on the positions of two circuits.

To derive Neumann's formula, let us recall that $\vec{B} = \nabla \times \vec{A}$, and then we find from equation (5.12)

$$\mathscr{E} = -\int_{A_\lambda} \vec{\dot{B}} \cdot d\vec{A_\lambda} = -\int_{A_\lambda} \nabla \times \vec{\dot{A}} \cdot d\vec{A_\lambda}$$

where A_λ is the area of circuit 2 bounded by a closed path λ. With the help of Stokes' theorem, this becomes

$$\mathscr{E} = -\oint \vec{\dot{A}} \cdot d\vec{\lambda} \tag{5.29}$$

where

$$\vec{A} = \frac{\mu_0}{4\pi} \oint \frac{I_1 ds}{r} \quad \text{and} \quad \vec{\dot{A}} = \frac{\mu_0}{4\pi} \frac{dI_1}{dt} \oint \frac{ds}{r} \tag{5.30}$$

because circuit 1 is considered fixed. Substitution of equation (5.30) into equation (5.29) gives

$$\mathscr{E} = -\frac{\mu_0}{4\pi} \frac{dI_1}{dt} \oint \oint \frac{d\vec{s} \cdot d\vec{\lambda}}{r}$$

Dividing by dI_1/dt, we have, from the definition of M in equation (5.28),

$$M = \frac{\mu_0}{4\pi} \oint \oint \frac{d\vec{s} \cdot d\vec{\lambda}}{r} \tag{5.31}$$

where $d\vec{s}$ and $d\vec{\lambda}$ are two elements of length and r is the distance between them (Figure 5.12). This is known as Neumann's formula.

5.10 Magnetic Energy

From Chapters 2 and 4, we can see a number of parallels between the magnetostatic and electrostatic fields. We cannot, however, extend these to comparisons of the energy density of the two fields. The reason for this is that we cannot introduce the concept of energy in magnetostatics because to set up a magnetic field from a steady current requires switching on the current so that there is an interval over which the field is established. The field during this interval is time dependent, and so induced *emf*s are present and these cause the current source to do work. We therefore delayed the introduction of the concept of magnetic energy until this chapter.

To obtain the energy of an electrostatic field, we calculated the work necessary to assemble the source charge distribution from infinitesimal volume elements. Similarly, to obtain the energy of a magnetic field, we calculate the work necessary to build up the source current distribution. When we build up a current in a circuit, we must do work against the back *emf* to get the current going. It represents energy latent in the circuit; as we shall see in a moment, it can be regarded as energy stored in the magnetic field.

For a circuit with a source of voltage $\mathscr{E}$, the work done by $\mathscr{E}$ in moving a charge increment $dq = Idt$ through the circuit is

$$dW = \mathscr{E}dq = \mathscr{E}Idt \tag{5.32}$$

The current is governed by the total *emf* in the circuit, and by Ohm's law we have

$$I = (\mathscr{E} + \mathscr{E}_{ind})/R \tag{5.33}$$

where $\mathscr{E}_{ind}$ is the induced *emf*:

$$\mathscr{E}_{ind} = -d\Phi/dt \tag{5.34}$$

Substituting equations (5.34) and (5.33) into equation (5.32), we obtain

$$dW = \left(I^2 R + I \frac{d\Phi}{dt} \right) dt \tag{5.35}$$

The first term on the right is the energy dissipated in the resistance as heat. The second term is the work done against the induced *emf* in the circuit. It is this part of work done of $\mathscr{E}_{ind}$ that is effective in altering the magnetic field structure. If there are no energy losses other than Joule heat loss (e.g., no hysteresis), then the work increment

$$dW = Id\Phi$$

is equal to the change in magnetic energy. If there are *n* circuits, then the work done against the induced *emf* is given by

$$dW = \sum_{i=1}^{n} I_i d\Phi_i \tag{5.36}$$

a. Magnetic Energy in Terms of Circuit Parameters

Equation (5.36) is a general expression; it is valid independently of how the flux increments $d\Phi_I$ are produced. We now apply it to *n* coupled circuits.

Then the flux changes are directed related with changes in the currents in the n circuits:

$$d\Phi_i = \sum_{j=1}^{n} \frac{d\Phi_{ij}}{dI_j} dI_j = \sum_{j=1}^{n} M_{ij} dI_j \tag{5.37}$$

For rigid and stationary circuits, no mechanical work is associated with the flux changes $d\Phi_i$, and then dW is just equal to the change in magnetic energy, dU, of the system. If the circuits move relative to one another, we cannot identify dU from dW.

The magnetic energy U of the system (of n rigid stationary circuits) can be obtained by integrating equation (5.36) from the zero flux situation (corresponding to all $I_i = 0$) to the final set of flux value. If the system contains only linear magnetic media, the Φ_i are linearly related to the currents in the circuit, and the magnetic energy is independent of the way in which the currents are brought to their final set of values. So we have freedom to choose a scheme for which W is easily calculated. Obviously, it is the one in which at any instant of time all currents are built up at the same fraction of their final values: $I_i' = \alpha I_i$, where I_i are the final values of the current. Then $d\Phi_i = \Phi_i d\alpha$. Integration of equation (5.36) gives

$$\int dW = \int_0^1 d\alpha \sum_{i=1}^{n} I_i' \Phi_i = \sum_{i=1}^{n} I_i \Phi_i \int_0^1 \alpha d\alpha = \frac{1}{2} \sum_{i=1}^{n} I_i \Phi_i$$

Thus, the magnetic energy of the system is

$$U = \frac{1}{2} \sum_{i=1}^{n} I_i \Phi_i = \frac{1}{2} \sum_{i=1}^{n} M_{ij} I_i I_j \tag{5.38}$$

(stationary circuits, linear media)

Note that for linear media, $M_{ij} = M_{ji}$ and $M_{ii} = I_i$. Equation (5.38) gives the magnetic energy of a current system in terms of circuit parameters, the inductances and currents. Because the currents and inductances are direct experimental measures, equation (5.38) is very useful.

b. Magnetic Energy in Terms of Field Vectors

From equation (5.38) we can derive an expression for the magnetic energy in terms of the field vectors $\vec{B}$ and $\vec{H}$. This alternative expression provides a picture in which energy is stored in the magnetic field itself.

For simplicity, we assume that each circuit consists of a single loop; then the flux Φ_i may be expressed as

$$\Phi_i = \int_S \vec{B} \cdot \hat{n} dS = \int_S \nabla \times \vec{A} \cdot \hat{n} dS = \oint_{C_i} \vec{A} \cdot d\vec{l}$$

where C is the path that encloses the surface S and $\vec{A}$ is the local vector potential; in the last equation we made use of Stokes' theorem. Substituting this into equation (5.38) we obtain

$$U = \frac{1}{2}\sum_{i=1}^{n} I_i \Phi_i = \frac{1}{2}\sum_i \oint_{C_i} I_i \vec{A} \cdot d\vec{l} \tag{5.39}$$

We now make this expression more general. Instead of current circuits defined by wires, we have each "circuit" as a closed path in the conducting medium that follows a line of current density. Equation (5.39) may be made to approximate this situation very closely by choosing a large number of contiguous circuits (C_i), replacing $I_i d\vec{l}_i$ by $\vec{j}dv$ and $\sum_i \oint_{C_i}$ by $\int_v$:

$$U = \frac{1}{2}\int_V \vec{j} \cdot \vec{A}dv \tag{5.40}$$

Because $\nabla \times \vec{B} = \mu_0 \vec{j}$ and $\vec{B} = \nabla \times \vec{A}$, it seems clear that we can work equation (5.40) into a different form. The trick is to use the vector identity

$$\nabla \cdot (\vec{A} \times \vec{B}) = \vec{B} \cdot \nabla \times \vec{A} - \vec{A} \cdot \nabla \times \vec{B} = \vec{B} \cdot \vec{B} - \mu_0 \vec{A} \cdot \vec{j}$$

Substituting this into equation (5.40), we obtain

$$U = \frac{1}{2\mu_0}\int_V \vec{B} \cdot \vec{B}dv - \frac{1}{2\mu_0}\int_V \nabla \cdot (\vec{A} \times \vec{B})\,dv$$

$$= \frac{1}{2\mu_0}\int_V \vec{B} \cdot \vec{B}dv - \frac{1}{2\mu_0}\int_S \vec{A} \times \vec{B} \cdot \hat{n}dS$$

where S is the surface that bounds the volume V. The integrations on the right are to be taken over the entire volume occupied by the current. We can take any region larger than this, for current density j is zero out there. The larger the volume we pick, the greater the contribution from the volume integral and smaller is that of the surface integral (because $\vec{B}$ falls off at least as fast as $1/r^2$ and $\vec{A}$ falls off at least as fast as $1/r$, and the surface area S is proportional to r^2). If S is moved out to infinity, the surface integral vanishes. Then the magnetic energy U is given by

$$U = \frac{1}{2\mu_0}\int_V \vec{B} \cdot \vec{B}dv = \frac{1}{2}\int_V \vec{H} \cdot \vec{B}dv \tag{5.41}$$

where $\vec{B} = \mu_0 \vec{H}$.

Equation (5.41) is also valid for systems that contain linear magnetic medium, with μ_0 replaced by μ the permeability of the medium. As in the electric case we may define the energy density in a magnetic field by

$$u = \frac{1}{2}\vec{H} \cdot \vec{B} \tag{5.42}$$

This reduces to, for the case of isotropic, linear, magnetic materials,

$$u = \frac{1}{2}\mu H^2 = \frac{1}{2}\frac{B^2}{\mu} \tag{5.43}$$

As an example, consider a portion of the long solenoid carrying a current I. With the usual notation, B is $\mu_0 NI$ everywhere inside and the volume is lA so that equation (5.43) gives the energy density $1/2\ \mu_0 N^2 lAI^2$, which is exactly equal to $1/2\ LI^2$.

For nonlinear magnetic materials such as ferromagnets, we have to modify the above result because for nonlinear materials, the L's and M's are not constant and $M_{ij} \neq M_{ji}$ in general. We put off such a discussion until we complete the discussion on magnetic properties of matter.

We now have two views on magnetic energy. Analogous to the interpretation of the electrostatic field, we may think of the magnetic energy as being either associated with the charges comprising the current or stored in the region containing the magnetic field resulting from the current. The first point of view says that the stored energy is in the conductor and in a kinetic form, whereas in the field point of view we think of the energy as traveling out into space while the field is being established and returning to the wire when the field collapses.

5.11 Forces and Torques on Linear Magnetic Materials

It is observed that currents exert forces on each other independently of any electrostatic forces that may be present. The magnetic force law is considerably difficult to discover for two reasons: (1) The force between two elements is not along the line joining them, and (2) the observations are carried out not on small isolated "point" elements but on extended circuits. The problem is made easier by the use of the energy method. Thus, we now see how the force, or torque, on one component of a system of current circuits may be calculated from knowledge of the magnetic energy.

Let us consider a system of current circuits and one component of the system is allowed to make a small displacement $d\vec{r}$ under the influence of the magnetic force $\vec{F}$ acting on it, while all currents remain constant. The work done by the force $\vec{F}$ is

$$dW = \vec{F} \cdot d\vec{r}$$

This work is done at the expense of the magnetic energy U. It contains another part: The work performed by an external energy source against the induced *emf*s to keep the current constant we denote by dW_b. Thus,

$$dW = dW_b - dU \tag{5.44}$$

To link U and the force on a part of the system, we eliminate dW_b from equation (5.44). This can be easily done for a system of rigid circuits in linear magnetic media. By equation (5.36), the electric work done against the induced *emf* is

$$dW_b = \sum_i I_i d\Phi_i$$

and

$$dU = \frac{1}{2} \sum_i I_i d\Phi_i$$

Thus, $dW_b = 2dU$. Combining this with equation (5.44), we find

$$dU = dW = \vec{F} \cdot d\vec{r}$$

or

$$\vec{F} = (\nabla U)_I \tag{5.45}$$

To use this equation, U, the magnetic energy, must be expressed as a function of the variable that is to be subject to the displacement.

If the circuit under consideration rotates about an axis by an angular displacement $d\vec{\theta}$, then

$$dW = \vec{\tau} \cdot d\vec{\theta} = \tau_1 d\theta_1 + \tau_2 d\theta_2 + \tau_3 d\theta_3$$

where $\vec{\tau}$ is the magnetic torque on the circuit. And equation (5.41) is replaced by

$$\tau_1 = \left(\frac{\partial U}{\partial \theta} \right)_I \tag{5.46}$$

and so on.

In some interesting cases, the flux through the circuits can be treated as constant; then $dW_b = \sum_i I_i d\Phi_i = 0$, and so

$$dW = -dU$$

and

$$\vec{F} = -(\nabla U)_I \tag{5.47}$$

5.12 The Transformer

Consider two coils with self-inductances L_1, L_2 and mutual inductance M, which inductively coupled (**Figure 5.14**). The flux of $\vec{B}$ through the first coil is $L_1 dI_1/dt + M dI_2/dt$, and then the net *emf* in this circuit is $\mathscr{E}_1 - L_1 dI_1/dt - M dI_2/dt$, and Ohm's law gives the differential equation

$$\mathscr{E}_1 - L_1 dI_1/dt - M dI_2/dt = R_1 I_1 \tag{5.48}$$

Similarly for circuit 2 we have

$$\mathscr{E}_2 - L_2 dI_2/dt - M dI_1 dt = R_2 I_2 \tag{5.49}$$

They can be solved easily if $\mathscr{E}_1$, $\mathscr{E}_2$ are known and the initial conditions are specified.

We now consider a case of special interest: $\mathscr{E}_1$ is alternating *emf* and both $\mathscr{E}_2$ and R_1 are zero. We call the first coil the primary coil or circuit and the second coil the secondary coil or circuit. The current in the secondary circuit is solely due to the alternating *emf* in the first. That is, the mutual inductance M provides an *emf* of a different numerical value from the original *emf* in the first circuit.

In practice, the two inductances L_1 and L_2 may consist of n_1 and n_2 turns,

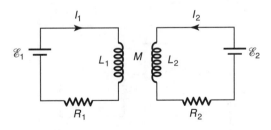

I FIGURE 5.14

Respectively, wound close together on the same iron core (**Figure 5.15**). When unit current flows in any one of these turns, a magnetic flux equal to Φ, say, crosses this particular turn of wire. This flux crosses all the other turns in both coils. Thus, unit current in the coil L_1 produces a total flux $n_1\Phi$ and this flux threads each of the n turns of the first coil and each of the n_2 turns of the second. Then from the definitions of L and M

$$L_1 = n_1{}^2\Phi, \qquad M = n_1 n_2 \Phi, \qquad L_2 = n_2{}^2\Phi \tag{5.50}$$

It follows that

$$L_1 L_2 = M^2 \tag{5.51}$$

This relation, sometimes called the transformer condition, is not totally satisfied in practice due to the leakage of lines of B, but if the leakage is negligible and condition (5.51) is satisfied, then we can eliminate I_1 from equations (5.48) and (5.49), with $\mathscr{E}_2 = 0$ and $R_1 = 0$, and obtain

$$\mathscr{E}_1 L_2 = -MR_2$$

Using equation (5.48), this becomes

$$R_2 I_2 = -(n_2/n_1)\,\mathscr{E}_1 \tag{5.52}$$

This states that the induced voltage across the ends of the second coil is equal to the constant ratio n_2/n_1 multiplied by $\mathscr{E}_1$. Thus, by varying n_2/n_1, we are able to step up or down a given alternating *emf*. For this reason a device of this kind is called a transformer.

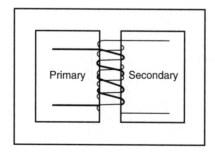

I FIGURE 5.15

We can obtain equation (5.52), without relying on the concept of inductance, using the fact that the sum of voltages around a closed loop of a circuit is zero (this is known as Kirchhoff's second rule). Thus

$$\mathcal{E}_1 - n_1(d\Phi/dt) = 0$$

and

$$R_2 I_2 - n_2(d\Phi/dt) = 0$$

Equation (5.50) follows immediately by eliminating $d\Phi/dt$ from these two equations.

The currents I_1 and I_2 flowing in the primary and secondary circuits are related by Ampere's law to the line integral of B around a circuit threading the coils:

$$\oint \vec{B} \cdot dl = \mu_0 I$$

Consider a circuit of length l lying entirely within the core. If the core has a uniform cross-section A, the magnetic field is nearly uniform over any cross-section and has the average magnitude $B = \Phi/A$. In applying Ampere's law to this circuit, we must remember that the secondary current is flowing in such a direction as to oppose the flux changes caused by the primary current. The two currents are therefore flowing in the opposite sense around the lines of B and

$$n_1 I_1 - n_2 I_2 = \frac{\oint \vec{B} \cdot d\vec{l}}{\mu\mu_0} = \frac{\Phi l}{\mu\mu_0 A} \approx 0$$

because μ is very large. Ignoring the small term on the right side, the above equation reduces to

$$I_2 = \frac{n_1}{n_2} I_1 \tag{5.43}$$

5.13 Eddy Currents and Magnetic Levitation

Currents that are induced in conductors such as conducting plates do not flow along well-defined paths like in a conducting wire; instead, they spread out over the whole conductor and are called eddy currents. It is very difficult to calculate eddy currents; the reader who is interested in the

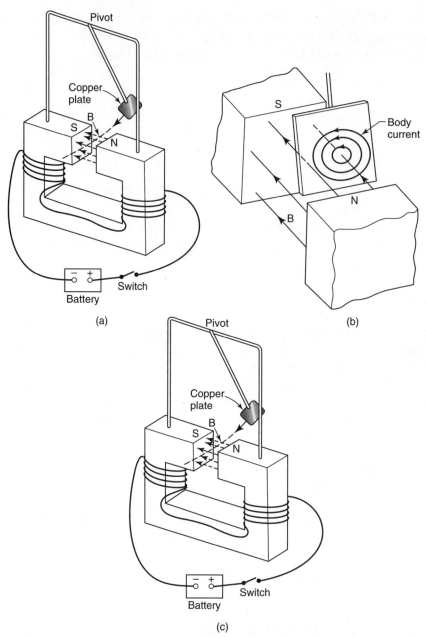

I FIGURE 5.16

calculation may find help from the article by W. M. Saslow (*American Journal of Physics*, **60**, 693, 1992). It is rather easy to demonstrate the effect of eddy currents. As shown in **Figure 5.16a,** a plate made of aluminum or copper is passed between the two poles of a magnet. When the plate enters the field, its motion is suddenly arrested and it then slowly settles to rest in the magnetic field. What happens is that when the plate enters into the magnetic field, the induced eddy currents in the plate (Figure 5.16b) act to oppose the change of flux through the plate and bring it to a stop. Then, as the currents die down, the plate slowly settles to rest in the magnetic field. This is an example of the "magnetic braking," which is used to stop circular motion of conducting plates in circular saws or high-speed trains. If we cut many slots in the plate (Figure 5.16c) to prevent the flow of large-scale currents, the plate now swings through the gap of the magnet with a small retarding force and can swing almost freely.

In addition to repulsive forces from eddy currents in a normal conductor, there can also be sidewise force. If we move a magnet sideways along a conducting surface, the induced eddy currents oppose the changing of the location of flux and so produce a force of drag, which is proportional to the velocity of the movement of the magnet and is like a kind of viscous force.

If the plate is made of a perfect conductor, as it enters the gap of the magnet, the induced currents would be so large that they would push the plate out. There is no resistance to current in a perfect conductor, and a slightest *emf* can generate a very large current, which can keep going forever. There are no perfect conductors at room temperature, but some materials become perfect conductors at low temperature. These materials are called superconductors, and their basic properties are discussed in Chapter 8.

If we place a conducting plate on the end of an electromagnet, when the magnet is turned on, the eddy current induced in the plate repels it from the magnet and tends to keep it suspended in the air. The magnetic levitation used for trains in Germany and Japan works on the same principle. A magnetic levitated train is almost friction free and also has much less noise. Recent development of high-temperature (~100 K) superconducting materials has greatly increased the magnetic levitation in mass transport. It is possible today to construct superconducting tracks with much lower refrigeration requirements compared with low-temperature superconductors, which are only a few degrees Kelvin.

Problems

1. Show that the total flux through any closed surface is zero.
2. Calculate the interaction energy between a long straight carrying current I_1 and a rectangular loop carrying current I_2 (**Figure 5.17**).
3. A single circular loop of wire is rotated in a uniform magnetic field B with a constant angular speed ω. The B field is in the z-direction and the axis of rotation of the loop is in the x-direction along a

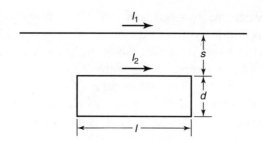

I FIGURE 5.17

diameter. Show that the induced *emf* varies with time in a sinusoidal way (**Figure 5.18**).

4. To measure magnetic fields, rotating coils are often used. A circular coil of radius 1 cm and with 100 turns is rotated at 60 cps in a magnetic field. The *emf* induced in the coil has a maximum value of 12.3 volts. Calculate the intensity of the magnetic field.

5. A metal airplane with a wing span of 40 meters flies at 355 m/sec in a region where the vertical component of the earth's magnetic field is 3×10^{-5} T, where T stands for tesla. Calculate the *emf* across the airplane.

6. Two conducting bars move outward with velocities $\vec{v}_1 = 12.5(-\hat{j})m/s$ and $\vec{v}_2 = 8.0\hat{j}m/s$ in the magnetic field $\vec{B} = 0.35\hat{k}$ tesla. Find the voltage of b with respect to c (**Figure 5.19**).

7. A long solenoid of n turns per unit length and cross-section A carries an alternating current $i = i_0 \sin\omega t$. Find the flux Φ through a closed curve that encircles the solenoid and calculate the *emf* around the curve.

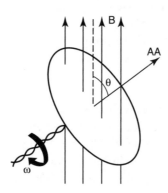

I FIGURE 5.18

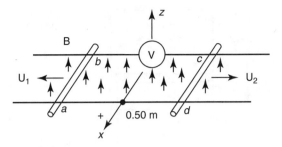

I FIGURE 5.19

8. A rigid slender rod OP of length R rotates in the x,y plane about the z-axis that passes through one end of the rod. A uniform magnetic field $\vec{B}$ is present over the area swept out by the rod. The field is directed along the z-axis. What is the induced *emf* in the rod if it rotates in the counterclockwise sense with an angular velocity ω? And what is the sign of the charge appearing on the free end P **(Figure 5.20)**.

9. Consider the fixed circular loop of Figure 5.4. Calculate the induced *emf* in the loop if there is a uniform time-varying magnetic field $\vec{B}$ directed at an angle θ with the normal to the plane of the loop. Assume that $\vec{B}$ is given by $B_0 \sin\omega t$, where B_0 and ω are constants.

10. Repeat the calculations in the preceding problem, but this time let $\vec{B}$ be nonuniform and time dependent, as given by

$$B = \left(1 - \frac{r}{R}\right) B_0 \sin\omega t$$

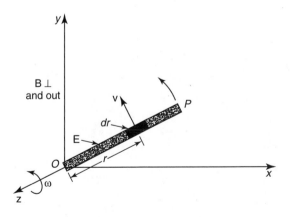

I FIGURE 5.20

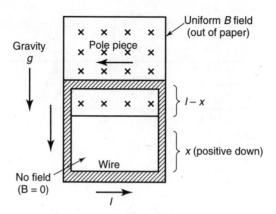

I FIGURE 5.21

11. There is a charge distribution ρ_0 in a medium of permittivity ε and conductivity σ. Show that this charge distribution decays exponentially.

12. Find the force on the rectangular loop in Problem 2.

13. **Figure 5.21** shows a square conducting ring of dimension l falling through the pole piece of a horizontal magnet. It is assumed that the **B** field is zero outside the pole piece. Show that the velocity of the falling ring is exponentially damped with a terminal velocity and a time constant that are directly proportional to the ring resistance R and inversely proportional to the square of the magnetic field B.

6 Maxwell Equations and Electromagnetic Waves in Vacuum

Until now, we have been studying electromagnetic theory in bits and pieces. In this chapter we come to the complete set of the four Maxwell equations; these four equations contain all the content of the electromagnetic fields that may be changing with time in any way. Let us first summarize the equations for the electric and magnetic fields that we have so far assembled. They are, in differential form,

Gauss' law

$$\nabla \cdot \vec{E} = \rho \varepsilon_0 \tag{6.1a}$$

Ampere's law

$$\nabla \times \vec{B} = \mu_0 \vec{j} \tag{6.1b}$$

Absence of magnetic monopole

$$\nabla \cdot \vec{B} = 0 \tag{6.1c}$$

Faraday's law of induction

$$\nabla \times \vec{E} + \frac{\partial \vec{B}}{\partial t} = 0 \tag{6.1d}$$

The continuity equation

$$\frac{\partial \rho}{\partial t} + \nabla \cdot \hat{j} = 0 \tag{6.1e}$$

These equations have been derived from various experiments performed in static situations where fields and current are not changing with time. Maxwell, who was largely responsible for putting these laws in the form of differential equations, contemplated the possibility that they were all valid even in time varying situations and realized that Ampere's law was inconsistent with the continuity equation. If we take divergence on both sides of Ampere's law and because $\nabla \cdot (\nabla \times \vec{B}) = 0$, we see immediately that $\nabla \cdot \vec{j} = 0$. This is, indeed, true if ρ does not change with time. But it is not true when ρ is changing with time. Maxwell suggested a way out of this unpleasant situation. Using Gauss' law, we can rewrite the continuity equation as

$$\nabla \cdot \vec{j} = -\frac{\partial \rho}{\partial t} = -\frac{\partial}{\partial t}\left(-\varepsilon_0 \nabla \cdot \vec{E}\right), \quad \text{or} \quad \nabla \cdot \left(\vec{j} + \varepsilon_0 \frac{\partial \vec{E}}{\partial t}\right) = 0$$

6.1 Displacement Current

Maxwell saw that if Ampere's law is modified by the addition of a new term, the time derivative term $\mu_0 \varepsilon_0 \partial \vec{E}/\partial t$ becomes

$$\nabla \times \vec{B} = \mu_0 \vec{j} + \mu_0 \varepsilon_0 \frac{\partial \vec{E}}{\partial t}$$

Then Ampere's law in this form is valid for steady-state phenomena and is also compatible with the equation of continuity for time-dependent fields. The term $\varepsilon_0 \partial \vec{E}/\partial t$ has the dimensions of current density and was called the "displacement current" by Maxwell, a term still in use today.

Evidently, the displacement current does not have the meaning of a current in the sense of being the motion of charges. Then what is exactly implied by the displacement current? To clarify this, let us consider a simple circuit such as a parallel plate capacitor connected to a battery. The current flowing in the wire is equal to the rate of change of charge on the plates,

$$I = dQ/dt$$

where Q is the charge on the positive plate of the capacitor. The field in the capacitor is equal to

$$E = \sigma/\varepsilon_0 = Q/\varepsilon_0 A$$

where A is the area of the plates. So

$$\varepsilon_0 \frac{\partial E}{\partial t} = \frac{1}{A}\frac{dQ}{dt} = \frac{I}{A}$$

Now I/A gives the current density. Hence, the quantity $\varepsilon_0 \partial E/\partial t$ can be interpreted as the density of the current that flows between a pair of plates of the capacitor when they are connected to a source. This description of current in terms of a changing electric field can be extended. Consider, for example, the charging of a sphere. At any point, distant r from the center of the sphere,

$$E = q/4\pi\varepsilon_0 r^2$$

instantaneously; hence,

$$\frac{\partial E}{\partial t} = \frac{1}{4\pi\varepsilon_0 r^2}\frac{dq}{dt}$$

or

$$\varepsilon_0\frac{\partial E}{\partial t} = \frac{I}{4\pi r^2} \quad \text{where} \quad I = \frac{dq}{dt}$$

Fixing attention on the surface of an imaginary sphere of radius r, we see therefore that the rate of change of field corresponds to an effective current density $\varepsilon_0\,\partial E/\partial t$, the displacement current density $\vec{j}_d$.

We needed the displacement current term to make Ampere's law consistent with the continuity equation, in the case of conduction currents changing in time. Does this new term imply the existence of a new induction effect in which a changing electric field is accompanied by a magnetic field? In any apparatus in which there are changing electric fields, there are present at the same time conduction currents, charges in motion. The **B** field, everywhere around the apparatus, is just about what we would expect the conduction currents to produce: It is almost exactly the field we would calculate if, ignoring the fact that the circuits may not be continuous, we used the "Biot-Savart" formula (see equation 4.45) to find the contribution of each conduction current element to the field at some point in space. We may include the entire displacement current distribution; its net effect is zero for relatively slowly varying fields. To see why this is so, let us consider a discharging parallel plate capacitor (**Figure 6.1**). Its electric field is practically an electrostatic field, except that it is slowly dying away. The vector function of the displacement current density $\vec{j}_d$ has the same form. The curl of $\vec{j}_d$ must be practically zero. Now let us take the curl of $\vec{j}_d$:

$$\nabla \times \vec{j}_d = \varepsilon_0 \nabla \times \left(\partial\vec{E}/\partial t\right) = \varepsilon_0\frac{\partial}{\partial t}\nabla\vec{E}$$

This becomes, with the help of Faraday's law of induction,

$$\nabla \times \vec{j}_d = -\varepsilon_0\frac{\partial^2\vec{B}}{\partial t^2}$$

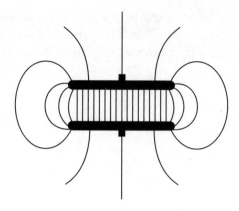

I FIGURE 6.1

This is negligible for sufficiently slow changes in field, the quasistatic fields.

In the quasistatic field, thus, the conduction currents alone are the only sources needed to account for the magnetic field. To see the induction effect by the displacement current, we need rapidly changing fields—changes to occur in the time it takes light to cross the apparatus. Thus, Faraday could not discover this new induction effect.

6.2 Maxwell Equations

Maxwell equations are the set of equations (6.1) with the displacement current density term included in Ampere's law:

Gauss' law

$$\nabla \cdot \vec{E}\,(\vec{r}, t) = \varepsilon_0 \rho\,(\vec{r}, t) \tag{6.2a}$$

Ampere's law with displacement current

$$\nabla \times \vec{B}\,(\vec{r}, t) - \mu_0 \varepsilon_0 \frac{\partial \vec{E}\,(\vec{r}, t)}{\partial t} = \mu_0 \vec{j}\,(\vec{r}, t) \tag{6.2b}$$

Absence of magnetic monopole

$$\nabla \cdot \vec{B}\,(\vec{r}, t) = 0 \tag{6.2c}$$

Faraday's law of induction

$$\nabla \times \vec{E}\,(\vec{r}, t) + \frac{\partial \vec{B}\,(\vec{r}, t)}{\partial t} = 0 \tag{6.2d}$$

The continuity equation is now redundant, because it is a consequence of equations (6.2a) and (6.2b).

Although displacement currents are not significant as the conduction current in the study of the continuous circuits, they have far reaching consequences in other respects. It is mainly due to the displacement current term in Ampere's law that a rich variety of new electromagnetic phenomena was uncovered.

If the electric and magnetic fields do not vary with time, Maxwell equations reduce to the equations of electrostatics and magnetostatics discussed in Chapters 2 and 3. These equations are as follows:

$$\nabla \cdot \vec{E} = \rho/\varepsilon_0, \quad \nabla \times \vec{E} = 0, \quad \text{and} \quad \nabla \cdot \vec{B} = 0, \quad \nabla \times \vec{B} = \mu_0 \vec{j}$$

For fields that vary rapidly with time, typically if they have frequency components of 100 kHz or higher, the displacement current has to be included. This gives rise to new solutions to Maxwell equations corresponding to electromagnetic radiation. Using spark discharges between two electrically charged spheres, Hertz in 1887 first detected electromagnetic radiation by inducing an electric field in a circuit some distance away from the spheres. Wave-like equations (from Maxwell equations) in free space are discussed later.

All four of Maxwell field equations (6.2) have the nature of linear differential equations. The linearity is manifest by the absence of terms involving any products or powers higher than one of the field quantities E, B, ρ, and J. This simple mathematical property has important physical consequences, which are usually summarized by the statement that the fields satisfy a *principle of superposition*. Whenever some solutions of Maxwell equations are known, this principle allows "new" solutions to be obtained from sums or superpositions of the known ones. More precisely, if the sets of fields E_1, B_1, ρ_1, J_1, and E_2, B_2, ρ_2, J_2 are known to separately satisfy Maxwell equations, then the fields $(E_1 + E_2)$, $(B_1 + B_2)$, $(\rho_1 + \rho_2)$, $(J_1 + J_2)$ also do so.

The principle of superposition has important practical application when some given charge and current distribution are seen to be a sum of distributions for each of which the associated E and B fields are either known or can be easily calculated. Then the total E and B fields are given as vector sums of the individual fields arising from each such distribution.

We can also cast Maxwell equations in integral forms:

Gauss' law

$$\oiint_S \vec{E} \cdot \vec{dS} = Q/\varepsilon_0 \tag{6.3a}$$

Modified Ampere's law

$$\oint_C \vec{B} \cdot d\vec{r} = \mu_0 \int_S \left(\vec{j} + \varepsilon_0 \frac{\partial \vec{E}}{\partial t} \right) \cdot d\vec{S} \tag{6.3b}$$

Absence of monopole

$$\oiint_S \vec{B} \cdot d\vec{S} = 0 \tag{6.3c}$$

Faraday's law

$$\oint_C \vec{E} \cdot d\vec{r} = -\frac{d}{dt} \int_S \vec{B} \cdot d\vec{S} \tag{6.3d}$$

6.3 Magnetic Monopoles

Gauss' law for magnetic field, $\nabla \cdot \vec{B} = 0$, is a statement that there is no magnetic charge or magnetic monopole. However, Dirac showed that the existence of monopoles would explain why electric charge is quantized. Because the quantization of electric charge is one of the most profound mysteries of the physical word, Dirac's idea has great appeal, and the search for a monopole is renewed whenever a new energy level is opened up in high-energy physics or new source of matter.

We now give a brief sketch to show that the existence of a magnetic charge implies the quantization of electric charge. If we assume a monopole of magnetic charge g at the origin, its magnetic field at a point r is

$$\vec{B} = \frac{\mu_0 g}{4\pi r^2} \hat{r} = -\frac{\mu_0 g}{4\pi} \nabla \left(\frac{1}{r} \right)$$

Now a particle of electric charge q passes the magnetic charge g at a distance b and with a velocity v. At a point r the charge e experiences a Lorentz force of

$$F = qvB = \frac{qg\mu_0 vb}{(b^2 + v^2 t^2)^{3/2}}$$

From this there results a momentum transfer of

$$\Delta p = \int F dt = \frac{2qg\mu_0}{b}$$

A change of momentum is also connected to a change of angular momentum:

$$\Delta L = b\Delta p = 2qg\mu_0$$

Quantum mechanically, the orbital angular momentum is quantized, that is,

$$L = n\hbar, \quad n = 0, 1, 2, 3, \ldots$$

where $\hbar = h/2\pi$ and h is the Planck's constant. From this there results the quantization of the electric charge e:

$$2qg\mu_0 = n\hbar$$

or

$$q = n\frac{\hbar}{2\mu_0 g}, \quad n = 0, \pm 1, \pm 2 \ldots$$

Because all electric charges of free particles are observed to be integral multiples of the electron charge magnitude e, we can identify $\hbar/2\mu_0 g$ with e. Thus, the above equation says that the magnetic charge of any monopole must be an integral multiple of the "unit"

$$g = \hbar/2\mu_0 e = 1.64 \times 10^{-9} A \cdot m$$

The grand unified theories of electromagnetic, weak, and strong interactions predict that monopoles exist with mass

$$m_{\text{monopole}} \approx 10^{16} m_{\text{proton}} \approx 10^{-8} \text{grams}$$

These monopoles might have been produced in the very early stages of the creation of the universe. The observation of such magnetic monopoles would support the unification of forces. These theories also predict that the proton is unstable, with a lifetime of about 10^{32} years, which is just at the level of present experimental limits on proton decay.

The existence of a magnetic charge could fit into electromagnetic theory nicely. In fact, the existence of a monopole gives Maxwell equations a pleasing symmetry. If ρ_m is the magnetic charge density, then equation (6.2c), Gauss' law for magnetic field, becomes

$$\nabla \cdot \vec{B} = \mu_0 \rho_m \tag{6.4}$$

We cannot modify Gauss' law for magnetic field without modifying Faraday's law, equation (6.2d). The appropriate form is

$$\nabla \times \vec{E} = -\mu_0 \vec{j}_m - \frac{\partial \vec{B}}{\partial t} \tag{6.5}$$

where $\vec{j}_m$ is the magnetic current density, and it is related to the magnetic charge density ρ_m by an equation of continuity

$$\nabla \cdot \vec{j}_m + \frac{\partial \rho_m}{\partial t} = 0 \tag{6.6}$$

Taking the divergence of equation (6.5) you get equation (6.6). Now, in the presence of magnetic charge, the four Maxwell equations take symmetric form:

$$\left.\begin{array}{llll} (i) & \nabla \cdot \vec{E} = \rho/\varepsilon_0, & (iii) & \nabla \times \vec{E} = -\mu_0 \vec{j}_m - \partial \vec{B}/\partial t \\[2mm] (ii) & \nabla \cdot \vec{B} = \mu_0 \rho_m, & (iv) & \nabla \times \vec{B} = \mu_0 \vec{j} + \mu_0 \varepsilon_0 \partial \vec{E}/\partial t \end{array}\right\} \tag{6.7}$$

In spite of diligent searches, no monopole has ever been detected. Thus, as far as classical electromagnetic theory is concerned, ρ_m is zero and so is $\vec{j}_m$ everywhere. All magnetic effects are attributed to charges in motion.

6.4 Decay of Free Charge

According to Maxwell's equations, a free charge should decay exponentially. To see this, we first replace $\vec{j}$ in equation (6.2b) by $\sigma\vec{E}$ and obtain

$$\nabla \times \vec{B} = \mu_0 \varepsilon_0 \frac{\partial \vec{E}}{\partial t} + \mu_0 \sigma \vec{E}$$

Now taking the divergence of each side and assuming that σ and ε_0 are constants, we have

$$\nabla \cdot \nabla \times \vec{B} = \mu_0 \varepsilon_0 \frac{\partial \nabla \cdot \vec{E}}{\partial t} + \mu_0 \sigma \nabla \cdot \vec{E}$$

from which we get, with the aid of equation (6.2a),

$$\frac{\partial \rho}{\partial t} + \frac{\sigma}{\varepsilon_0} \rho = 0$$

Integration results

$$\rho = \rho_0 e^{-t/\tau} \tag{6.8}$$

where

$$\tau = \varepsilon_0/\sigma \tag{6.8a}$$

is known as the relaxation time. The relation shows that any original distribution of charge decays exponentially at a rate that is independent of any other electromagnetic disturbances that may be taking place.

6.5 Electromagnetic Potentials and Wave Equations in Vacuum

We saw in Chapter 5 $\vec{E}$ and $\vec{B}$ can be expressed in terms of the electromagnetic potentials $\vec{A}$ and ϕ:

$$\vec{B} = \nabla \times \vec{A} \tag{6.9}$$

$$\vec{E} = \nabla\phi - \partial\vec{A}/\partial t \tag{6.10}$$

The electromagnetic potentials satisfy the gauge transformations:

$$\vec{A}' = \vec{A} + \nabla\chi \tag{6.11}$$

$$\phi' = \phi - \partial\chi/\partial t \tag{6.12}$$

where χ is an arbitrary function. The two sets of the electromagnetic potentials $(\vec{A}, \phi)$ and $(\vec{A}', \phi')$ yield the same $\vec{E}$ and $\vec{B}$. Equations involving electromagnetic potentials $\vec{A}$ and ϕ must be gauge invariant.

Substituting equation (6.10) in equation (6.2a), we obtain

$$\nabla \cdot (-\nabla\phi - \partial\vec{A}/\partial t) = \rho/\varepsilon_0$$

or

$$\nabla^2\phi + \frac{\partial}{\partial t} (\nabla \cdot \vec{A}) = -\rho/\varepsilon_0 \tag{6.13}$$

Now substituting equations (6.9) and (6.10) into equation (6.2b), we obtain

$$\nabla \times (\nabla \times \vec{A}) - \mu_0\varepsilon_0 \frac{\partial}{\partial t}\left(-\nabla\phi - \frac{\partial\vec{A}}{\partial t}\right) = \mu_0\vec{j}$$

Using the vector identity $\nabla \times (\nabla \times \vec{A}) = \nabla (\nabla \cdot \vec{A}) - \nabla^2\vec{A}$, this equation becomes

$$-\nabla^2\vec{A} + \nabla\left(\nabla \cdot \vec{A} + \mu_0\varepsilon_0\frac{\partial\phi}{\partial t}\right) + \mu_0\varepsilon_0\frac{\partial^2\vec{A}}{\partial t^2} = \mu_0\vec{j} \tag{6.14}$$

We can simplify this equation by imposing some conditions on $\vec{A}$ and ϕ in such a way that it does not change the physics. In other words, it must be consistent with the transformations (6.11) and (6.12) so that $\vec{B}$ and $\vec{E}$ remain unaffected. The Coulomb gauge, $\nabla \cdot \vec{A} = 0$, used in magnetostatics does not help us very much. Another choice is to impose a condition on $\vec{A}$ and ϕ such that

$$\nabla \cdot \vec{A} + \mu_0 \varepsilon_0 \frac{\partial \phi}{\partial t} = 0 \qquad (6.15)$$

This is known as the Lorentz gauge condition, which is not quite as arbitrary as it appears at first sight. We can relate it to the first principle of electromagnetic theory: Coulomb's law, Biot-Savart's law, and the principle of conservation of energy.

With the Lorentz gauge condition, equation (6.13) and equation (6.15) become

$$\nabla^2 \phi - \mu_0 \varepsilon_0 \frac{\partial^2 \phi}{\partial t^2} = -\rho / \varepsilon_0 \qquad (6.16)$$

$$\nabla^2 \vec{A} - \mu_0 \varepsilon_0 \frac{\partial^2 \vec{A}}{\partial t^2} = -\mu_0 \vec{j} \qquad (6.17)$$

In space where ρ and $\hat{j}$ vanish, we then have two homogeneous equations

$$\nabla^2 \vec{A} - \mu_0 \varepsilon_0 \frac{\partial^2 \vec{A}}{\partial t^2} = 0 \qquad (6.18)$$

$$\nabla^2 \phi - \mu_0 \varepsilon_0 \frac{\partial^2 \phi}{\partial t^2} = 0 \qquad (6.19)$$

They are wave equations. In free space where both ρ and $\hat{j}$ vanish, the electric and magnetic fields also satisfy the homogeneous wave equation

$$\left(\nabla^2 - \mu_0 \varepsilon_0 \frac{\partial^2}{\partial t^2} \right) \begin{cases} \vec{E} \\ \vec{B} \end{cases} = 0 \qquad (6.20)$$

Equations (6.2b) and (6.2d) show how this occurs: A changing $\vec{E}$ creates some new $\vec{B}$, and a changing $\vec{B}$ creates some new $\vec{E}$. Thus, the fact of changing fields leads to new (changing) fields, and the wave, the electromagnetic wave, propagates itself. We thus have the remarkable result that in space the electric and magnetic fields $\vec{E}$ and $\vec{B}$ as well as the scalar and vector potential all satisfy the same basic wave equation:

$$\nabla^2 U - \frac{1}{v^2}\frac{\partial^2 U}{\partial t^2} = 0 \tag{6.21}$$

Equation (6.20) says that the changes of the fields propagate through empty space as waves with a constant speed given by $1/\sqrt{\mu_0\varepsilon_0}$, where the constant μ_0 is the permeability of the vacuum and it has, by definition, the exact value $4\pi \times 10^{-7}$ henries per meter (H/m); the constant ε_0 is the permittivity of the vacuum. The value of ε_0 must be determined by measurement and is found to be 8.854×10^{-12} farads per meter (F/m). Thus, the electromagnetic waves propagate through empty space with a speed equal to

$$c = 1/\sqrt{\mu_0\varepsilon_0} = 3 \times 10^8\,\text{m/s}$$

Maxwell identified this with the speed of light c. Interference and diffraction permit very exact determinations of wavelength λ. For long wavelengths it is sometimes possible to measure the frequency ν precisely by comparison with oscillators. The basic idea of wave propagation is conformed by the agreement of the values of $\nu\lambda$ with c.

Light is a form of electromagnetic radiation. Gamma rays, x-rays, ultraviolet, infrared radio, and microwave radiations are all electromagnetic radiations, differing only in wavelengths. If we use rectangular coordinates and resolve the vector wave equation (6.19) into components, each component of $\vec{E}$ and $\vec{B}$ satisfies the general scalar wave equation

$$\frac{\partial^2 U}{\partial x^2} + \frac{\partial^2 U}{\partial y^2} + \frac{\partial^2 U}{\partial z^2} = \frac{1}{v^2}\frac{\partial^2 U}{\partial t^2} \tag{6.22}$$

where the quantity U stands for any one of the field components E_x, E_y, E_z, B_x, B_y, B_z. Equation (6.22) is well known and its general solution is a superposition of an infinite set of one-dimensional waves traveling in all possible directions. We can obtain the basic information we need by studying the equation for one spatial dimension and the time:

$$\frac{\partial^2 U}{\partial z^2} = \frac{1}{v^2}\frac{\partial^2 U}{\partial t^2} \tag{6.23}$$

Suppose that at $t = 0$ its solution is an arbitrary function $f(z)$. It travels along the z direction at speed v, so that at time t it has traveled a distance vt in space and it is not distorted. The solution is therefore of the form $U(z, t) = f(z - vt)$; you can check this by direct substitution.

U is any periodic function of $(z - vt)$; a possible case would be a sine or a cosine harmonic function. **Figure 6.2** shows a plane wave $U(z,t)$ traveling along Ox at speed v in the time t. $\Delta z = vt$ is the distance between any two

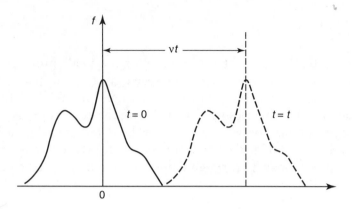

I **FIGURE 6.2**

points of corresponding phase. This is the reason why v is called the phase velocity.

It is also possible for the wave to travel in the opposite direction $-z$. Hence, a general solution of equation (6.23) is of the form

$$U(z, t) = f(z - vt) + f(z + vt)$$

Electromagnetic waves differ from the simple waves such as waves on a string, because the quantity that varies in space and time is a vector field and can exist in a vacuum.

6.6 Plane Electromagnetic Waves

We first show that electromagnetic waves are transverse. As shown above, a solution of the wave equation (6.20) that corresponds to a traveling wave in the Oz-direction, with plane wave fronts parallel to the Oxy-plane, is of the form

$$\vec{E}(z - ct) = [f(z - ct), \quad g(z - ct), \quad h(z - ct)]$$

where f, g, h are any functions of $(z - ct)$, and $c = (\mu_0 \varepsilon_0)^{1/2}$.

The waves must satisfy Maxwell equations. Now the equation $\nabla \cdot \vec{E} = 0$ gives

$$\nabla \cdot \vec{E} = \frac{\partial f}{\partial x} + \frac{\partial g}{\partial y} + \frac{\partial h}{\partial z} = \frac{\partial h}{\partial z} = 0$$

giving h = constant. A constant component of the field is of no interest here, so set it to be zero, and then

$$\vec{E} = [f(z - ct), g(z - ct), 0] \tag{6.24}$$

Then the Maxwell equation (6.1d), Faraday's law of induction, gives

$$\partial\vec{B}/\partial t = -\nabla \times \vec{E} = -(\partial/\partial x, \partial/\partial y, \partial/\partial z) \times [f(z - ct), g(z - ct), 0]$$

or

$$\partial\vec{B}/\partial t = [g'(z - ct), -f'(z - ct), 0]$$

From which we obtain, by integrating with respect to t,

$$c\vec{B} = [-g(z - ct), f(z - ct), 0] = \hat{k} \times \vec{E}$$

Thus, the electric wave is accompanied by a magnetic wave, and $\vec{E}$ and $\vec{B}$ are mutually perpendicular. Furthermore, $\vec{E} \times \vec{B}$ is in the direction of wave propagation, $\vec{E}$, $\vec{B}$ and the direction of propagation $\hat{k}$ form a right-handed set of orthogonal vectors. It is also worth noting that $|\vec{E}| = c|\vec{B}|$ everywhere; this means that $|\vec{B}|$ is much smaller numerically than $|\vec{E}|$.

We showed earlier that the solution of equation (6.23) is any periodic function of $(z - ct)$. By direct substitution it is easy to verify that the function $U = |\vec{E}|$ where

$$\vec{E} = \hat{i}E_0 \sin k(z - ct) = \hat{i}E_0 \sin(kz - \omega t) \tag{6.25}$$

is a solution of equation (6.23) provided that the ratio of ω and k equal to the speed c:

$$k = 2\pi/\lambda = 2\pi f/\lambda f = \omega/c \tag{6.26}$$

$k = 2\pi/\lambda$ is called the wave number or the propagation constant. For $\vec{B}$ field we can write, for generality,

$$\vec{B} = \hat{j}B_y = \hat{j}B_0 \sin(\pm kz - \omega t + \varphi)$$

To find which sign should be used, we apply it to Faraday's law and Ampere's law:

$$\frac{\partial E_x}{\partial z} = -\frac{\partial B_y}{\partial t}, \quad \frac{\partial B_y}{\partial z} = -\varepsilon_0\mu_0\frac{\partial E_x}{\partial t}$$

From these we get

$$kE_0 \cos(kz - \omega t) = \omega B_0 \cos(\pm kz - \omega t + \varphi) \tag{6.27}$$

$$\varepsilon_0 \mu_0 \omega E_0 \cos(kz - \omega t) = \pm kB_0 \cos(\pm kz - \omega t + \varphi) \tag{6.28}$$

Dividing the equations we get

$$\frac{k}{\varepsilon_0 \mu_0 \omega} = \pm \frac{\omega}{k}$$

Because $c = \omega/k$, only the solution with the plus sign makes sense,

$$\vec{B} = \hat{k} B_y = \hat{k} B_0 \sin(kz - \omega t + \varphi) \tag{6.29}$$

What is the value of the phase factor φ? To answer this question, let us rewrite equation (6.27) as

$$\frac{2\pi}{\lambda} E_0 \cos 2\pi \left(\frac{z}{\lambda} - ct\right) = 2\pi c B_0 \cos 2\pi \left(\frac{z}{\lambda} - ct + \frac{\varphi}{2\pi}\right) \tag{6.27a}$$

which is true for all values of t and z. At the point $z/\lambda - ct = 1/4$, we have

$$\frac{2\pi}{\lambda} E_0 \cdot 0 = 2\pi c B_0 \cos 2\pi \left(\frac{\pi}{2} + \varphi\right) = 2\pi c B_0 \sin \varphi$$

which gives

$$\sin \varphi = 0 \quad \text{or} \quad \varphi = 0 \text{ or } \pi$$

To determine which value should be used, let us apply equation (6.27a) at the point $z/\lambda - ct = 0$ and obtain

$$\frac{2\pi}{\lambda} E_0 \cdot 1 = 2\pi c B_0 \cos \varphi$$

Because λ, E_0, c, and B_0 are all positive, φ must be equal to 0. Equation (6.29) becomes

$$\vec{B} = \hat{j} B_z = \hat{j} B_0 \sin(kz - \omega t) \tag{6.30}$$

and

$$B_0 = E_0/c \tag{6.31}$$

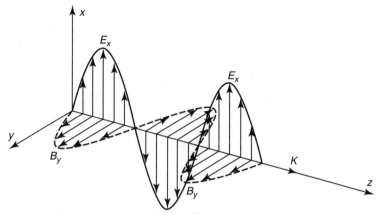

I FIGURE 6.3

The physical restrictions on electromagnetic waves can now be summarized below (**Figure 6.3**):

1. The $\vec{B}$ wave is perpendicular to $\vec{E}$ wave; they are always in phase.
2. The direction of travel of the wave is $\vec{E} \times \vec{B}$.
3. $B_0 = E_0/c$, $c^2 = 1/\varepsilon_0\mu_0$.

Because one of the waves determines the other, we need only discuss one of them explicitly. We usually choose $\vec{E}$.

6.7 Polarization

The plane wave described by equation (6.25) is said to be linearly polarized because the electric field exists in one transverse direction (the x-axis) only as it propagates. The electric vector in a plane wave could also be circularly or elliptically polarized. To discuss the latter two cases, we now let the electric field in the xy plane instead along the x-axis:

$$\vec{E} = \vec{E}_0 \exp i(kz - \omega t) \tag{6.32}$$

where $\vec{E}_0$ is a vector in the xy plane normal to the direction of propagation Oz. To see this, let us first recall that E_z is constant:

$$\frac{\partial E_z}{\partial z} = 0$$

Now equation (6.32) gives

$$\frac{\partial \vec{E}}{\partial z} = k\vec{E}$$

The only value of E_z that satisfies the last two equations is

$$E_z = 0$$

Therefore, $\vec{E}_0$ is a vector in the xy plane normal to the direction of propagation Ox:

$$\vec{E}_0 = E_{0y}\hat{j} + E_{0x}\hat{i} \qquad (6.33)$$

where E_{0y} and E_{0x} are constants. The electric field exists in one transverse direction as it propagates, given by

$$\tan \theta = E_{0y}/E_{0x}$$

so the plane wave described by equation (6.32) is linearly polarized. Although E_{0y} *and* E_{0x} can be of different magnitudes, they are always in phase and so θ is constant.

A plane wave composed of x and y components with constant phase differences and different amplitudes is elliptically polarized; if the amplitudes are the same, the wave is circularly polarized. These can be seen from the following example:

$$\vec{E} = \hat{j}E_{0y}\exp\left(kz - \omega t + \phi_y\right) + \hat{i}E_{0x}\exp\left(kz - \omega t + \phi_x\right)$$

If $\phi_y - \phi_x = \pi/2$, the real part is

$$\vec{E} = \hat{j}E_{0y}\cos\left(kz - \omega t + \phi_y\right) - \hat{i}E_{0x}\sin\left(kz - \omega t + \phi_y\right)$$

Thus, $\vec{E}$ rotates in the xy plane with direction

$$\tan\left(kz - \omega t + \phi_y\right) = - E_{0y}/E_{0x}$$

which trace out a circle when $E_{0y} = E_{0x}$ and an ellipse when $E_{0y} \neq E_{0x}$, as shown in **Figure 6.4**: (a) linearly polarization, (b) circularly polarization, and (c) elliptically polarization. Plane waves are the simplest to describe mathematically and are used extensively to study propagation in dielectrics and conductors in later chapters.

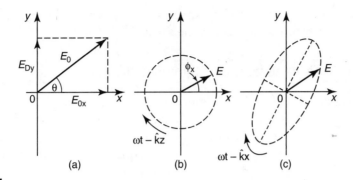

I FIGURE 6.4

Maxwell's theory was put forward in 1865, but it was not until 1888 that H. Hertz demonstrated the production of radiation by recognizably electrical means. He produced a current distribution by inducing a discharge between two spheres to which were attached large metal sheets (**Figure 6.5**). This generated a plane-polarized wave with its electric vector parallel to the gap. Hertz detected the radiation with a broken loop of wire, which acted as a resonator. The electric field of the waves generated a sufficiently strong potential gradient across the detector gap to break down its insulation and cause a spark. Hertz found further that the detector failed to respond if its gap was perpendicular to the source gap and that its sensitivity was greatest when the two gaps were parallel. The state of polarization of the radiation was thus clearly shown. With Hertz's arrangement, the wavelength of the radiation was about 5 meters, that is, 10^7 times longer than ordinary light,

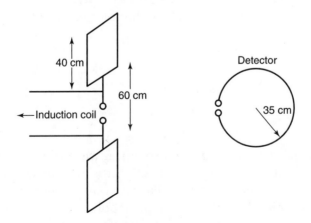

I FIGURE 6.5

and it was a clear tribute to the success of Maxwell's theory that it could account for both of them in the same terms.

6.8 Waves in Three Dimensions, Spherical Waves

Plane waves are the simplest to describe mathematically, and we use them to study wave propagation in later chapters. In free space they propagate without attenuation in amplitude. In three dimensions they are spherical waves, and so it is instructive to rewrite the wave equation in spherical polar coordinates and consider its spherical symmetric solution.

The Laplacian ∇^2 in spherical polar coordinates with spherical symmetric is

$$\nabla^2 U = \frac{1}{r^2} \frac{\partial}{\partial r}\left(r^2 \frac{\partial U}{\partial r}\right) = \frac{1}{r^2}\left(2r \frac{\partial U}{\partial r} + r^2 \frac{\partial^2 U}{\partial r^2}\right)$$

This can be rewritten in a more compact form:

$$\nabla^2 U = \frac{1}{r}\left[\frac{\partial^2}{\partial r^2}(rU)\right]$$

Then wave equation (6.21) becomes

$$\frac{1}{r}\left[\frac{\partial^2}{\partial r^2}(rU)\right] = \frac{1}{c^2}\frac{\partial^2 U}{\partial t^2}$$

or

$$\frac{\partial^2}{\partial r^2}(rU) = \frac{1}{c^2}\frac{\partial^2}{\partial t^2}(rU)$$

This equation tells us that the function rU satisfies the one-dimensional wave equation in the variable r. There the solution is of the same form

$$rU(r,t) = f(r - ct)$$

or

$$U(r,t) = \frac{1}{r}f(r - ct)$$

Such a function represents a general spherical wave traveling outward from the origin at the speed c. Because of the factor $1/r$, we see that a spherical wave decays in amplitude as $1/r$, unlike the constant amplitude plane wave. Obviously, the spherical solution cannot apply at the origin where $r = 0$. There is another difference. We saw that for plane wave traveling in the opposite direction, $g(z + ct)$ also satisfies the plane wave equation, but $(1/r)g(r + ct)$ implies a spherical wave collapsing to a source point. We see in Chapter 13 on the generation of electromagnetic waves this inward solution as well as the outgoing solution are not valid near the origin when $r \ll \lambda$. More detailed discussion on spherical waves may be found in *Foundations of Electromagnetic Theory*, third edition, by J. R. Ritz, F. J. Milford, and R. W. Christy (Addison-Wesley, Reading, MA, 1979).

6.9 Energy and Momentum in Electromagnetic Waves

It was shown in Chapter 2 that the work done in setting up, in vacuum, a static charge distribution from a state in which the charges are infinity dispersed can be expressed in alternative form

$$\frac{1}{2}\int \rho V d\tau = \frac{1}{2}\varepsilon_0 \int E^2 d\tau$$

and that right-hand expression embodies of field energy density $\varepsilon_0 E^2/2$. Likewise, in Chapter 5, we introduced the concept of magnetic energy density $\mu_0 H^2/2$ for the vacuum magnetic field of steady current. We now establish that the Maxwell equations yield the conclusion that the same expressions can legitimately be interpreted as energy densities in the general time varying case when both electric and magnetic fields are present. That is, we show that electromagnetic waves carry energy, and the energy density u is calculated as

$$u = 1/2 \ \varepsilon_0 E^2 + 1/2 \ \mu_0 B^2 \tag{6.34}$$

For the simple plane waves this becomes

$$u = \varepsilon_0 E_0^2 \sin^2(kx - \omega t) \tag{6.34a}$$

Averaged over time and space we get

$$\langle u \rangle_{av} = 1/2 \ \varepsilon_0 E_0^2 \tag{6.34b}$$

To show electromagnetic waves carrying energy, we go back to the vector $\vec{E} \times \vec{B}$ that points in the direction of propagation. Let us take divergence of this vector

$$\nabla \cdot (\vec{E} \times \vec{B}) = \vec{B} \cdot (\nabla \times \vec{E}) - \vec{E} \cdot (\nabla \times \vec{B}) \tag{6.35}$$

Now $\nabla \times \vec{E} = -\partial \vec{B}/\partial t, \quad \nabla \times \vec{B} = \varepsilon_0 \mu_0 (\partial \vec{E}/\partial t)$

Substituting these into equation (6.35) we obtain

$$\nabla \cdot (\vec{E} \times \vec{B}) = -\vec{B} \cdot (\partial \vec{B}/\partial t) - \vec{E} \cdot \varepsilon_0 \mu_0 (\partial \vec{E}/\partial t)$$

$$= -\frac{\partial}{\partial t} \left(\frac{\varepsilon_0 \mu_0 E^2}{2} + \frac{B^2}{2} \right)$$

or

$$\nabla \cdot \left(\frac{\vec{E} \times \vec{B}}{\mu_0} \right) = -\frac{\partial}{\partial t} \left(\frac{\varepsilon_0 E^2}{2} + \frac{B^2}{2\mu_0} \right)$$

We recognize that the quantity inside the parentheses on the right is the energy density in an electromagnetic field. Now integrate over a finite volume τ of area a, and then apply the divergence theorem on the left:

$$\int_a \frac{\vec{E} \times \vec{B}}{\mu_0} \cdot d\vec{a} = -\frac{\partial}{\partial t} \int_\tau \frac{1}{2} (\varepsilon_0 E^2 + B^2/\mu_0) \, d\tau \tag{6.36}$$

a. Poynting Theorem and Poynting Vector

The term on the right of equation (6.36), with its negative sign, gives the decrease in the total electromagnetic energy inside the volume τ per unit time. The integral on the left must then give the rate at which electromagnetic energy flows out of τ. Equation (6.36) is simply a statement of the conservation of energy in an electromagnetic field. It is known as the *Poynting theorem*, after John Poynting (1852–1914), and the vector

$$\vec{S} = \frac{1}{\mu_0} (\vec{E} \times \vec{B}) \tag{6.37}$$

is called the Poynting vector, and it gives the energy per unit time, per unit area, transported by the fields, that is, it specifies the power flux intrinsic to an electromagnetic field. For monochromatic plane wave propagating in the z direction,

$$S = c\varepsilon_0 E_0^2 \sin^2(kz - \omega t) \, \hat{k} = cu\hat{k}. \tag{6.37a}$$

In many situations we are only interested in time-average values of the energy flux. Denoting time averages by bracket < >, equation (6.37) becomes

$$\left\langle \vec{S} \right\rangle = \frac{1}{\mu_0} (\vec{E} \times \vec{B}) \tag{6.37b}$$

There is a degree of arbitrariness in the expression (6.37). Obviously, the addition of any solenoidal vector to (6.37) cannot be rejected, because it would not affect equation (6.36). More generally, the addition of a scalar function f to the energy density u [equation (6.34)] and a vector function $\vec{F}$ to equation (6.37) would not affect equation (6.36) provided only that

$$\nabla \cdot \vec{F} + f = 0$$

Hence, alternative expressions for the energy density and the Poynting vector have been proposed, but the reasons for advocating a departure from the conventional expressions (6.34) and (6.37) are still too inclusive.

The Poynting theorem, equation (6.36), is for the special case where there is no motion of charges (and, hence, currents). We now examine the general case that allows the possibility of moving charges. We begin with Faraday's law and Ampere's law and rewrite them in the following form, respectively:

$$\frac{\partial \vec{B}}{\partial t} = -\nabla \times \vec{E}$$

$$\varepsilon_0 \mu_0 \frac{\partial \vec{E}}{\partial t} = \nabla \times \vec{B} - \mu_0 \vec{j}$$

Next, we take the scalar product of $\vec{B}$ with the first of these equations and the scalar product of $\vec{E}$ with the second and add to obtain

$$\vec{B} \cdot \frac{\partial \vec{B}}{\partial t} + \varepsilon_0 \mu_0 \vec{E} \cdot \frac{\partial \vec{E}}{\partial t} = -\mu_0 \vec{j} \cdot \vec{E} - (\vec{B} \cdot \nabla \times \vec{E} - \vec{E} \cdot \nabla \times \vec{B})$$

This becomes, with the help of the identity $\nabla \cdot (\vec{E} \times \vec{B}) = \vec{B} \cdot \nabla \times \vec{E} - \vec{E} \cdot \nabla \times \vec{B}$

$$\frac{1}{2} \frac{\partial}{\partial t} (B^2 + \varepsilon_0 \mu_0 E^2) = -\mu_0 \vec{j} \cdot \vec{E} - \nabla \cdot (\vec{E} \times \vec{B})$$

or

$$\frac{1}{2} \frac{\partial}{\partial t} (\mu_0^{-1} B^2 + \varepsilon_0 E^2) = -\vec{j} \cdot \vec{E} - \nabla \cdot (\mu_0^{-1} \vec{E} \times \vec{B}) \tag{6.38}$$

Note that the quantity $\mu_0^{-1}(\vec{E} \times \vec{B})$ is the Poynting vector $\vec{S}$ in free space. Integrating this equation over a fixed volume τ, we obtain

$$\frac{d}{dt} \int_\tau \frac{1}{2} (\varepsilon_0 E^2 + B^2/\mu_0)\, d\tau = -\int_\tau \vec{j} \cdot \vec{E}\, d\tau - \int_\tau \nabla \cdot \vec{S}\, d\tau$$

where we have moved the partial derivative $\partial/\partial t$ out of integral sign and changed it to a total derivative d/dt, because both the volume τ and the bounding surface a are fixed. By transforming the last term on the right to a surface integral by means of the divergence theorem, we obtain

$$\frac{d}{dt}\int_\tau \frac{1}{2}\,(\varepsilon_0 E^2 + B^2/\mu_0)\,d\tau = -\int_\tau \vec{j}\cdot\vec{E}\,d\tau - \oint_a \vec{S}\cdot d\vec{a} \tag{6.39}$$

What is the meaning of the $\vec{j}\cdot\vec{E}$ term on the right side? The current distribution represented by $\vec{j}$ can be considered as made up of various charges q_α moving with velocities $\vec{v}_\alpha$. Then the volume integral of $\vec{j}\cdot\vec{E}$ may be replaced by

$$\int_\tau \vec{j}\cdot\vec{E}\,d\tau \rightarrow \sum_\alpha q_\alpha \vec{v}_\alpha \cdot \vec{E}_\alpha$$

where $\vec{E}_\alpha$ is the electric field at the position of the charge q_α.

The work done per unit time on the charge q_α by the electromagnetic force is

$$dW_\alpha/dt = \vec{F}_\varepsilon \cdot \vec{v}_\varepsilon$$

where the electromagnetic force on the charge q_α is given by the Lorentz expression

$$\vec{F}_\alpha = q_\alpha(\vec{E}_\alpha + \vec{v}_\alpha \times \vec{B}_\alpha)$$

Thus, $dW_\alpha/dt = \vec{F}_\varepsilon \cdot \vec{v}_\varepsilon = q_\alpha \vec{v}_\alpha \cdot (\vec{E}_\alpha + \vec{v}_\alpha \times \vec{B}_\alpha)$.

But the magnetic force is always perpendicular to the velocity and so can do no work on a charged particle:

$$\vec{v}_\alpha \cdot (\vec{v}_\alpha \times \vec{B}) = (\vec{v}_\alpha \times \vec{v}_\alpha) \cdot \vec{B} = 0$$

so dW_α/dt reduces to

$$dW_\alpha/dt = q_\alpha \vec{v}_\alpha \cdot \vec{E}_\alpha$$

Equation (6.39) may now be written as

$$\frac{dW}{dt} = -\frac{d}{dt}\int_\tau \frac{1}{2}\,(\varepsilon_0 E^2 + \frac{1}{\mu_0} B^2)\,d\tau - \oint_a \vec{S}\cdot d\vec{a} \tag{6.40}$$

where $W = \sum_\alpha W_\alpha$. The first integral on the right is the total energy stored in the electromagnetic field

$$U_{EB} = \int \frac{1}{2}\,(\varepsilon_0 E^2 + B^2/\mu_0)\,d\tau \tag{6.41}$$

and the second integral is the energy flow-out across its boundary surface a per unit time. The Poynting theorem now states that

> The work done on the charges by the electromagnetic force is equal to the decrease in energy stored in the field, less the energy that flowed out through the surface.

The work done on the charges by the field increases their mechanical energy U_m:

$$dW/dt = dU_m/dt$$

Substituting these into equation (6.40) we have

$$-\frac{dU_{EB}}{dt} = \frac{dU_m}{dt} + \oint_a \vec{S} \cdot d\vec{a} \tag{6.42}$$

Equation (6.42) states that the decrease in field energy is equal to the work done on the charges by the field and the energy flowing out the volume through surface a.

If τ is the whole space and the charges are distributed in the finite region, the fields then tend to be zero at boundary surface a, which is infinitely far from the charges. This means that the surface integral on the right side of equation (6.42) is zero, and we have

$$\frac{d}{dt}(U_{EB} + U_m) = 0, \quad \text{or} \quad U_{EB} + U_m = \text{constant}. \tag{6.43}$$

That is, neither the mechanical energy of the charges nor the field energy is conserved. Only their sum is conserved.

The energy transport in time-varying fields is readily measurable, and in all cases the Poynting vector gives the correct distribution and amount. But electric and magnetic fields are decoupled under static conditions, so it is quite artificial to link them together in the Poynting vector, and rather strange results can be obtained.

b. Electromagnetic Field Momentum

Evidently, a unidirectional flow of energy implies a momentum, and it is gratifying this is so. Otherwise, momentum conservation would not be valid in electrodynamics. In the following we show that the field momentum density is given by

$$\vec{p}_{EB} = \varepsilon_0 \mu_0 \vec{S} = \vec{S}/c^2 \tag{6.44}$$

where $\vec{S}$ is the Poynting vector. To show this, we consider some volume τ, in which there is an electromagnetic field interacting with charges enclosed in this volume. The total electromagnetic force on the charges in τ is given by

$$\vec{F} = \int_\tau (\rho\vec{E} + \vec{j} \times \vec{B})\, d\tau \tag{6.45}$$

where $\hat{j} = \rho\vec{v}$. We can eliminate ρ and $\hat{j}$ by using Gauss' law and Ampere's law

$$\rho = \varepsilon_0 \nabla \cdot \vec{E}, \quad \hat{j} = \mu_0^{-1} \nabla \times \vec{B} - \varepsilon_0 \partial\vec{E}/\partial t$$

and obtain

$$\vec{F} = \int_\tau \left[\varepsilon_0 (\nabla \cdot \vec{E})\vec{E} + \left(\mu_0^{-1} \nabla \times \vec{B} - \varepsilon_0 \partial\vec{E}/\partial t \right) \times \vec{B} \right] d\tau$$

Now

$$\frac{\partial\vec{E}}{\partial t} \times \vec{B} = \frac{\partial}{\partial t}(\vec{E} \times \vec{B}) - \vec{E} \times \frac{\partial\vec{B}}{\partial t} = \frac{\partial}{\partial t}(\vec{E} \times \vec{B}) + \vec{E} \times (\nabla \times \vec{E})$$

In the last step Faraday's law was used. Now using the identity

$$\nabla(\vec{A} \cdot \vec{B}) = \vec{A} \times (\nabla \times \vec{B}) + \vec{B} \times (\nabla \times \vec{A}) + (\vec{A} \cdot \nabla)\vec{B} + (\vec{B} \cdot \nabla)\vec{A}$$

we find

$$\nabla(E^2) = \nabla(\vec{E} \cdot \vec{E}) = 2\vec{E} \times (\nabla \times \vec{E}) + 2(\vec{E} \cdot \nabla)\vec{E}$$

or

$$\vec{E} \times (\nabla \times \vec{E}) = \frac{1}{2}\nabla(E^2) - (\vec{E} \cdot \nabla)\vec{E}$$

The same goes for $\vec{B}$. Now the total force $\vec{F}$ is transformed into

$$\vec{F} = \int_\tau \left\{ \varepsilon_0 \left[(\nabla \cdot \vec{E})\vec{E} + (\vec{E} \cdot \nabla)\vec{E} \right] + \frac{1}{\mu_0} \left[(\nabla \cdot \vec{B})\vec{B} + (\vec{B} \cdot \nabla)\vec{B} \right] \right.$$
$$\left. - \frac{1}{2}\nabla(\varepsilon_0 E^2 + \mu_0^{-1} B^2) - \varepsilon_0 \frac{\partial}{\partial t}(\vec{E} \times \vec{B}) \right\} d\tau \tag{6.46}$$

We can shorten this long expression by introducing the Maxwell stress tensor:

$$\vec{T} = \left[\varepsilon_0 \vec{E}\,\vec{E} + \mu_0^{-1}\vec{B}\,\vec{B} - \frac{1}{2}(\varepsilon_0 E^2 + \mu_0^{-1}B^2)\vec{I} \right] \tag{6.47}$$

where $\overrightarrow{I} = \hat{i}\hat{i} + \hat{j}\hat{j} + \hat{k}\hat{k}$ and $\hat{i}, \hat{j}, \hat{k}$ are unit vectors along x, y, z-axes. In component we have

$$T_{ij} = (\varepsilon_0 E_i E_j + \mu_0^{-1} B_i B_j) - \frac{1}{2}\delta_{ij}(\varepsilon_0 E^2 + \mu_0^{-1} B^2) \tag{6.47a}$$

The indices i, j, k refer to the coordinates x, y, z, and so E_1, E_2, ..., B_1, B_2, ... are E_x, E_y ..., B_x, B_y, ..., respectively. Similarly, T_{11}, T_{12} ... stand for T_{xx}, T_{xy}, ...; there are nine of them. δ_{ij} is the Kronecker delta symbol: $\delta_{ij} = 1$ if $i = j$ and $\delta_{ij} = 0$ if $i \neq j$. We can form dot product of $\overrightarrow{T}$ with a vector $\overrightarrow{A}$:

$$(\overrightarrow{A} \cdot \overrightarrow{T})_j = \sum_i A_i T_{ij} \tag{6.48}$$

In terms of Maxwell stress tensor, equation (6.44) takes a very simple form:

$$\overrightarrow{F} = \int_\tau \left[\nabla \cdot \overrightarrow{T} - \varepsilon_0 \mu_0 \frac{\partial \overrightarrow{S}}{\partial t} \right] d\tau \tag{6.49}$$

where $\overrightarrow{S}$ is the Poynting vector. It is easy to show that equation (6.49) is the same as equation (6.45). We note that

$$\nabla \cdot (\overrightarrow{E}\,\overrightarrow{E}) = (\nabla \cdot \overrightarrow{E})\overrightarrow{E} + (\overrightarrow{E} \cdot \nabla)\overrightarrow{E}$$

and

$$\nabla \cdot \left(\frac{1}{2} E^2 \overrightarrow{I} \right) = \nabla \cdot \left[\frac{1}{2} E^2 (\hat{i}\hat{i} + \hat{j}\hat{j} + \hat{k}\hat{k}) \right]$$

which can be rewritten as

$$\nabla \cdot \left(\frac{1}{2} E^2 \overrightarrow{I} \right) = \left(\nabla \cdot \frac{1}{2} E^2 \hat{i} \right)\hat{i} + \left(\nabla \cdot \frac{1}{2} E^2 \hat{j} \right)\hat{j} + \left(\nabla \cdot \frac{1}{2} E^2 \hat{k} \right)\hat{k}$$

$$= \frac{\partial}{\partial x}\left(\frac{1}{2} E^2 \right)\hat{i} + \frac{\partial}{\partial y}\left(\frac{1}{2} E^2 \right)\hat{j} + \frac{\partial}{\partial z}\left(\frac{1}{2} E^2 \right)\hat{k}$$

$$= \hat{i}\left[\frac{\partial \overrightarrow{E}}{\partial x} \cdot \overrightarrow{E} \right] + \hat{j}\left[\frac{\partial \overrightarrow{E}}{\partial y} \cdot \overrightarrow{E} \right] + \hat{k}\left[\frac{\partial \overrightarrow{E}}{\partial z} \cdot \overrightarrow{E} \right] = (\nabla \overrightarrow{E}) \cdot \overrightarrow{E}$$

Hence,

$$\nabla \cdot (\overrightarrow{E}\,\overrightarrow{E} - E^2 \overrightarrow{I}/2) = (\overrightarrow{E} \cdot \nabla)\overrightarrow{E} + (\nabla \cdot \overrightarrow{E})\overrightarrow{E} - (\nabla \overrightarrow{E}) \cdot \overrightarrow{E}$$

$$= (\nabla \cdot \overrightarrow{E})\overrightarrow{E} + [(\overrightarrow{E} \cdot \nabla)\overrightarrow{E} - (\nabla \overrightarrow{E}) \cdot \overrightarrow{E}] \tag{6.50}$$

$$= (\nabla \cdot \overrightarrow{E})\overrightarrow{E} + (\nabla \times \overrightarrow{E}) \times \overrightarrow{E}$$

where in the last step we have used the identity

$$(\vec{\alpha} \cdot \nabla)\vec{\beta} - (\nabla\vec{\beta}) \cdot \vec{\alpha} = (\nabla \times \vec{\beta}) \times \vec{\alpha}$$

Similarly, we have

$$\nabla \cdot (\vec{B}\vec{B} - B^2 \vec{I}/2) = (\nabla \cdot \vec{B})\vec{B} + (\nabla \times \vec{B}) \times \vec{B} \tag{6.51}$$

Adding equations (6.50) and (6.51) together, we see that equation (6.49) is the same as equation (6.46).

It is easy to show that

$$\int_{\tau}(\nabla \cdot \vec{T})\,d\tau = \oint_{a}\vec{T} \cdot d\vec{a} \tag{6.52}$$

where a is the surface bounding τ. To show this you first write

$$\vec{T} = T_{11}\hat{i}\hat{i} + T_{12}\hat{i}\hat{j} + T_{13}\hat{i}\hat{k} + T_{21}\hat{j}\hat{i} + \ldots = \vec{T_1}\hat{i} + \vec{T_2}\hat{j} + \vec{T_3}\hat{k}$$

where $\vec{T_1} = T_{11}\hat{i} + T_{21}\hat{j} + \vec{T_{31}}\hat{k}$, and so on. The proof then follows naturally. Equation (6.52) is not valid if $\vec{T}$ is not symmetric. Setting equation (6.52) into equation (6.49) we obtain

$$\vec{F} = \oint_{a} d\vec{a} \cdot \vec{T} - \varepsilon_0\mu_0 \frac{d}{dt}\int_{\tau}\vec{S}\,d\tau \tag{6.53}$$

Note that we have replaced $\partial/\partial t$ by d/dt and take it out of the integral sign, because τ is fixed. Physically, $\vec{T}$ is the stress acting on the surface: diagonal elements T_{xx}, T_{yy}, and T_{zz} represent pressures, and off-diagonal elements are shears.

Now the force acting on the charges is equal to the rate of change of momentum of the charges, $d\vec{P}/dt$. Thus, the last equation can be rewritten as

$$-\varepsilon_0\mu_0 \frac{d}{dt}\int_{\tau}\vec{S}\,d\tau = \frac{d\vec{P}}{dt} - \oint_{a} d\vec{a} \cdot \vec{T} \tag{6.54}$$

If we identify the integral on the left side as the momentum of the electromagnetic field in τ,

$$\vec{P}_{EB} = \varepsilon_0\mu_0 \int_{\tau}\vec{S}\,d\tau \tag{6.55}$$

Equation (6.54) is similar in structure to Poynting theorem and admits an analogous interpretation: The decrease in field momentum per unit time is equal to the momentum increase of the charges per unit time plus the momentum flowing out of τ per unit time through the boundary surface a. If τ is all of space, then no momentum flows in or out (this is equivalent to make the surface integral on the right to be zero), and the total momentum $(\vec{P} + \vec{P}_{EB})$ is conserved.

We can rewrite equation (6.54) in terms of the density of mechanical momentum $\vec{p}$ and the density of the field momentum $\vec{p}_{EB}$, where

$$\vec{p}_{EB} = \varepsilon_0 \mu_0 \vec{S} = \vec{S}/c^2 \tag{6.56}$$

In terms of these densities, equation (6.54) is equivalent to

$$\nabla \cdot (-\overleftrightarrow{T}) = -\frac{\partial}{\partial t}(\vec{p} + \vec{p}_{EB}). \tag{6.57}$$

This shows that $(-\overleftrightarrow{T})$ is the momentum density, playing the role of $\vec{j}$ in the continuity equation.

We now make a brief summary. The Poynting vector $\vec{S}$ is the energy per unit area, per unit time, transported by the electromagnetic field, whereas $\vec{S}/c^2$ is the field momentum per unit volume, $\overleftrightarrow{T}$ is the electromagnetic stress acting on a surface, and $\overleftrightarrow{T} \cdot d\vec{a}$ is the radiation pressure. Kepler, in about 1600, suggested quite correctly that the pressure of light caused the tails of comets to point away from the sun. But the matter was not tested quantitatively until Lebedev (1900) and Nichols and Hull (1903) carried out laboratory experiments on the pressure of light.

Through the transport of energy and momentum, the electromagnetic field acquires properties more usually associated with material particles. In quantum mechanics this association of wave and particle properties becomes a fundamental feature of the theory. According to Planck's hypothesis, a classic electromagnetic wave of angular frequency ω consists of a flow of *photons*, each carrying a quantum unit of energy, $\hbar\omega$, where $\hbar = h/2\pi$ is Planck's constant. The linear momentum of a photon then has a magnitude given by $\hbar\omega/c = h/\lambda$ (see Problem 12) entirely in accordance with de Broglie's hypothesis.

6.10 Angular Momentum

As electromagnetic fields carry energy and momentum, we expect electromagnetic fields to also carry angular momentum. Bethe confirmed this expectation in 1936. To discuss field angular momentum and its conservation, we start with the force exerted on charges by the field again:

$$dF = \rho(\vec{E} + \vec{v} \times \vec{B})\, d\tau$$

Then the torque about the origin of coordinates that is exerted by the field on the charge is just

$$d\vec{\Im} = \vec{r} \times d\vec{F} = \vec{r} \times \rho(\vec{E} + \vec{v} \times \vec{B})\, d\tau \tag{6.58}$$

If $\vec{L}_m$ is the angular momentum of the charges, then $d\vec{L}_m/dt$ is equal to the total torque acting on the charges. Thus,

$$\frac{d\vec{L}_m}{dt} = \int_\tau \vec{r} \times \rho(\vec{E} + \vec{v} \times \vec{B})\, d\tau \tag{6.59}$$

According to equation (6.49) the force density $\rho(\vec{E} + \vec{v} \times \vec{B})$ can be written as

$$\rho(\vec{E} + \vec{v} \times \vec{B}) = \nabla \cdot \vec{T} - \varepsilon_0 \mu_0 \frac{\partial}{\partial t}(\vec{E} \times \vec{B}) = \nabla \cdot \vec{T} - \frac{1}{c^2} \frac{\partial \vec{S}}{\partial t}$$

Substituting this into equation (6.59) we obtain

$$\frac{d\vec{L}_m}{dt} = -\frac{1}{c^2} \frac{d}{dt} \int_\tau \vec{r} \times \vec{S}\, d\tau + \int_\tau \vec{r} \times (\nabla \cdot \vec{T})\, d\tau$$

which can be rewritten as

$$\frac{d\vec{L}_m}{dt} = -\frac{1}{c^2} \frac{d}{dt} \int_\tau \vec{r} \times \vec{S}\, d\tau - \int_\tau \nabla \cdot (\vec{T} \times \vec{r})\, d\tau \tag{6.60}$$

where we have made use of the relation $\vec{r} \times (\nabla \cdot \vec{T}) = -\nabla \cdot (\vec{r} \times \vec{T})$. Applying the divergence theorem to the second integral on the right, equation (6.60) becomes

$$\frac{d\vec{L}_m}{dt} = -\frac{d}{dt} \int_\tau \frac{1}{c^2} \vec{r} \times \vec{S}\, d\tau - \oint (\vec{T} \times \vec{r}) \cdot d\vec{a} \tag{6.61}$$

Similar to previous discussion on energy and momentum conservation, we can identify $c^{-2}\vec{r} \times \vec{S}$ as the angular momentum density of the electromagnetic field, and

$$\vec{L}_{EB} = \int_\tau \frac{1}{c^2} \vec{r} \times \vec{S}\, d\tau \tag{6.62}$$

is the angular momentum of the field in volume τ.

Equation (6.61) now can be written as

$$-\frac{d\vec{L}_{EB}}{dt} = \frac{d\vec{L}_m}{dt} + \oint_a (\vec{r} \times \vec{T}) \cdot d\vec{a} \tag{6.63}$$

The second term on the right represents the angular momentum flux of the field.

Equation (6.62) is a statement of angular momentum conservation of the electromagnetic system (field and the charges).

In quantum theory, photons are the quanta of the quantized electromagnetic field. Classically, the electromagnetic field brings about the interaction between charged particles. Quantum mechanically, we can think of interaction between charges as due to an exchange of photons. The angular momentum of the electromagnetic field $\vec{L}_{EB}$ as given by equation (6.62) contains $\vec{r}$; it seems that $\vec{L}_{EB}$ only represents the orbital angular momentum. But it contains both orbital and spin angular momentum. It can be shown that $\vec{L}_{EB}$ can be decomposed into two terms, with one term describing spin. When this spin term combines wth the photon concept, we can obtain the spin of photon to be $1\hbar$, where $\hbar = h/2\pi$.

Problems

1. Derive displacement current from charge conservation and Biot-Savart law.

$$\nabla \cdot \vec{j} + \partial\rho/\partial t = 0 \text{ (charge conservation)}$$

$$\vec{B}(\vec{r}) - \frac{\mu_0}{4\pi} \int \frac{\vec{j}(\vec{r_1}) \times (\vec{r} - \vec{r_1}) dv_1}{|\vec{r} - \vec{r_1}|^3} \text{ (Biot-Savart law)}$$

2. A point charge q moves uniformly along a straight line with a velocity v that is nonrelativistic. Find the vector of the displacement current density at a point P lying at a distance r from the charge on a straight line (1) coinciding with its trajectory and (2) perpendicular to its trajectory and passing through the charge.

3. A parallel-plate capacitor is formed by two discs, the space between which is filled with a homogeneous poorly conducting medium. The capacitor was charged and then disconnected from a power source. Ignoring edge effects, show that the magnetic field inside the capacitor is absent.

4. A point charge q moves in a vacuum uniformly and rectilinearly with a nonrelativistic velocity v. Using's Maxwell equation for the

circulation of vector B, obtain the expression for B at a point P whose position relative to the charge is characterized by radius vector r (**Figure 6.6**).

5. Protons having the same velocity v form a beam of a circular cross-section with current I. Find the direction and magnitude of Poynting's vector outside the beam at a distance r from its axis.

6. A current flowing through the winding of a long straight solenoid is being increased. Show that the rate of increase in the energy of the magnetic field in the solenoid is equal to the flux of Poynting's vector through its lateral surface.

7. Assuming that Coulomb's law for magnetic charges (q_m) takes the form

$$\vec{F} = \frac{\mu_0}{4\pi} \frac{q_{m_1} q_{m_2}}{r^2} \hat{r}$$

work out the force law for a magnetic charge (monopole) q_m moving with velocity $\vec{v}$ through electric and magnetic fields $\vec{E}$ and $\vec{B}$.

8. Show that, in free space, both electric field $\vec{E}$ and magnetic field $\vec{B}$ satisfy the homogeneous wave equation

$$\left(\nabla^2 - \mu_0 \varepsilon_0 \frac{\partial^2}{\partial t^2} \right) \begin{Bmatrix} \vec{E} \\ \vec{B} \end{Bmatrix} = 0$$

9. We have seen that the electric field $\vec{E}$ and the magnetic field $\vec{B}$ are invariant under the gauge transformations (6.10) and (6.11):

$$\vec{A}' = \vec{A} + \nabla\chi, \quad \phi' = \phi - \frac{\partial \chi}{\partial t}$$

Show that, to preserve the Lorentz condition, the gauge parameter χ satisfies the equation

$$\nabla^2 \chi - \mu_0 \varepsilon_0 \frac{\partial^2 \chi}{\partial t^2} = 0.$$

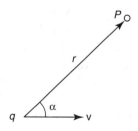

I FIGURE 6.6

10. In the discussion of polarization, we take the electric field in a plane electromagnetic wave as the real part of

$$\vec{E} = \vec{E}_0 \exp i(kx - \omega t) \tag{6.64}$$

and showed that $E_z = 0$ and so $\vec{E}_0$ is a vector in the yz plane normal to the direction of propagation:

$$\vec{E}_0 = E_{0u}\hat{j} + E_{0z}\hat{k} \tag{6.65}$$

where E_{0y} and E_{0z} are constants. In a similar way, the magnetic field in a plane electromagnetic wave is given by the real part of

$$\vec{B} = \vec{B}_0 \exp i(kx - \omega t) \tag{6.66}$$

Show that, in a similar argument for E_z,

$$B_z = 0$$

and so $\vec{B}_0$ is also a vector in the yz plane normal to the direction of propagation. $\vec{E}$ and $\vec{B}$ are normal to each other and both normal to the direction. To see this, first setting (1) and (2) in Faraday's law of induction, $\nabla \times \vec{E} = -\partial \vec{B}/\partial t$. Show that this substitution yields

$$\left(-ikE_{0z}\hat{j} + ikE_{0y}\hat{k}\right) \exp i(kx - \omega t) = -\partial \vec{B}/\partial t$$

and

$$-\frac{1}{c}E_{0z}\hat{j} + \frac{1}{c}E_{0y}\hat{k} = \vec{B}_0$$

where $c = \omega/k$. The last expression shows that

$$\vec{B} = \frac{1}{c}(\hat{i} \times \vec{E})$$

That is, $\vec{E}$ and $\vec{B}$ are normal to each other and both normal to the direction.

11. Using the Maxwell stress tensor, find the net force that the southern hemisphere of a uniformly charged sphere exerts on the northern hemisphere.

12. In a plane wave, the energy density U moves at the speed c in the direction of wave propagation given by the unit vector $\hat{i} = \vec{E} \times \vec{B}/EB$.

The flow of energy through unit area per unit time at any point can be specified in magnitude and direction by an energy flow density vector $\vec{S}$, the Poynting vector: $\vec{S} = cU\hat{i}$. Show that if the energy density is given by equation (6.40), the $\vec{S}$ can be rewritten as equation (6.36). Then the field momentum density $\vec{p}_{EB} = \vec{S}/c^2$ can be expressed as $\vec{p}_{EB} = U/c\,\hat{i}$.

13. A long coaxial cable, of length l, consists of an inner conductor (radius a) and an outer conductor (radius b). It is connected to a battery at one end and a resistor at the other (**Figure 6.7**). The inner conductor carries a uniform charge per unit length λ and a steady current l to the right; the outer conductor has the opposite charge and current. What is the electromagnetic momentum stored in the fields?

14. Using energy conservation, show that Maxwell equations have a unique solution.

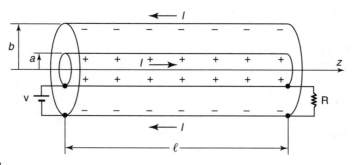

I FIGURE 6.7

7 Electrostatics of Dielectric Media

Materials, in terms of electric properties, are divided into two broad categories: conductors and insulators (or dielectrics). Conductors contain charges that are free to move within the material; by contrast, an ideal dielectric has no free charges, and so it is not capable of yielding a steady flow of current when under the influence of a potential gradient. In this chapter, dielectrics are our concerns. Macroscopic aspects of dielectric polarization are discussed first, and then we study microscopic aspects of dielectrics.

Between the conductors and insulators lie the semiconductors. A discussion of semiconductors requires more than classical electrodynamics. Hence, we do not discuss them here. Many substances have been observed to behave as superconductors under low temperatures. Recently, superconducting behavior at temperatures of around 80 K has been claimed for some organic conductors. We discuss superconductivity briefly in Chapter 8.

A. Macroscopic Aspects of Dielectric Polarization

An ideal dielectric material is one that has no free charges. Nevertheless, atoms of the dielectric are affected by the presence of an electric field, with positive particles being pushed in the direction of the field and negative particles oppositely. Thus, the positive and negative parts of each atom are displaced from their equilibrium positions in opposite directions, and the dielectric is said to be polarized.

A polarized dielectric, even though it is electrically neutral on the average, produces an electric field at exterior points and inside the dielectric as well. As a result, we are confronted with what appears to be an awkward problem. The main purpose of this section is to develop general methods for handling this problem.

7.1 Induced Dipoles

We may take as a crude model of a dielectric two interpenetrating arrays of charge, as shown in **Figure 7.1**, where the representation is in two dimensions for simplicity. We can think of the positive array as being due to the nuclei of the atoms making up the solid and the negative array as due to the average positions of the electrons associated with the atoms. In the absence of an external field, we should expect the two arrays to be superposed, but the relative displacement of the charges under an applied field produces a separation of charge, as shown in the lower drawing. This model gives a crude idea of the situation within the polarized solid, though it cannot give exact information on an atomic scale.

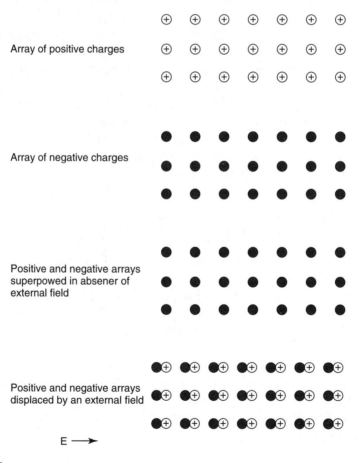

Array of positive charges

Array of negative charges

Positive and negative arrays superpowed in absener of external field

Positive and negative arrays displaced by an external field

E ⟶

I **FIGURE 7.1**

As a result of the polarization, each atom becomes a tiny dipole whose strength depends on $\vec{E}$. If the electron cloud is displaced by a distance δ, the vector

$$\vec{p} = q\vec{\delta}$$

pointing from the center of the cloud toward the nucleus can be defined as the dipole moment of the polarized atom, where $q = Ze$ for an element with atomic number Z. For a dielectric material consisting of one kind of element only, in a uniform electric field the center of the electron cloud in each atom moves the same distance away from the nucleus in a direction opposite to $\vec{E}$. Each atom therefore acquires a dipole moment of magnitude $q\delta$ in the direction of the field $\vec{E}$. The measurement we make is over regions containing many thousands of atoms. If there are N atoms in unit volume, we can define a dipole moment per unit volume, $\vec{P}$ with

$$\vec{P} = Nq\vec{\delta} \tag{7.1}$$

If the field $\vec{E}$ is not too large, the strength of each atomic dipole is proportional to $\vec{E}$ and we may write $\vec{p}$ as

$$\vec{p} = \alpha\vec{E} \tag{7.2}$$

The constant of proportionality α is called the atomic polarizability.

For molecules of the dielectric, the situation is not so simple, because some molecules are themselves permanent dipoles, such as water. In the absence of a field these permanent dipoles are randomly oriented, but when a field is applied, a torque is established that acts to align them. Consequently, there is a resultant moment in the direction of the field. Let us suppose that this effect also has been included in the value of α. The total moment induced by an electric field $\vec{E}$ is $N\alpha\vec{E}$ per unit volume. Obviously, this is not exact but an approximation, because dipoles in the immediate neighborhood of any one dipole will themselves contribute to the field that polarizes the original dipole. We often write the polarization $\vec{P}$ for isotropic materials in terms of a scalar quantity denoted by χ:

$$\vec{P} = N\alpha\vec{E} = \varepsilon_0\chi\vec{E} \tag{7.3}$$

The parameter χ is called the electric susceptibility of the material and depends on the density of the material and on the polarizability α, both of which may vary with temperature. The susceptibility χ should be properly written as $\chi(E)$, but in practice over a wide range of field intensities χ is sensibly field independent. To make χ dimensionless, we have included the factor ε_0 in equation (7.3).

For anisotropic dielectric materials such as certain crystals or plasma in a magnetic field, $\vec{P}$ is not parallel to $\vec{E}$. Piezoelectric crystals become electrically polarized when they are mechanically strained. Because a ferroelectric crystal has a spontaneous polarization $\vec{P}(T)$ at a temperature below its characteristic ferroelectric "Curie temperature," T_c, the crystal is easy to polarize and hence has a very high susceptibility. This property of ferroelectrics is of some use technically in the construction of capacitors.

7.2 Electric Field at an Exterior Point

We see above that the effect of an electric field $\vec{E}$ on a dielectric is to create the polarization $\vec{P}$, with its associated dipole moment $\vec{P}d\vec{r}$ in each volume element $d\vec{r}$. In calculating the potential for such a polarized dielectric system, we may now replace the dielectric itself by the volume polarization $\vec{P}d\vec{r}$. From equation (2.34) we know that the potential at any point due to a single dipole $\vec{p}$ is $(1/4\pi\varepsilon_0)(\vec{p} \cdot \vec{r}/r^3)$ for a dipole centered at the origin. More generally, for a dipole placed at some point $\vec{r}'$,

$$V(\vec{r}) = \frac{1}{4\pi\varepsilon_0} \frac{\vec{p}\,(\vec{r} - \vec{r}')}{|\vec{r} - \vec{r}'|^3} \tag{7.4}$$

Now $(\vec{r} - \vec{r}')/|\vec{r} - \vec{r}'|^3 = -\nabla(|\vec{r} - \vec{r}'|^{-1}) = \nabla'(|\vec{r} - \vec{r}'|^{-1})$

where ∇ operates only on r and ∇' operates on the primed coordinates r' (**Figure 7.2**). We may now rewrite equation (7.4) in a more appropriate form:

$$V(\vec{r}) = \frac{1}{4\pi\varepsilon_0}\, \vec{p} \cdot \nabla'\left(\frac{1}{|\vec{r} - \vec{r}'|}\right)$$

To apply this formula to our polarized dielectrics, we simply replace the point dipole by a volume polarization, that is, to replace $\vec{p}$ by $\vec{P}(\vec{r}')d\vec{r}'$. Then the potential arising from the induced polarization is

$$V(\vec{r}) = \frac{1}{4\pi\varepsilon_0} \int \vec{P}(\vec{r}') \cdot \nabla'\left(\frac{1}{|\vec{r} - \vec{r}'|}\right) d\vec{r}' \tag{7.5}$$

Because

$$\nabla'\left(\frac{\vec{P}(\vec{r}')}{|\vec{r} - \vec{r}'|}\right) = \vec{P}(\vec{r}') \cdot \nabla'\left(\frac{1}{|\vec{r} - \vec{r}'|}\right) + \frac{\nabla' \cdot \vec{P}(\vec{r}')}{|\vec{r} - \vec{r}'|}$$

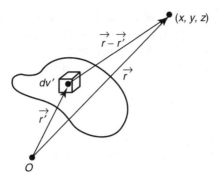

I FIGURE 7.2

equation (7.5) now becomes

$$V(\vec{r}) = \frac{1}{4\pi\varepsilon_0} \int \nabla' \left(\frac{\vec{P}(\vec{r}')}{|\vec{r} - \vec{r}'|} \right) d\vec{r}' - \frac{1}{4\pi\varepsilon_0} \int \frac{\nabla' \cdot \vec{P}(\vec{r}')}{|\vec{r} - \vec{r}'|} d\vec{r}'$$

$$= \frac{1}{4\pi\varepsilon_0} \int \frac{\vec{P}(\vec{r}') \cdot d\vec{S}'}{|\vec{r} - \vec{r}'|} - \frac{1}{4\pi\varepsilon_0} \int \frac{\nabla' \cdot \vec{P}(\vec{r}')}{|\vec{r} - \vec{r}'|} d\vec{r}'$$

(7.6)

where we have used divergence theorem in transform the volume integral into a surface integral; the surface integral is taken over the boundary of the dielectric.

7.3 The Polarization Charge Densities ρ_P and σ_P

Equations (7.6) and (2.13) have the same mathematical structure. (Note that the element volume in equation (2.13) is denoted by dv' and the element surface by da'.) This shows that the first term in the potential arising from the induced polarization is from a surface distribution of charge $\sigma_P = \vec{P} \cdot \hat{n}$ (where $\hat{n}$ is the normal unit vector to dS) on the boundary of the dielectric, and the second term is from a volume distribution of charge $\rho_P = -\nabla \cdot \vec{P}$ throughout the dielectric. These charges

$$\rho_P = -\nabla \cdot \vec{P}, \quad \sigma_P = \vec{P} \cdot \hat{n}$$

(7.7)

are referred to as polarization (or bound) charge densities. The dielectric may be replaced by these bound charges without affecting the potential (and so the field) outside the dielectric.

If the polarization is constant throughout the medium, the charge per unit volume in its interior is zero because each of the derivative in $\nabla \cdot \vec{P}$ vanishes.

It is intuitively obvious that the total polarization charge is always zero. This follows simply from the divergence theorem:

$$Q_P = \int_S \sigma_P \, dS + \int_V \rho_P \, d\vec{r} = \int_S \vec{P} \cdot \hat{n} \, dS + \int_V (-\nabla \cdot \vec{P}) \, d\vec{r} = 0 \tag{7.8}$$

This result is found because the polarization charges merely represent the effects of the atomic or molecular dipoles. The bound surface charge can be visualized simply as the ends of the dipoles appearing at surface.

Previously we found the polarization charges σ_P and ρ_P in the course of mathematical manipulations of equation (7.5). Alternatively, we could find them from purely physical grounds. For simplicity, again consider an isotropic dielectric material consisting of the same element placed in a uniform electric field (**Figure 7.3**). As a result of polarization, each atom becomes a tiny dipole. There is no net charge density within the slab as all the individual dipoles are aligned parallel to the field and, hence, negative charge of one dipole is next to the positive charge of the next dipole. Thus, the average density of the positive and negative charges displaced under the external field is the same, the net charge being zero. But the neutrality at the surface of the dielectric is not preserved. There is a positive charge on one surface and a negative charge on the other.

Consider a small area dS on one of the surfaces of the slab. If σ_P is the surface density of the polarization charge, the total charge on dS is $\sigma_P dS$. This charge is also equal to the net charge in a box with cross-section dS and length δ equal to the separation of positive and negative charges.

$$\sigma_P dS = Nq\delta dS$$

from which we find

$$\sigma_P = Nq\delta = Np = P$$

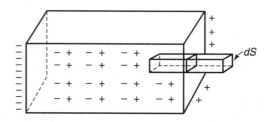

I FIGURE 7.3

where P is the magnitude of the polarization vector $\vec{P}$. This result generally holds, even for surfaces that are not normal to the electric field. In **Figure 7.4** the surface element dS is at an angle θ to the field, and its projected area along the field is $dS \cos\theta$. The net charge acquired by the box after polarization is

$$Nq\delta dS\cos\theta = Npd S\cos\theta = PdS\cos\theta = \vec{P} \cdot \hat{n} dS$$

i.e., $\sigma_P = \vec{P} \cdot \hat{n} = P_n$
same as equation (7.7).

If the polarization is not uniform, we may expect some net charge in the body of a dielectric. We now calculate the average density of charge in a small volume V. Evidently, this charge density vanishes if the medium is unpolarized. So our problem is to calculate the excess of the charge leaving V over that entering V when the bound electrons in each atom become displaced by an impressed electric field. The total charge displaced out of volume V by the polarization is the integral of the outward normal of $\vec{P}$ over the surface S that bounds V. An equal excess charge of the opposite sign is left behind. If we denote the net charge within V by ΔQ_P, we can then write

$$\Delta Q_P = -\int_S \vec{P} \cdot \hat{n} dS$$

We can attribute ΔQ_P to a volume distribution of charges with the density ρ_P, and so

$$\Delta Q_P = \int_V \rho_P dV$$

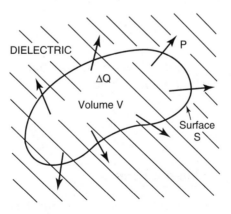

I FIGURE 7.4

Combining the two equations yields

$$\int_V \rho_P \, dV = -\int_S \vec{P} \cdot \hat{n} \, dS$$

Using divergence theorem we can transform the surface integral on the right-hand side into a volume integral; the result is

$$\int_V \rho_P \, dV = -\int_S \nabla \cdot \vec{P} \, dV \quad \text{or} \quad \int_V (\rho_P + \nabla \cdot \vec{P}) \, dV = 0$$

Because this is valid for any arbitrary volume V, the integrand must vanish identically, and we get

$$\rho_P = -\nabla \cdot \vec{P}$$

same as equation (7.7).

Example 1

Consider a dielectric cylinder polarized uniformly in the direction of its length, $\vec{P} = P_0 \hat{k}$. Find the polarization charge densities ρ_P and σ_P.

Solution

Because $\nabla \cdot \vec{P} = 0$, $\rho_P = 0$ inside the dielectric cylinder. $\sigma_P = 0$ on the cylinder surface, P_0 on the positive end surface and $-P_0$ on the negative end surface. The problem reduces to a pair of uniformly charged disks.

Example 2

Consider a spherical cavity in a uniformly polarized dielectric, assuming that the polarization in the remaining material remains unchanged. Find the surface charge density σ_P.

Solution

$\sigma_P = \vec{P} \cdot \hat{n}$, where $\hat{n}$ is a unit normal vector directed out of the material as shown in **Figure 7.5**. Hence, we have a point located at angle θ from the direction of $\vec{P}$

$$\sigma_P = -P \cos \theta$$

7.4 Electric Field at an Interior Point

Often, we need to know the electric potential and the electric field at points inside a dielectric. For example, we may want to know the potential difference between two conductors separated by a dielectric.

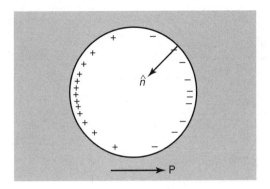

I FIGURE 7.5

The electric field intensity at a point inside a dielectric changes rapidly both in time and position. The macroscopic electric field $\vec{E}$ that we want to know is the average electric field in a small region of the dielectric that nevertheless contains a large number of molecules. This microscopic electric field is a slowly varying function of the coordinates and is independent of the time in the static case. The part of this macroscopic field that originates from the dipoles within the dielectric can be calculated from polarization charge densities σ_P and ρ_P discussed above, exactly as in equation (7.6) for an exterior point.

To proof this we use a simple but elegant argument. (A proof worked out in great detail can be found in the book *Electromagnetic Fields and Waves*, second edition, by Paul Lorrain and Dale Corson, 1970, W. H. Freeman.)

The electric field in a dielectric must have the same basic properties that we found applied to electric field in vacuum; in particular, the electric field $\vec{E}$ is a conservative field and it is derivable from a scalar function. $\nabla \times \vec{E} = 0$, or its equivalent $\oint \vec{E} \cdot d\vec{l} = 0$, is true both in vacuum and in dielectric. We now cut a needle-shaped cavity in the dielectric (**Figure 7.6**) and $\oint \vec{E} \cdot d\vec{l} = 0$ to the path $ABCD$ lies partly in the needle-shaped cavity and partly in the dielectric. Because the segments AD and BC may be made arbitrarily small, the line integral reduces to

$$\vec{E}_0 \cdot \vec{l} - \vec{E}_d \cdot \vec{l} = 0, \quad E_{0t} = E_{dt} \tag{7.9}$$

where the subscripts 0 and d refer to vacuum and dielectric, respectively, and the subscript t stands for tangential component. Equation (7.9) is true regardless the orientation of the needle-shaped cavity. But our needle-shaped cavity is parallel to $\vec{E}$, so $E_{dt} = E_d$. Also, by symmetry, the field in

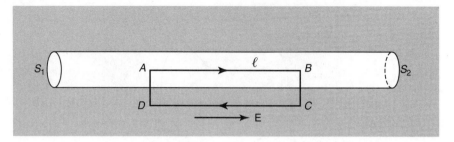

I FIGURE 7.6

the cavity is along the direction of the needle, so $E_{0t} = E_0$. What do we see now? It is very important: For isotropic dielectrics, the electric field in a dielectric is equal to the electric field inside a needle-shaped cavity in the dielectric, with the cavity axis oriented parallel to the direction of the electric field.

Now the problem of calculating the electric field inside a dielectric reduces to calculating the electric field inside a needle-shaped cavity in the dielectric. But the electric field in the cavity is an external field and hence may be determined by means of the results of equation (7.6). That is, we assume here that the polarization of the dielectric is a given function $\vec{P}(\vec{r}')$, and we calculate the potential and electric field arising from this polarization. Taking the field point $\vec{r}$ at the center of the cavity and using equation (7.6), we obtain for the potential

$$V(\vec{r}) = \frac{1}{4\pi\varepsilon_0} \int_{\tau_0 - \tau_1} \frac{\rho_P(\vec{r}') \, d\vec{r}'}{|\vec{r} - \vec{r}'|} + \frac{1}{4\pi\varepsilon_0} \int_{S_0 + S'} \frac{\sigma_P(\vec{r}') \, dS'}{|\vec{r} - \vec{r}'|} \tag{7.10}$$

where $\tau_0 - \tau_1$ is the volume of the dielectric excluding the needle-shaped cavity, S_0 is the exterior surface of the dielectric, and $S' = S_1 + S_2 + S_c$ are the needle surfaces. But from Figure 7.6 it is seen that $\sigma_P = 0$ on the cylindrical surface S_c of the needle. Furthermore, the needle may be made arbitrarily thin so that the surfaces S_1 and S_2 have negligible area. Thus, only the exterior surfaces of the dielectric contribute, and the surface integral of equation (7.10) becomes identical in form to the surface integral of equation (7.6). The volume integral of equation (7.10) excludes the cavity. However, volume τ_1 of the needle may be made arbitrarily small by making the cavity thin, and so the contribution of the cavity to this integral is negligible. Thus, we do not need to exclude the volume τ_1. The charge density ρ_P is bounded; the quantity $d\vec{r}'/|\vec{r} - \vec{r}'|$ does not diverge at the field point (i.e., when $\vec{r}' = \vec{r}$)

because the volume of a point is a higher order zero than the $\lim |\vec{r} - \vec{r}'|$. Finally, equation (7.10) becomes similar in form to equation (7.6). In other words, equation (7.10) gives the potential $V(\vec{r})$ regardless of whether the point $\vec{r}$ is located inside or outside the dielectric. The electric field $\vec{E}(\vec{r})$ may be calculated as minus the gradient of equation (7.10).

The calculations indicated in equations (7.6) and (7.10) are straightforward for cases in which the polarization vector $\vec{P}$ is a known function of position. In most cases, however, the polarization arises in response to an electric field that has been imposed on the dielectric medium. That is, $\vec{P}$ is a function of the total macroscopic electric field, and under these conditions, the situation is much more complicated. First, although it is necessary to know the functional form of $\vec{P}(\vec{E})$, this form is known experimentally in most cases and hence is not a source of difficulty. The real complication arises because $\vec{P}$ depends on the total electric field, including the contribution from the dielectric itself, and it is this contribution that we are in the process of evaluating. Thus, we cannot determine $\vec{P}$ because we do not know $\vec{E}$, and vice versa.

In the following we discuss Gauss' law in the presence of polarization charges.

7.5 Gauss' Law for Charges in Dielectric

Consider a dielectric that contains free charges $q_1, q_2, q_3, \ldots$. Describe any closed surface S that may lie wholly or partially within the dielectric and surround some of the immersed free charges. Gauss' law states that the electric flux through this Gaussian surface S is equal to $1/\varepsilon_0$ times the total charge (free and polarization charges) enclosed:

$$\int_S \vec{E} \cdot \hat{n} \, dS = \frac{1}{\varepsilon_0} \left(\sum q_i + \Delta Q_P \right)$$

where $\hat{n}$ is the normal unit vector to dS and ΔQ_P is the polarization or bound charge inside S. But as shown in the preceding section, ΔQ_P is equal to

$$\Delta Q_P = -\int_S \vec{P} \cdot \hat{n} \, dS$$

Gauss' law becomes, by rearranging terms to combine the two surface integrals,

$$\int_S \left(\varepsilon_0 \vec{E} + \vec{P} \right) \cdot \hat{n} \, dS = \sum q_i \tag{7.11}$$

In the early days of electricity, the atomic mechanism of polarization was not known and the existence of ρ_P was not appreciated. To write

Maxwell equations in a simple form, a new field vector $\vec{D}$, the electric displacement vector, was defined as the vector sum of $\varepsilon_0\vec{E}$ and $\vec{P}$:

$$\vec{D} = \varepsilon_0\vec{E} + \vec{P} \tag{7.12}$$

Then Gauss' law, equation (7.11), becomes

$$\int_S \vec{D} \cdot \hat{n}dS = \sum q_i \tag{7.13}$$

If we apply Gauss' law to a small region in which all the free charges enclosed are distributed as a charge density ρ_f, then

$$\int_S \vec{D} \cdot \hat{n}dS = \int_V \rho_f d\vec{r}$$

from which we obtain

$$\nabla \cdot \vec{D} = \rho_f \tag{7.14}$$

This is Gauss' law in differential form. Don't forget that $\nabla \times \vec{E}$ is always zero for electrostatic fields.

Example 3: Plasma Oscillations

A simple application of the material developed above is to electric oscillations in plasma. Plasma is a gaseous body of ions and electrons of sufficiently low density, and it is normally neutral and free of electric field in its interior. But the electrons and ions may oscillate about their mean positions.

Electrons are much more mobile than ions. Now if a sudden disturbance, such as an external electric field applied for a short time, displaces each electron by an average amount ξ in the x-direction, the resulting polarization $\vec{P}$ is given by

$$\vec{P} = -ne\xi$$

where n is the number of electrons per unit volume. All the electrons and ions can be counted as bound charges; thus, the free charge density ρ_f is zero and so equation (7.14) reduces to

$$\nabla \cdot \vec{D} = 0$$

But $\vec{D} = 0$ before the disturbance, and after the disturbing field has gone, $\vec{D}$ is still zero. Thus, we have from equation (7.12)

$$\varepsilon_0\vec{E} + \vec{P} = 0 \quad \text{or} \quad \vec{E} + \vec{P}/\varepsilon_0 = 0$$

From this we obtain

$$eE_x = ne^2\xi/\varepsilon_0$$

This is the principal force acting on electrons, and the equation of motion for each electron is, to a good approximation,

$$-eE_x = m_e\frac{d^2\xi}{dt^2}$$

or

$$m_e\frac{d^2\xi}{dt^2} = -\frac{Ne^2}{\varepsilon_0}\xi \qquad (7.15)$$

We recognize that this is the equation for simple harmonic oscillations, and its solution can be written as

$$\xi = \xi_0 \sin(\omega_p t + \phi)$$

where the angular frequency ω_p is the so-called plasma frequency and is given by

$$\omega_p = (ne^2/\varepsilon_0 m_e)^{1/2}$$

The plasma oscillations here were assumed to be induced by pulse-like disturbance; imposing alternating external fields can induce them as well.

Example 4

A long straight wire, carrying uniform line charge q, is surrounded by insulation material of radius R. Find the electric fields outside and inside the insulation material.

Solution

We first find the electric displacement vector $\vec{D}$. To this end, we draw a cylindrical Gaussian surface of radius r and length l (**Figure 7.7**), and then Gauss law, equation (7.14), gives

$$D(2\pi rl) = ql$$

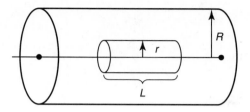

I FIGURE 7.7

and so

$$\vec{D} = \frac{q}{2\pi r}\hat{r}$$

which holds both inside and outside the insulation. Outside the insulation, the polarization $\vec{P}$ charge is zero, and so

$$\vec{E}_{out} = \frac{1}{\mathring{a}_0}\vec{D} = \frac{q}{2\partial \mathring{a}_0 r}\hat{r}, \quad \text{for } r > R$$

Inside the insulation, we do not know $\vec{P}$, and therefore the electric field cannot be determined.

7.6 Electric Displacement Vector $\vec{D}$

The electric displacement vector $\vec{D}$ deserves some special attention, because it plays a very important role in electrodynamics.

Maxwell introduced the field vector $\vec{D}$ and called it the electric displacement. In isotropic materials (such as glass or paraffin) for which equation (7.4) holds

$$\vec{D} = \varepsilon_0\vec{E} + \vec{P} = \varepsilon_0\vec{E} + \varepsilon_0\chi\vec{E} = \varepsilon_0(1+\chi)\vec{E} \tag{7.16}$$

If we write

$$\varepsilon = \varepsilon_0(1+\chi) \tag{7.17}$$

then

$$\vec{D} = \varepsilon\vec{E} \tag{7.18}$$

where ε is called the permittivity of the dielectric materials. Because the dielectric susceptibility α is field dependent, so too is the permittivity $\varepsilon = \varepsilon(E)$. In an anisotropic dielectric, such as crystalline quartz or plasma in magnetic fields, the permittivity is a tensor ε_{ij}.

It is often convenient to introduce a dimensionless quantity K to characterize the electric behavior of materials, where

$$K = \varepsilon / \varepsilon_0 = 1 + \chi \tag{7.19}$$

and it is called the dielectric constant (or relative permittivity). (The values of dielectric constant K can be found from reference books, such as the *Handbook of Chemistry and Physics*, CRC Press, Inc., Cleveland, Ohio.)

Now, in terms of K, we can write $\vec{D}$ as

$$\vec{D} = \varepsilon \vec{E} = \varepsilon_0 K \vec{E} \tag{7.20}$$

As an application of these relations, let us consider the electric field in an isotropic dielectric surrounding a spherical charge Q in which the charge density is a function of distance from the center only. We can find the electric field by means of Gauss' law, equation (7.12), at once. First, let us rewrite equation (7.12), with the aid of equations (7.16) and (7.19), as

$$\int_S \varepsilon_0 K \vec{E} \cdot \hat{n} dS = \sum q_i \quad \text{or} \quad \int_S \varepsilon \vec{E} \cdot \hat{n} dS = \sum q_i / \varepsilon_0 K$$

This differs from Gauss' law for charges in empty space only in appearance of the factor K in the denominator on the right. Thus, when the charge is immersed in a dielectric, the electric field must be given by

$$\vec{E} = \frac{Q}{4\pi\varepsilon_0 K r^2}$$

Why $\vec{E}$ is reduced in the ratio 1 to K by the dielectric is made clearer if we examine the polarization charges in the medium. The polarization $\vec{P}$ can be written by eliminating χ from equations (7.4) and (7.19) as

$$\vec{P} = \varepsilon_0 (K - 1) \vec{E}$$

and making use of the last expression for $\vec{E}$,

$$\vec{P} = \frac{K - 1}{K} \frac{Q}{4\pi r^2}$$

If a is the radius of the charge Q, the polarization charge per unit area of the cavity in the dielectric in which Q lies is

$$\sigma_P = -(P)_{r=a} = -\frac{K-1}{K}\frac{Q}{4\pi a^2}$$

and the total polarization charge on the surface of the cavity is

$$Q_P = 4\pi a^2 \sigma_P = -\frac{K-1}{K}Q$$

The effective charge producing the field is the sum of Q and the polarization charge Q_P, that is,

$$Q + Q_P = Q - \frac{K-1}{K}Q = \frac{Q}{K}$$

Consequently, $\vec{E}$ in the dielectric is only $1/K$ times that which would be produced by Q alone.

We may also introduce lines of electric displacement and flux of displacement. The flux of displacement through the surface dS is defined as the product of the component of the displacement $\vec{D}$ normal to the surface by the area of the surface; the lines of electric displacement may be drawn so as to have everywhere the direction of $\vec{D}$, and they are given by the differential equations

$$dx/D_x = dy/D_y = dz/D_z \tag{7.21}$$

Moreover, these lines can originate and end only on free charges, whereas the lines of $\vec{E}$ begin or end on either free or bound charges. This does not mean that the value of $\vec{D}$ at a particular point depends only on the free charge; the quantity that depends on the free charge alone is ($\nabla \cdot \vec{D}$, as indicated by equation (7.14). Equation (7.14) looks like equation (2.17), with the total charge density ρ replaced by the free charge density ρ_f and $\vec{E}$ by $\vec{D}$. This may be a good scheme to remember Gauss' law for $\vec{D}$ in differential form; however, one should never conclude that $\vec{D}$ is just like $\vec{E}$ (apart from the factor ε), except that its source is the free charge density ρ_f. This is an erroneous conclusion!

To clarify the concepts involved in the preceding discussion, let us consider the special and important example of two charged plane parallel metallic plates separated by a dielectric.

Example 5: Parallel Plate Separated by a Dielectric

Given a parallel plate capacitor between the plates of which there is placed a slab of dielectric, what is the capacity of this capacitor?

Solution

When the intervening space is free of dielectric slab, equal and opposite free surface charge densities, ±σ, are present on the plates (**Figure 7.8a**), and the electric field $\vec{E_0}$ is uniform and has the value σ/ε_0. If A is the area of plate, then the capacitance C is

$$C = \frac{Q}{V} = \frac{\sigma A}{\sigma d/\varepsilon_0} = \frac{\varepsilon_0 A}{d}$$

where $V = E_0 d = \sigma d/\varepsilon_0$ and d is the separation of the two plates. Keeping the charge on the plates unchanged, let a dielectric slab be inserted between the two plates. Under the action of $\vec{E_0}$, induced dipoles are formed. As a result, a positive charge appears on the surface of the slab that faces the negative plate and a negative charge appears on the other side of the slab facing the positive plate. These induced charges produce a macroscopic field inside the slab that is opposed to the field of the plates, effectively reducing it. This macroscopic field is called the depolarizing field. The effective field inside the capacitor containing the dielectric slab is reduced to

$$E = E_0 - E_i = (\sigma - P)/\varepsilon_0$$

as shown in Figure 7.8c. Now $P = \varepsilon_0(K - 1)E$. Substituting this in the last expression and solving for E,

$$E = \frac{\sigma}{\varepsilon_0 K}$$

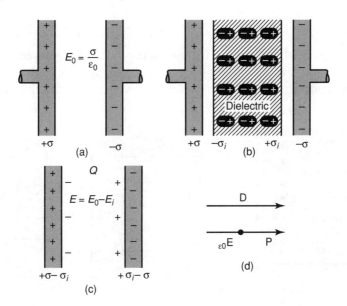

I **FIGURE 7.8a, b, c, d**

We see that the field is reduced by a factor $1/K$. Now

$$V = Ed = \frac{\sigma}{\varepsilon_0 K} d \quad \text{and} \quad C = \frac{Q}{V} = K \frac{\varepsilon_0 A}{d}$$

The capacity is increased by a factor K. This explains Faraday's discovery: The capacity of a capacitor could be increased greatly by inserting a dielectric into it. **Figure 7.9** shows a mapping of the fields of $\vec{D}$, $\vec{E}$, and $\vec{P}$.

This result is true for a capacity of any shape filled with a *uniform linear dielectric*. Without the dielectric, the equations to be solved are

$$\nabla \cdot \vec{E_0} = \rho_f / \varepsilon_0, \quad \nabla \times \vec{E_0} = 0$$

With the dielectric present we have

$$\nabla \cdot (K\vec{E}) = \rho_f / \varepsilon_0, \quad \nabla \times \vec{E} = 0$$

Because K is constant, $\nabla \times \vec{E} = 0$ may be written as $\nabla \times (K\vec{E}) = 0$. And the last two equations become

$$\nabla \cdot (K\vec{E}) = \rho_f / \varepsilon_0, \quad \nabla \times (K\vec{E}) = 0$$

Comparing these with the two equations without the dielectric present, we see that

$$K\vec{E} = \vec{E_0}, \quad \text{or} \quad \vec{E} = \vec{E_0}/K$$

The field $\vec{E}$ is smaller than $\vec{E_0}$ by the factor $1/K$. The voltage V is reduced by the same factor. As $C = Q/V$, the charge Q on the electrodes of the capacity is taken constant; hence, the capacity C is increased by the factor K.

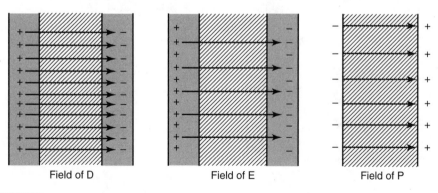

Field of D   Field of E   Field of P

I FIGURE 7.9

The measurement of the capacitance of a suitable capacitor with or without a dielectric provides a convenient method for measuring a dielectric constant K.

7.7 Poisson's Equation for Dielectrics

Now let us combine equation (7.18) with equation (7.15):

$$\nabla \cdot \vec{D} = \nabla \cdot (\varepsilon\vec{E}) = \nabla \cdot (\varepsilon_0 K \vec{E}) = \rho_f$$

For constant K we have

$$\nabla \cdot \vec{E} = \rho_f / \varepsilon_0 K$$

Because $\vec{E} = -\nabla V$, it follows that

$$\nabla^2 V = -\rho_f / \varepsilon_0 K = -\rho_f / \varepsilon \qquad (7.22)$$

This is Poisson's equation, which is true for regions over which K is constant.

7.8 Boundary Conditions for $\vec{D}$ and $\vec{E}$

For solving electrostatic problems involving more than one dielectric medium, it is important to know the boundary conditions for $\vec{D}$ and $\vec{E}$ at the surfaces separating the different materials.

Let K_1 and K_2 be two dielectrics in contact (**Figure 7.10**). We draw a pillbox-shaped surface $ABCD$ about an area Δs of the surface of separation, the height BC of the pillbox being very small compared with the diameter of the bases. Let D_{1n} and D_{2n} be the normal components of the displacement $\vec{D}$ in the two media. Then, as there is no free charge inside the pillbox, Gauss' law requires that

$$D_{2n}\Delta s - D_{1n}\Delta s = 0$$

or

$$D_{2n} = D_{1n} \qquad (7.23)$$

Therefore, the normal component of the displacement is the same on both sides of the surface of separation.

Next consider rectangle $EFGH$ of length Δl and negligible height. Let E_{1t} and E_{2t} be the tangential components of the electric intensity in the two media. Then the work done in taking a unit charge around the rectangle is

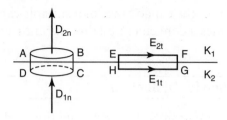

I FIGURE 7.10

$$E_{2t}\,\Delta l - E_{1t}\,\Delta l$$

Because the electric field is conservative, no net work is done (or, as this represents the drop in potential around a closed path, it must vanish). Hence,

$$E_{2t} = E_{1t} \tag{7.24}$$

and the tangential component of the electric intensity is the same on both sides of the surface.

When two dielectrics are in contact, both lines of force and lines of displacement are bent in passing from one to the other. We may use equations (7.23) and (7.24) to calculate the law of refraction of these lines. As shown in **Figure 7.11**, θ_1 and θ_2 are the angles that the lines of displacement make with the normal to the surface of separation of the two dielectrics. From equation (7.23) we have

$$\varepsilon_1 E_1 \cos\theta_1 = \varepsilon_2 E_2 \cos\theta_2$$

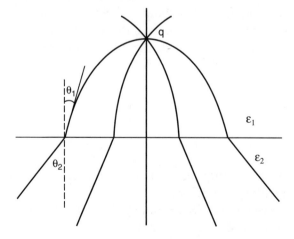

I FIGURE 7.11

and from equation (7.24) we have

$$\varepsilon_1 \sin\theta_1 = \varepsilon_2 \sin\theta_2$$

Dividing one by the other yields

$$\frac{\tan\theta_1}{\tan\theta_2} = \frac{\varepsilon_1}{\varepsilon_2} \qquad (7.25)$$

In passing from empty space into a dielectric, the lines of displacement are bent away from the normal to the surface, as shown in Figure 7.10, where a point charge q is placed in space above a dielectric.

Example 6

Given $\vec{E}_1 = 2\hat{i} - 3\hat{j} + 5\hat{k}$ V/m at the charge-free dielectric interface of Figure 7.12, find $\vec{D}_2$ and the angles θ_1 and θ_2.

Solution

By equations (7.24) and (7.23) we have

$$\vec{E}_1 = 2\hat{i} - 3\hat{j} + 5\hat{k}, \quad \vec{E}_1 = 2\hat{i} - 3\hat{j} + E_{z2}\hat{k}$$

$$\vec{D}_1 = \varepsilon_0 K_1 \vec{E}_1 = 4\varepsilon_0\hat{i} - 6\varepsilon_0\hat{j} + 10\varepsilon_0\hat{k}, \quad \vec{D}_2 = D_{x2}\hat{i} + D_{y2}\hat{j} + 10\varepsilon_0\hat{k}.$$

The unknown components can be found from $\vec{D}_2 = \varepsilon_0 K_2 \vec{E}_2$:

$$D_{x2}\hat{i} + D_{y2}\hat{j} + 10\varepsilon_0\hat{k} = 2\varepsilon_0 K_2\hat{i} - 3\varepsilon_0 K_2\hat{j} + \varepsilon_0 K_2 E_{z2}\hat{k}$$

from which

$$D_{x2} = 2\varepsilon_0 K_2 = 10\varepsilon_0, \quad D_{y2} = -3\varepsilon_0 K_2 = -15\varepsilon_0, \quad E_{z2} = 10/K_2$$

The angles made with the plane of the interface are found from (**Figure 7.12**)

$$\vec{E}_1 \cdot \hat{k} = |E_1| \cos(90^0 - \theta_1), \quad \vec{E}_2 \cdot \hat{k} = |E_2| \cos(90^0 - \theta_2)$$

$$5 = \sqrt{38} \sin\theta_1, \quad 2 = \sqrt{17} \sin\theta_2$$

$$\theta_1 = 54.2^0, \quad \theta_2 = 29.0^0$$

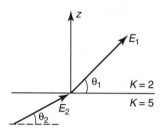

I FIGURE 7.12

Example 7: Dielectric Sphere in Uniform Field

An uncharged dielectric sphere of dielectric constant K is placed in a uniform electric field E_0. Find the potential at any point in space exterior to the dielectric sphere.

Solution

This is essentially Example 2 of Chapter 3, with the conducting sphere replaced by a dielectric sphere. In this case the surface of the sphere no longer needs to be an equipotential surface, but instead it is necessary that the normal components of the electric displacement and the tangential components of the electric intensity should be the same on both sides of the sphere.

   As in the case of the conducting sphere, the potential outside the sphere is of the form

$$V_0 = A_1 r \cos\theta + \frac{B_1 \cos\theta}{r^2}$$

To satisfy the boundary condition at infinity, we find again $A_1 = -E_0$. The components of the electric field intensity are

$$(E_r)_0 = -\frac{\partial V_0}{\partial r} = E_0 \cos\theta + \frac{2B_1 \cos\theta}{r^3},$$

$$(E_\theta)_0 = -\frac{\partial V_0}{r\partial\theta} = -E_0 \sin\theta + \frac{B_1 \sin\theta}{r^3}$$

If there is no volume distribution of free charge inside the sphere, the potential inside the sphere must be a solution of Laplace's equation. We try a solution of the same form as that for outside:

$$V_i = A_i r \cos\theta + \frac{B_i \cos\theta}{r^2}$$

Evidently, the second term on the right becomes infinite at the origin, so we must set $D_1 = 0$, and we get

$$V_i = A_i r \cos\theta$$

The components of the electric field intensity inside the sphere are

$$(E_r)_i = -\frac{\partial V_i}{\partial r} = -A_i \cos\theta,$$

$$(E_\theta)_i = -\frac{\partial V_i}{r\partial\theta} = A_i \sin\theta$$

Now $V_i = V_0$ at the surface of the sphere. So if a is radius of the sphere, then

$$A_i a \cos\theta = -E_0 a \cos\theta + \frac{B_1 \cos\theta}{a^2}$$

The continuity of the normal component of the displacement vector at the surface gives another relation:

$$-KA_i \cos\theta = E_0 \cos\theta + \frac{2B_1 \cos\theta}{a^3}$$

Solving the last equations we get

$$A_i = -\frac{3}{K+2} E_0, \quad B_1 = \frac{K-1}{K-2} a^3 E_0$$

Consequently, we finally have

$$V_0 = -\left[1 - \frac{K-1}{K+2}\frac{a^3}{r^3}\right] E_0 r \cos\theta,$$

$$V_i = -\frac{3}{K+2} E_0 r \cos\theta = -\frac{3}{K+2} E_0 z$$

The electric field intensity inside the sphere has the constant value

$$E_i = -\frac{\partial V_i}{\partial z} = \frac{3}{K+2} E_0$$

Evidently, the electric displacement inside the sphere also has the constant value

$$D_i = \varepsilon_0 K E_i = \varepsilon_0 \frac{3K}{K+2} E_0$$

and the lines of displacement inside the sphere are parallel to the z-axis. Moreover, because the dielectric constant K is always greater than unity, the lines of displacement are crowed together inside the sphere, as shown in **Figure 7.13**. That is, the lines are "pulled" into the sphere.

The last equation also shows that $E_i < E_0$. If the dielectric constant K is less than unity (i.e., $\varepsilon < \varepsilon_0$, as for a spherical cavity in a dielectric), the lines of displacement are "pushed" out from the sphere and $E_i > E_0$.

One more observation, namely, the results of this example for a dielectric sphere in a uniform field reduce to the results for a conducting sphere in a uniform field when $K \to \infty$. In electrostatic problems, a conductor may be considered as a dielectric of infinite permittivity.

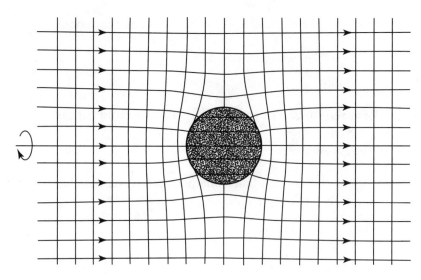

I FIGURE 7.13

7.9 Energy of the Electrostatic Field

Obviously, the expression obtained in Chapter 2 for the energy of the electric field needs to be modified when dielectrics are present. We may start from equation (2.22), because its proof is unaffected by the presence of dielectrics:

$$W = \frac{1}{2}\int \rho V d\vec{r} + \frac{1}{2}\int \sigma V dS$$

However, we must now substitute $\nabla \cdot \vec{D} = \rho$. Then

$$\frac{1}{2}\int \rho V d\vec{r} = \frac{1}{2}\int V(\nabla \cdot \vec{D})\, d\vec{r} = \frac{1}{2}\int \left\{ \nabla \cdot (V\vec{D}) - \vec{D} \cdot \nabla V \right\} d\vec{r}$$

where we have used the vector identity

$$\nabla \cdot (V\vec{D}) = V\nabla \cdot \vec{D} + \vec{D} \cdot \nabla V$$

With the help of the divergence theorem, we can transform the first integral on the right-hand side into a surface integral and obtain

$$\frac{1}{2}\int \rho V d\vec{r} = \frac{1}{2}\int V\vec{D} \cdot d\vec{S} + \frac{1}{2}\int \vec{D} \cdot \vec{E}\, d\vec{r}$$

The surface integral must be taken over all medium surfaces that cover conductors in our system and a closed surface bounding our system from out-

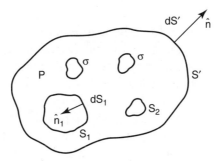

I FIGURE 7.14

side (**Figure 7.14**). We may choose to locate the latter surface at infinity, and then the surface integral over it vanishes. We have now

$$\frac{1}{2}\int \rho V d\vec{r} = \frac{1}{2}\int_S V\vec{D}\cdot d\vec{S} + \frac{1}{2}\int \vec{D}\cdot\vec{E}d\vec{r}$$

where S is the surfaces of all conductors. Substituting this into the expression (2.22) for W we obtain

$$W = \frac{1}{2}\int \sigma V dS + \frac{1}{2}\int_S V\vec{D}\cdot d\vec{S} + \frac{1}{2}\int \vec{D}\cdot\vec{E}d\vec{r}$$

The two surface integrals on the right-hand side cancel out, because the second surface integration is over surface of the medium, the vector $\hat{n}_i$ is drawn outward from the medium and so into the conductor as shown in Figure 7.14. Thus, D_n is $-\sigma$ and

$$\frac{1}{2}\int_S V\vec{D}\cdot d\vec{S} = -\frac{1}{2}\int \sigma V dS$$

which cancels the first surface integral on the right-hand side. We are left with

$$W = \frac{1}{2}\int \vec{D}\cdot\vec{E}d\vec{r} \tag{7.26}$$

We may regard the electrostatic energy as being distributed over the field with density being given by $\frac{1}{2}\vec{D}\cdot\vec{E}$. For isotropic dielectrics we may write this as $\varepsilon E^2/2$ or $D^2/2\varepsilon$. In the parallel plate capacitor of Example 4, the energy stored when it is filled with dielectric is $\frac{1}{2}CV^2 = \frac{1}{2}\varepsilon_0 K(A/d)V^2$. Dividing this by the volume of the capacitor (Ad), we obtain the energy density

$$\frac{1}{2}\varepsilon_0 K\left(\frac{V}{d}\right)^2 = \frac{1}{2}\varepsilon_0 KE^2 = \frac{1}{2}DE$$

A word of caution should be added here. In assembling the system of charges piece by piece to arrive at equation (7.26), we can begin with unpolarized dielectric with no free charges and then bring in the free charge from infinity, a bit at a time. In arriving at equation (7.26) we also have tacitly assumed the polarization to be proportional to the electric field. This is always true except in some very uncommon materials, where nonlinear effects do occur; then the external work done may depend on the precise path by which the final state of the system is approached. Energy is of course always conserved, but sometimes it is not possible to make a unique distinction between potential energy and heat. We meet the equivalent problem in magnetism, where nonlinear materials are of practical importance.

7.10 Forces on Dielectrics

A dielectric material placed in an electric field is subject to forces that arise from the interaction of the electric field with the dipoles in the dielectric. This force depends on the variation of the electric field. **Figure 7.15** shows a dipole with the negative charge at the origin. Set the field intensity at the origin to be

$$\vec{E} = E_x \hat{i} + E_y \hat{j} + E_z \hat{k}$$

then the x-component of the force on the dipole is

$$
\begin{aligned}
F_x &= (-q)\,E_x + q\,(E_x + \vec{l} \cdot \nabla E_x) \\
&= q\,(x\hat{i} + y\hat{j} + z\hat{k}) \cdot \left(\frac{\partial E_x}{\partial x}\,\hat{i} + \frac{\partial E_x}{\partial y}\,\hat{j} + \frac{\partial E_x}{\partial z}\,\hat{k} \right) \\
&= p_x \frac{\partial E_x}{\partial x} + p_y \frac{\partial E_y}{\partial y} + p_z \frac{\partial E_z}{\partial z} = \vec{p} \cdot \nabla E_x, \quad (p_x = qx, etc)
\end{aligned}
$$

Similarly,

$$F_y = \vec{p} \cdot \nabla E_y, \quad F_z = \vec{p} \cdot \nabla E_z$$

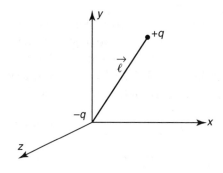

I FIGURE 7.15

And so

$$\vec{F} = \vec{p} \cdot \nabla E_x \hat{i} + \vec{p} \cdot \nabla E_y \hat{j} + \vec{p} \cdot \nabla E_z \hat{k}$$
$$= \left(p_x \frac{\partial}{\partial x} + p_y \frac{\partial}{\partial y} + p_z \frac{\partial}{\partial z} \right) (E_x \hat{i} + E_y \hat{j} + E_z \hat{k})$$
$$= (\vec{p} \cdot \nabla) \vec{E}$$

The force per unit volume of the dielectric is then given by

$$\vec{F} = N(\vec{p} \cdot \nabla) \vec{E} = (\vec{P} \cdot \nabla) \vec{E} \tag{7.27}$$

But

$$\vec{P} = \vec{D} - \varepsilon_0 \vec{E} = (\varepsilon - \varepsilon_0) \vec{E} = \varepsilon_0 \chi \vec{E}$$

Substituting this in equation (7.27)

$$\vec{F} = \varepsilon_0 \chi (\vec{E} \cdot \nabla) \vec{E} \tag{7.28}$$

This can be further simplified. Let us first examine F_x:

$$F_x = \varepsilon_0 \chi \left(E_x \frac{\partial E_x}{\partial x} + E_y \frac{\partial E_x}{\partial y} + E_z \frac{\partial E_x}{\partial z} \right)$$

From $\nabla \times \vec{E} = 0$ we have

$$\frac{\partial E_x}{\partial y} = \frac{\partial E_y}{\partial x}, \quad \frac{\partial E_x}{\partial z} = \frac{\partial E_z}{\partial x}$$

Substituting this in the preceding expression, we obtain

$$F_x = \varepsilon_0 \chi \left(E_x \frac{\partial E_x}{\partial x} + E_y \frac{\partial E_y}{\partial x} + E_z \frac{\partial E_z}{\partial x} \right) = \frac{1}{2} \varepsilon_0 \chi \frac{\partial}{\partial x} E^2$$

similarly,

$$F_y = \frac{1}{2} \varepsilon_0 \chi \frac{\partial}{\partial y} E^2, \quad F_z = \frac{1}{2} \varepsilon_0 \chi \frac{\partial}{\partial z} E^2$$

Finally, we have

$$\vec{F} = \frac{1}{2} \varepsilon_0 \chi \nabla E^2 = \frac{1}{2} \varepsilon_0 (K - 1) \nabla E^2 \tag{7.29}$$

The force is proportional to the gradient of E^2 and therefore pulls the dielectric in the direction of increasing field strength, as anticipated.

Equation (7.27) is also valid in alternating electric field, provided $\nabla \times \vec{E} = 0$ still holds. What does this mean? From $\nabla \times \vec{E} = -\partial \vec{B}/\partial t$, we have, the x-component

$$\frac{\partial E_y}{\partial z} - \frac{\partial E_z}{\partial y} = \frac{\partial B_x}{\partial t}$$

Thus, $\nabla \times \vec{E} = 0$ approximately holds whenever

$$\left| \frac{\partial E_y}{\partial z} \right| \cong \left| \frac{\partial E_z}{\partial y} \right| \gg \left| \frac{\partial B_x}{\partial t} \right|, \text{ etc.}$$

Example 8

A slab of linear dielectric material is partially inserted between the plates of a parallel-plate capacitor (**Figure 7.16**). Calculate the force on the slab.

Solution

According to equation (7.29) the force on the slab is

$$F = \frac{1}{2} \varepsilon_0 \chi \int \frac{\partial}{\partial x} E^2 \, dx \, dy \, dz$$

By symmetry, the y- and z-components integrate out. Now

$$\int \frac{\partial}{\partial x} E^2 \, dx = E_0^2 - E_i^2$$

where E_0 denotes the field at the right end of the slab, outside the capacitor, and E_i is the field at the left end of the slab, so it is inside the capacitor. E_0 is zero as long as the slab extends well outside so that the fringe effect is neg-

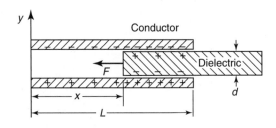

I FIGURE 7.16

ligible. $E_i = V/d$ in the uniform field region. The y and z integrals give d and W (the width of the plate), respectively. Putting these all together, we obtain

$$F = -\frac{1}{2}\varepsilon_0\chi\frac{W}{d}V^2$$

The minus sign indicates that the force is in the negative x-direction, so the slab is pulled into the capacitor.

Alternatively, we can calculate the force by examining what happens to the stored energy in the capacitor when a dielectric is introduced. We shall learn that the stored energy becomes less if the charge is hold fixed while we add a dielectric medium. This follows from the expression

$$U = \frac{1}{2}QV = \frac{Q^2}{2C}$$

Where does the energy go? It goes to the work done by the electric force on the slab. As shown in Figure 7.16, the dielectric slab is partially inserted. There is a force driving the slab in, which is related to non-uniformities in the field around the edges of the dielectric and the plates. The electric force on the slab is

$$F_x = -dU/dx$$

where U is energy of the capacitor, which can be written as

$$U = \frac{1}{2}CV^2$$

The capacitance is the ratio of the total free charge on the plates to the voltage between the plates. Now the total free charge Q on the plate is

$$Q = \frac{K\varepsilon_0 V}{d}(L - x)W + \frac{\varepsilon_0 V}{d}xW$$

then the capacitance C is

$$C = \frac{\varepsilon_0 W}{d}[(1 - K)x + KL] = \frac{\varepsilon_0 W}{d}[-\chi x + KL]$$

So the force on the slab is

$$F_x = -\frac{dU}{dx} = \frac{1}{2}\frac{Q^2}{C^2}\frac{dC}{dx} = -\frac{1}{2}\varepsilon_0\chi\frac{W}{d}V^2$$

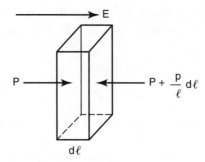

I FIGURE 7.17

which is the same result we obtained with equation (7.29). The energy method can often be used to avoid enormous complications in determining the forces on dielectric materials.

a. Stresses in a Dielectric

We now return to equation (7.29). If we designate the pressure in the dielectric by p, we see that the pressure over the right-hand face of the rectangular parallelepiped of unit cross-section and thickness dl pictured in **Figure 7.17** must be greater than that over the left-hand face by an amount just sufficient to balance the electrical force given by equation (7.29), that is,

$$\frac{\partial p}{\partial l} dl = \frac{\varepsilon_0(K-1)}{2} \frac{\partial}{\partial l}(E^2) dl$$

Integrating we have

$$p = \frac{\varepsilon_0(K-1)}{2} E^2 \tag{7.30}$$

except for a constant of integration that is of no significance because it represents merely a constant pressure that is everywhere the same. Equation (7.30) shows that the pressure in a dielectric is proportional to the square of the electric intensity, being large near charges, where E is great, and small at great distances from charges, where the field is weak.

b. Stress on Surface of Dielectric

At the surface of the dielectric there is a rise in pressure that can be understood with the aid of **Figure 7.18**. The electric field intensity is assumed to be normal to the surface. The bound charge σ' resides in a thin layer of

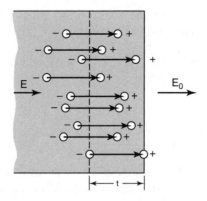

I FIGURE 7.18

thickness t inside which the electric field intensity rises from a value E in the interior of the dielectric to E_0 outside. The dipole whose positive charges constitute the surface layer experience the field E on their negative ends and a field that is greater than E on their positive ends. The resultant force per unit area, $\mathscr{I}$, is directed to the right, and

$$\mathscr{I} = \sigma' \frac{(E_0 + E)}{2} - \sigma'E = \frac{\sigma'}{2}(E_0 - E)$$

Now E_0 is larger than the field E inside the dielectric by a factor equal to the dielectric constant K

$$E_0 = KE$$

and so F becomes

$$\mathscr{I} = \frac{\sigma'}{2}(K - 1)E \tag{7.31}$$

Now

$$\sigma' = \vec{P} \cdot \hat{n} = \varepsilon_0(K - 1)E_n$$

where E_n is the component of the field intensity normal to the dielectric surface, and in our case $E_n = E$. Thus we may write equation (7.31) as

$$\mathscr{I} = \frac{\varepsilon_0(K - 1)^2}{2}E^2 \tag{7.32}$$

To get the total stress $\mathscr{T}_t$ on the surface, we must add the pressure in the homogeneous body of the dielectric given by (7.30):

$$\mathscr{T}_t = \frac{\varepsilon_0(K-1)^2}{2} E^2 + \frac{\varepsilon_0(K-1)}{2} E^2$$

After simplification we have

$$\mathscr{T}_t = \frac{\varepsilon_0 K(K-1)}{2} E^2 = \frac{\varepsilon_0}{2} \frac{K-1}{K} E_0^2 \tag{7.33}$$

As equation (7.33) for the stress involves only the square of the field strength, it represents a positive stress or tension whether the electric intensity is directed along the outward normal to the surface, as we have supposed, or in the opposite sense. This tension tends to stretch a solid dielectric in the direction of the lines of force of the field in which it is placed, an effect known as electrostriction. The result (7.33) has been verified by using a liquid dielectric with an air bubble above it; the increase in pressure of the air as the field is increased is observed to accord with formula (7.33).

Example 9

A charged parallel-plate capacitor is immersed in a liquid dielectric. Find the net force pulling the capacitor plates together.

Solution

Using equations (7.31) and (7.33), we can calculate the total force per unit area with which the liquid dielectric pushes against the capacitor plates. There is one rise in pressure in passing through the fringe field into the region between the plates and another rise in pressure in passing out of the dielectric through the surface layer. From equations (7.31) and (7.33) the total pressure at the surface of the capacitor plate is

$$p_t = \frac{\varepsilon_0(K-1)^2 E^2}{2} + \frac{\varepsilon_0(K-1) E^2}{2} + p_0$$

where p_0 is the hydrostatic pressure on the outside surface of the capacitor plate, $p_t - p_0$ is the excess pressure tending to push the capacitor plates apart, and

$$p_t - p_0 = \frac{\varepsilon_0}{2} K(K-1) E^2$$

Because the field intensity within the dielectric is given by

$$E = \sigma/K\varepsilon_0$$

where σ is the surface charge density of the capacitor plate. Now we can rewrite the expression for $p_t - p_0$ in terms of the surface charge density σ

$$p_t - p_0 = \frac{\sigma^2}{2\varepsilon_0} \frac{K-1}{K}$$

The capacitor plates are charged conductors, however; thus the electric field intensity acting on the charges on the plates produces a force that tends to pull the plates together. The magnitude of this force per unit area is

$$\frac{dF}{da} = \frac{\sigma^2}{2\varepsilon_0}$$

which is greater than the pressure $p_t - p_0$ that tends to push the capacitor plates apart. Then the net force per unit area with which the plates are pulled together is

$$F = \frac{\sigma^2}{2\varepsilon_0} - \frac{\sigma^2}{2\varepsilon_0} \frac{K-1}{K} = \frac{\sigma^2}{2K\varepsilon_0}$$

Thus, the attractive force on the capacitor plates is reduced by the factor K over what it would be if the plates carried the same charge density in free space.

Calculations of the forces between charged conductors immersed in a liquid dielectric always show that the force is reduced by the factor K. There is a tendency to think of this as representing a reduction in the electrical forces between the charges on the conductors, as though Coulomb's law for the interaction of two charges should have the dielectric constant included in its denominator. This is incorrect, however. The strictly electric forces between charges on the conductors are not influenced by the presence of the dielectric medium. The medium is polarized, however, and the interaction of the electric field with the polarized medium results in an increased fluid pressure on the conductors that reduces the net forces acting on them.

B. Microscopic Aspects of Dielectrics

So far we have limited our discussion on the macroscopic aspects of dielectric polarization. In many cases we could understand the properties of electric systems with dielectrics once we appreciated the polarization induced by an electric field. However, one basic question remains to be answered, namely what determines the value of the polarization $\vec{P}$ at any point inside a dielectric? Or stated differently, what determines the dipole moment of a given molecule? It is not the macroscopic electric field that we talked about in the above sections. The macroscopic field that has been smoothed out over a large region containing many molecules is not necessarily the same as the local field (or the molecular field) that polarizes one molecule. Our task now is to examine what this local field is and how is it related to the macroscopic electric field. Moreover, we shall see that it is possible to understand, on the basis of a simple molecular model, the linear behavior that is characteristic of a large class of dielectric.

7.11 The Local Field $\vec{E}_{loc}$

The local field at a molecular position in the dielectric is the electric field generated by all charges besides the particular molecule under consideration, that is, by all external sources and by all polarized molecules in the dielectric with the exception of the molecule under consideration.

To calculate the local field, let us suppose that we have a linear and isotropic dielectric sample that has been polarized by placing it in the uniform field between two parallel plates that are oppositely charged. Now imagine a small spherical surface constructed within the material (**Figure 7.19**). Its volume is chosen so that it is small compared with the entire sample of the material and large enough so that all dipoles beyond it can be accounted for via the induced polarization surface charges. But inside the imaginary sphere we treat the dipoles individually. The imaginary sphere divides the dielectric sample into two volumes, V outside the sphere and V- inside. The electric field at point P has several contributions: (1) the external applied field, (2) all dipoles outside the sphere, and (3) local contribution of the set of dipoles in the sphere.

We first consider all dipoles outside the sphere. Their contributions can be divided into two parts, $\vec{E}_d + \vec{E}_S$. $\vec{E}_d$ is just the depolarization field caused by the surface polarization charges on the dielectric. In our case, $\vec{E}_d = -\sigma_P/\varepsilon_0 = -\vec{P}/\varepsilon_0$, $(\sigma_p = P_n = P)$. In general, it is given by $\vec{E}_d = -LP/\varepsilon_0$,

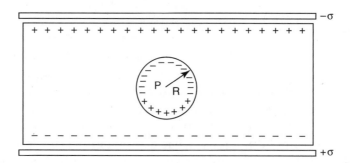

I FIGURE 7.19

where L is the depolarization factor, depending on the specimen geometry. The term $\vec{E}_S$ is also a depolarizing field. It arises from those charges at the imaginary spherical boundary that are the near ends of the dipoles just outside of the boundary. They correspond closely to the polarization surface charges that appear whenever a cavity is cut inside a dielectric medium. Because these charges are on surfaces with opposite symmetry from those at the outside of the specimen, they have opposite signs, and $\vec{E}_S$ is opposed to $\vec{E}_d$. We show a moment later that $\vec{E}_s$ is given by

$$\vec{E}_s = \vec{P}/3\varepsilon_0 \tag{7.34}$$

We are now left with the task of evaluating the local contribution of the set of dipoles in the sphere, $\vec{E}_{in}$. It turns out that if we take the rather common case of lattice having cubic symmetry, the contributions of these nearby dipoles cancel out and $\vec{E}_{in} = 0$. This is shown later. We may therefore write for the local field

$$\vec{E}_{loc} = \vec{E}_{ex} + \vec{E}_d + \frac{1}{3}\frac{\vec{P}}{\varepsilon_0} \tag{7.35}$$

where $\vec{E}_{ex}$ is the external field, due to the charged parallel plate in our case. The local field is often written as

$$\vec{E}_{loc} = \vec{E} + \frac{1}{3\varepsilon_0}\vec{P} \tag{7.36}$$

where $\vec{E}$ stands for the fields in the dielectric:

$$\vec{E} = \vec{E}_{ex} + \vec{E}_d \tag{7.37}$$

7.12 Evaluation of the Depolarizing Field $\vec{E}_s$

The depolarizing field $\vec{E}_s$ arises from those charges at the imaginary spherical boundary and can be written as

$$\vec{E}_s = \frac{1}{4\pi\varepsilon_0}\int_{S'}\frac{\sigma_P ds'}{r^2}\hat{r} \tag{7.38}$$

where r is the distance from the element of surface dS' to the point P and $\hat{r}$ is the unit vector in the direction of r. The axis z is drawn parallel to the polarization vector $\vec{P}$ (**Figure 7.20**). At the point A, where the normal to the surface S' is at an angle θ to $\vec{P}$, the polarization charge on an element of surface ds' is $\sigma_P(\theta)ds'$, where

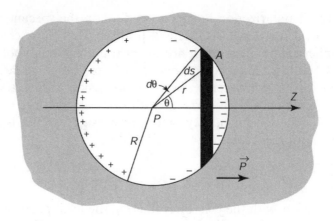

I FIGURE 7.20

$$\sigma_P(\theta)\,ds' = \vec{P} \cdot \vec{ds'} = -P\cos\theta\,ds'$$

The minus sign arises because the outward normal to the dielectric is the inward normal to the sphere. By symmetry the net field generated at the center P by the polarization charge is directed along the axial direction

$$d\vec{E}_{S'}'' = -\frac{\vec{P}\cos\theta}{4\pi\varepsilon_0 r^2}\cos^2\hat{r}\,ds'$$

the area of the band on the surface lying between θ and $(\theta + d\theta)$ is $2\pi R^2 \sin\theta d\theta$. The field at P due to the complete surface is then given by

$$\vec{E}_{S'}'' = -\frac{\vec{P}}{2\varepsilon_0}\int_\pi^0 \cos^2\theta\sin\theta d\theta = \frac{1}{3\varepsilon_0}\vec{P} \qquad (7.39)$$

7.13 Evaluation of $\vec{E}_{in}$ Due to the Dipoles in the Sphere

The field $\vec{E}_{in}$ can be obtained by summing over the fields due to the dipoles within the sphere. We have the potential of an individual dipole from Chapter 2

$$V = \frac{1}{4\pi\varepsilon_0}\frac{\vec{p}\cdot\vec{r}}{r^3}$$

and the field at a distance r from it is

$$\vec{E} = -\nabla V = \frac{-1}{4\pi\varepsilon_0}\left[\frac{\vec{p}}{r^3} - \frac{3(\vec{p}\cdot\vec{r})\vec{r}}{r^5}\right]$$

Summing over all the dipoles within the sphere is equivalent to taking the spatial average over one of the components of the field, say the x-component:

$$\sum E_x = \frac{-1}{4\pi\varepsilon_0}\left[\frac{p_x}{r^3} - \frac{3\,(p_x x^2 + p_y xy + p_z xz)}{r^5}\right]$$

Because we are only dealing with isotropic dielectric, the x-, y-, z-directions are equivalent, and

$$\langle x^2 \rangle = \langle y^2 \rangle = \langle z^2 \rangle = \langle r^2 \rangle/3, \quad \langle xy \rangle = \langle yz \rangle = \langle zx \rangle = 0$$

Hence, $\vec{E}_{in} = 0$ for isotropic dielectric.

It is evident that equation (7.36) for the local field is at best approximate, due to the approximations that have gone into it. Fortunately, our dielectric theory only demands that the expression for the local field to be of the form

$$\vec{E}_{loc} = \vec{E} + \frac{b}{\varepsilon_0}\vec{P} \tag{7.36a}$$

where b is a constant that depends on the nature of the dielectric, but it is of the order of 1/3.

Equation (7.36) is also valid for a gas. Under normal conditions, the polarization in a gas is small compared with the electric field, so $\vec{E}_{loc} \cong \vec{E}$. However, if the gas is under very high pressure, then its density may be comparable with that of a liquid and the difference between $\vec{E}_{loc}$ and $\vec{E}$ cannot be neglected.

There are several mechanisms of dielectric response to a local field. The basic mechanism is that electrons in each atom are pushed one way and the nuclei the other way by the local field. This is known as electronic polarization. In polyatomic molecules, the shifting of the electron orbits leads to shifts in the nuclear positions. It is more complicated to calculate this ionic polarizability, but in general the resulting polarizability is comparable with the electronic term. The ionic polarization is the major one in ionic crystals.

Electronic and ionic polarization can occur in all classes of dielectric materials. If these are the only mechanisms present, the dielectric is called nonpolar.

Some molecules have permanent dipole moments. Molecules consisting of two or more different species of atoms exhibit permanent dipole moments. During molecule formation, some of the electrons may be completely or partially transferred from one atomic species to the other, with the resulting electronic arrangement being such that positive and negative charge centers do not coincide in the molecule. Typical examples are *HCL*, *H₂O*, and *CO*, all of which have asymmetric structures.

In a gas, the molecules are far apart except during occasional collisions and are thus free to rotate without interference from their neighbors. If there is no electric field, thermal agitation (collisions) completely randomizes the orientations. So without an electric field, a macroscopic piece of the polar dielectric is not polarized.

7.14 Linear Dielectrics and the Clausius-Mossotti Equation

In linear and isotropic dielectrics, the molecular dipole moment is directly proportional to, and in the same direction as, the local field. Thus we write

$$\vec{p} = \alpha \vec{E}_{loc} = \alpha \left(\vec{E} + \vec{P}/3\varepsilon_0 \right) \tag{7.40}$$

where the constant α is known as the molecular polarizability.

The polarization (the dipole moment per unit volume) is then given by

$$\vec{P} = N\vec{p} = N\alpha \vec{E}_{loc} = N\alpha \left(\vec{E} + \vec{P}/3\varepsilon_0 \right) \tag{7.41}$$

where N is again the number of molecules per unit volume. Solving for $\vec{P}$:

$$\vec{P} = \frac{N\alpha}{1 - N\alpha/3\varepsilon_0} \vec{E} \tag{7.41a}$$

People usually write this in terms of the electric susceptibility χ

$$\vec{P} = \varepsilon_0 \chi \vec{E} \tag{7.42}$$

Then we have

$$\frac{N\alpha}{1 - N\alpha/3\varepsilon_0} = \varepsilon_0 \chi \tag{7.43}$$

Solving for α:

$$\alpha = \frac{3\varepsilon_0 \chi}{N(3 + \chi)} = \frac{3\varepsilon_0 (K - 1)}{N(K + 2)} \tag{7.44}$$

where we have used the relation

$$K = \varepsilon/\varepsilon_0 = 1 + \chi$$

Table 7-1 **Dielectric constant of some substances at normal temperature and pressure**

Substance	Gas			Liquid				
	κ (exp)	$N\alpha$	Density	Density	Ratio*	$N\alpha$	κ (predict)	κ (exp)
CS_2	1.0029	0.0029	0.00339	1.293	381	1.11	2.76	2.64
O_2	1.000523	0.000523	0.00143	1.19	832	0.435	1.509	1.507
CCl_4	1.0030	0.0030	0.00489	1.59	325	0.977	2.45	2.24
A	1.000545	0.000545	0.00178	1.44	810	0.441	1.517	1.54

*Ratio = density of liquid/density of gas.

Equation (7.44) is known as the Clausius-Mossotti equation, which relates the measurable property K to the microscopic property α and the density N. N can be written as $(\rho/M)N_A$, ρ is the mass density, M is the molecular weight, and N_A is Avogadro's number. K may be measured by recording the change in C as we fill a capacitor with the dielectric. **Table 7.1** lists some experimental data.

The Clausius-Mossotti equation is also known, in its application to optics, as the Lorentz-Lorenz equation.

7.15   Polar Molecules and the Langevin Equation

We now discuss the dielectric behavior of molecules that have permanent dipole moments. When the polar dielectric is subjected to an electric field, the individual dipoles experience torques, which tend to align them with the field. If the field is strong enough, the dipoles may be completely aligned, and the polarization achieves the saturation value

$$\vec{P}_S = N\vec{p}_m \qquad (7.45)$$

N is the number of molecules per unit volume, and $\vec{p}_m$ is the electric dipole moment of a single molecule. At field strengths normally encountered, the polarization of a polar dielectric usually does not reach its saturation value. This orientation effect is in addition to the induced dipole effects, which are also usually present. Let us ignore the induced dipole contribution at the moment and add its effect later.

In gaseous or liquid polar dielectrics, thermal agitation causes collisions between molecules that tend to destroy the alignment of molecular dipoles with the local field. But the local field exerts a restoring force between collisions and there remains some partial alignment and, consequently, on the average, a net dipole moment $\vec{P}$ per unit volume.

The calculation of this net polarization is a simple problem of statistical mechanics. Let us consider a unit volume containing N molecules. Without an external electric field, the dipoles are oriented at random—all orientations of dipoles are equally probable. If, at any moment, dN dipoles are oriented at angles lying between θ and $\theta + d\theta$ with respect to a given direction, then the fraction dN/N is equal to the ratio of the solid angle $d\Omega$ corresponding to the angular interval $d\theta$ to the total solid angle 4π. Now the solid angle in the range of orientation between θ and $\theta + d\theta$ is (**Figure 7.21**)

$$d\Omega = \frac{dA}{r^2} = \frac{2\pi r^2 \sin\theta\, rd\theta}{r^2} = 2\pi\sin\theta\, d\theta$$

and we have

$$\frac{dN}{N} = \frac{d\Omega}{4\pi} = \frac{1}{2}\sin\theta\, d\theta \tag{7.46}$$

If now a field $\vec{E}_{loc}$ is applied, the dipoles making angle θ with the field will gain an additional potential energy $W = -pE_{loc}\cos\theta$. It is shown in statistical mechanics that if a large number of molecules are in statistical equilibrium, the number possessing a particular energy W is proportional to $\exp(-W/K_B T)$, where K_B is the Boltzmann constant 1.38×10^{-23} joule/Kelvin and T is the absolute temperature in Kelvin. The number of molecules per unit volume with their dipoles in the range between θ and $d\theta$ is

$$dN = N(\theta)\, d\Omega = C\exp(\vec{P}\cdot\vec{E}_{loc}/K_B T)\, d\Omega$$
$$= C\exp(u\cos\theta)\sin\theta\, d\theta \tag{7.47}$$

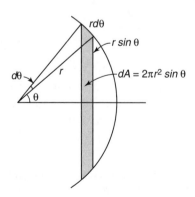

I FIGURE 7.21

where we have set

$$u = pE_{loc}/K_BT \tag{7.48}$$

and where the constant C must be chosen so that the total number of molecules in a unit volume is N: $N = C \int_0^\pi \exp(u\cos\theta)\sin\theta d\theta$, or

$$C = N/\int_0^\pi \exp(u\cos\theta)\sin\theta d\theta \tag{7.49}$$

The effective dipole moment of a molecule dipole is its component along the field direction:

$$dP = pdN\cos\theta = pN \frac{\exp(u\cos\theta)\sin\theta\cos\theta \, d\theta}{\int_0^\pi \exp(u\cos\theta)\sin\theta \, d\theta}$$

To obtain the dipole moment per unit volume P, we integrate the numerator from $\theta = 0$ to $\theta = \pi$:

$$P = pN \frac{\int_0^\pi \exp(u\cos\theta)\sin\theta\cos\theta \, d\theta}{\int_0^\pi \exp(u\cos\theta)\sin\theta \, d\theta}$$

Now let $t = u\cos\theta$, then

$$P = pN \frac{\int_{-u}^u te^t \, dt}{\int_{-u}^u te^t \, dt} = \frac{pN}{u} \frac{\left[te^t - e^t\right]_u}{e^t|_{-u}^u}$$

We finally have

$$P = pN(\coth u - 1/u) = pN\left(\coth\frac{pE_{loc}}{K_BT} - \frac{K_BT}{pE_{loc}}\right) \tag{7.50}$$

where

$$\coth u = \frac{e^u + e^{-u}}{e^u - e^{-u}}, \quad e^u = 1 + u + u^2/2! + u^3/3! + \ldots.$$

The function on the right is called the Langevin function and has the form shown in **Figure 7.22**. For large u $(= pE_{loc}/K_BT)$, that is, for large fields and low temperatures, P approaches Np. The dipoles are all aligned with the field, and the polarization reaches maximum.

However, u is normally smaller than unity. A typical dipole is of the order of 10^{-30} coulomb-meter, and at room temperature $K_BT \sim 4 \times 10^{-21}$

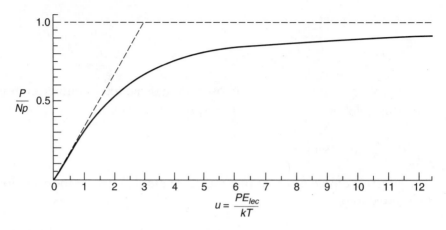

I FIGURE 7.22

joule, so even with a local field of 10^7 volts/meter, u is about of the order of 2×10^{-3}. For such small u, we can expand the exponentials in equation (7.50) and let us retain only the terms up to u^3:

$$P \approx pN \left[\frac{2+u^2}{2u\left(1+u^2/6\right)} - \frac{1}{u} \right]$$

$$\approx pN \left[\frac{1}{2u}(2+u^2)(1-u^2/6) - \frac{1}{u} \right] \approx \frac{Npu}{3}$$

But $u = pE_{loc}/K_B T$ and hence

$$P \approx \frac{Npu}{3} = \frac{Np^2}{3K_B T} E_{loc} \tag{7.51}$$

which may be written in vector form

$$\vec{P} = \frac{Np^2}{3K_B T} \vec{E}_{loc}, \text{ or } \frac{1}{N}\vec{P} = \frac{p^2}{3K_B T} \vec{E}_{loc} \tag{7.52}$$

Hence, when $pE_{loc} \ll K_B T$, the polarization P in a polar dielectric is proportional to the local field and the dielectric is linear; P (and so the susceptibility and the dielectric constant) is also inversely proportional to T. This $1/T$ dependence is called Curie's law. From a practical point of view, the temperature dependence of the dielectric constant distinguishes a polar from a nonpolar dielectric.

When the molecules are crowded close together the situation gets very complicated. In solids the rotation may be limited to a number of discrete positions by the crystal structure. In liquids the rotation is still largely unhindered, but each dipole so strongly affects its neighbors that the theory based on average values is not adequate any more. In both cases a polarizability proportional to u $(= pE_{loc}/K_BT)$ still obtained, but the proportionality constant is different from $1/3$ and depends on structural details.

7.16 The Debye Equation

Equation (7.52) has been derived by neglecting the induced dipole moment; it represents what is often termed orientation polarizability. Besides this, the polar molecules also acquire an induced dipole moment along the field, irrespective of the direction of the permanent dipole moment, and give rise to what is often termed deformation polarizability, α_0. In the general case, then, the total or net molecular polarizability is

$$\alpha = \alpha_0 + \frac{p^2}{3K_BT} \tag{7.53}$$

Combining this with equation (7.41),

$$\vec{P} = N\left(\alpha_0 + \frac{p^2}{3K_BT}\right)\vec{E}_{loc} \tag{7.54}$$

where

$$\vec{E}_{loc} = \vec{E} + \frac{\vec{P}}{3\varepsilon_0}$$

Let us rewrite $\vec{E}_{loc}$ in terms of $\vec{E}$ only. From equation (7.43)

$$\vec{P} = \varepsilon_0\chi\vec{E} = \varepsilon_0(K-1)\vec{E}$$

then

$$\vec{E}_{loc} = \vec{E} + \frac{1}{3\varepsilon_0}\varepsilon_0(K-1)\vec{E} = \frac{K+2}{3}\vec{E} \tag{7.55}$$

Accordingly, equation (7.54) becomes

$$\varepsilon_0(K-1)\vec{E} = N\left(\alpha_0 + \frac{p^2}{3K_BT}\right)\frac{K+2}{3}\vec{E}$$

From which we have

$$\frac{K-1}{K+2} = \frac{N}{3\varepsilon_0}\left(\alpha_0 + \frac{p^2}{3K_BT}\right) \qquad (7.56)$$

Multiplying by the molecular weight M and dividing by the mass density ρ, we get a new expression for the molecular polarization that is valid for polar dielectrics:

$$\alpha_M = \frac{M}{\rho}\frac{K-1}{K+2} = \frac{N_A}{3\varepsilon_0}\left(\alpha_0 + \frac{p^2}{3K_BT}\right) \qquad (7.57)$$

where the quantity α_M

$$\alpha_M = \frac{M}{\rho}\frac{K-1}{K+2} \qquad (7.58)$$

is called the molar polarization. Equation (7.57) is known as the *Debye equation*. The plot of α_M versus $1/T$ gives a straight line, as shown in **Figure 7.23**, where intercept on the vertical axis is $N_A\alpha_0/3\varepsilon_0$ and whose slope is $N_Ap^2/9\varepsilon_0K_BT$. Thus, in principle, the Debye equation can be used to determine both the molecular polarizability α_0 and the permanent dipole moment p of a molecule. But in practice it provides a reliable way of measuring α_0 and p for gases only.

The Debye equation is only an approximation, for at least two reasons. First, we have assumed that the field $\vec{E}_{loc} = \vec{E} + \vec{P}/3\varepsilon_0$ is responsible both for the induced and the oriented polarization. We saw that the factor 1/3 is only approximate. Second, as a permanent dipole turns, part of the local field turns with it and plays no part in orienting the dipole. This comes about because the field of the dipole polarizes the surrounding molecules, and these in turn produce a field at the dipole that is in the same direction as its moment. Furthermore, molecular associations in liquids and in solids complicate the problem. The result is that dielectric behavior in liquids, and

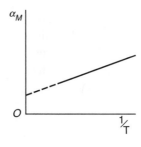

I FIGURE 7.23

especially in crystalline solids, is more complicated than the simple picture provided by the Debye equation. The molar polarization of water, for example, shows little temperature dependence, even though water has a large permanent dipole moment of 6.2×10^{-30} coulomb-meter. A precise determination of molecular polarizability and permanent dipole moment must rest on an improved treatment.

7.17 Permanent Polarization: Ferroelectricity

We have seen in preceding sections that it is the local electric field responsible for the polarization of the individual molecule, and the local field is related to the macroscopic electric field by equation (7.36):

$$\vec{E}_{loc} = \vec{E} + \frac{1}{3\varepsilon_0}\vec{P}$$

In most cases, the polarization vector $\vec{P}$ is proportional to the macroscopic field $\vec{E}$, so that the local field $\vec{E}_{loc}$ vanishes when $\vec{E}$ is zero. Under certain conditions, a permanent polarization $\vec{P}_0$ exists when $\vec{E}$ is set equal to zero

$$\vec{E}_{loc} = \frac{1}{3\varepsilon_0}\vec{P}_0 \tag{7.59}$$

Then, what is the condition under which a permanent polarization can exist? If there are N molecules per unit volume, we have

$$\vec{P}_0 = N\alpha\vec{E}_{loc} = \frac{N\alpha}{3\varepsilon_0}\vec{P}_0$$

The above relation is satisfied when either $\vec{P}_0 = 0$ or

$$N\alpha/3\varepsilon_0 = 1 \tag{7.60}$$

This is the condition for a permanent polarization.

We can look at this from a different angle. If we substitute equation (7.60) into equation (7.43), then there is an infinite susceptibility, so that $\vec{P}$ can have a nonzero value even in zero field, and we have a spontaneous polarization. Materials with such properties are called ferroelectrics. Barium titanate ($BaTiO_3$) is a good example of a ferroelectric material; it exhibits a spontaneous dipole moment at temperatures below 120°C (the Curie temperature of the material).

The polarized state of a ferroelectric material is a relatively stable one that can persist for long periods of time. A ferroelectric is also stable against a reversed electric field provided this electric field is not too large. **Figure 7.24**

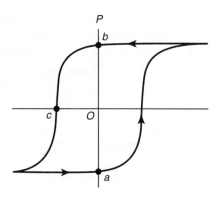

I FIGURE 7.24

shows the complete curve of polarization versus electric field. It is evident that for low fields there are two values of P for each value of E. A curve such as that in Figure 7.24 is called a *hysteresis loop*. Hysteresis means "to lag behind," and it is apparent that the polarization vector lags the electric field vector.

A ferroelectric slab between two parallel plates may serve as the basic elements of a memory device. This is related to the fact that there is no depolarizing field in a polarized ferroelectric specimen. Suppose we have a specimen that is polarized by placing it between parallel conducting plates that subsequently have a large potential difference applied to them. In this process, the *free* charge from the plates is, to a large extent, neutralized by surface polarization charges. If the parallel plates are now brought to the same potential by short-circuiting them, the polarized state of the ferroelectric slab is still energetically favorable, so the *free charge stays in place*, still neutralizing the polarization charge. If a large potential difference of the opposite sign is now applied to the plates surrounding the polarized ferroelectric slab, the specimen changes its polarization, and the free charge of the opposite sign flows to the plates from the external circuit, sufficient not only to neutralize the free charge already there, but also to neutralize the new polarization charge. Thus, a ferroelectric slab between two parallel plates may serve as the basic element of a memory device. It is capable of storing charges on the two plates surrounding the ferroelectric as ± or ∓ (they correspond to points b and a in Figure 7.24). The "number" ± or ∓ may be read by applying a potential difference across the specimen. If the applied field is in the direction of the original polarization, no charge will pass through the external circuit: If the potential difference is opposite to the original polarization, a charge will flow through the external circuit as the polarization of the ferroelectric changes its direction.

7.18 Frequency-Dependent Linear Response

So far we have tacitly assumed that the external field was static in time. We now remove this restriction and assume that

$$\vec{E} = \vec{E_0}\cos\omega t \tag{7.61}$$

This is a reasonable assumption, because any arbitrary time variation can be expressed as a Fourier series. So we need to consider only harmonic variation. When such a field is first applied, the initial response is very complicated as the polarization builds up. Eventually, the transients die away, and the polarization P varies sinusoidally at the same frequency of the field but may not be in phase with the field. This means that in general we have

$$P = P_1\cos\omega t + P_2\sin\omega t \tag{7.62}$$

P is no longer proportional to E. It is more convenient to work with exponential form, so we rewrite equations (7.61) and (7.62) as

$$E = Re(E_0 e^{-i\omega t}) \quad \text{and} \quad P = Re(P_0 e^{-i\omega t}) \tag{7.63}$$

where Re means that we take the real part of the complex quantity in the parentheses. For simplicity, we drop Re, but it is understood that we always take the real part of the final answer. Moreover, P_0 is equal to

$$P_0 = P_1 + iP_2 \tag{7.64}$$

Note that P_1 is in phase with E and P_2 lags behind by 90 degrees of lagging; we can plot them in a complex plane as in **Figure 7.25**. It is easy to see that

$$|P_0| = \sqrt{P_1^2 + P_2^2}, \quad \text{and} \quad \tan\phi = P_2/P_1 \tag{7.65}$$

Because we express the field and the polarization in complex form, the susceptibility and the dielectric constant must be complex. Moreover, they are functions of frequencies. To see this, we consider a simple model, in which a point charge Ze is surrounded by a spherically symmetric cloud of charge $(-Ze)$ and the density is uniform up to the atomic radius. When an electric field E is applied, the nucleus is displaced by a distance x in the

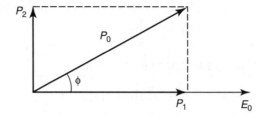

I FIGURE 7.25

direction of the field (because the nucleus is almost 2000 times more massive than the negative charge cloud, the center of the negative charge cloud probably does the moving). A restoring force between the nucleus and the charge cloud tends to restore the initial configuration. The equation of motion of the nucleus is

$$m\ddot{x} = -Kx - eE_0 e^{-i\omega t} \tag{7.66}$$

where $-Kx$ is the restoring force. This equation has a solution of the form

$$x = x_0 e^{-i\omega t}$$

Substituting this in equation (7.66) we obtain

$$(-m\omega^2 + K)x_0 e^{-i\omega t} = -eE_0 e^{-i\omega t}$$

or

$$x_0 = \frac{eE_0}{m\omega^2 - K} \tag{7.67}$$

Now $p_m = -ex = \alpha E$, and from this we find

$$\alpha = -ex/E = \frac{e^2/m}{(K/m) - \omega^2} \tag{7.68}$$

For this simple model, we see the polarizability dependents on frequency. When $\omega^2 = K/m$, the polarizability becomes infinite. The quantity $(K/m)^{1/2}$ is called the resonant frequency ω_r. In real physical systems there is always some frictional force that prevents the system from reaching the resonant frequency. Very often this frictional force can be represented by a term $-b\dot{x}$

on the right side of the equation of motion (7.66), which becomes the equation of motion for a forced damped oscillator:

$$m\ddot{x} = -Kx - b\dot{x} - eE_0 e^{-i\omega t}$$

or

$$\ddot{x} = -\omega_r^2 x - \beta\dot{x} + \frac{qE_0}{m} e^{-i\omega t} \qquad (7.69)$$

where

$$\omega_r = \sqrt{K/m}, \quad q = -e, \quad \text{and} \quad \beta = b/m$$

On assuming a solution of the form $x = x_0 e^{-i\omega t}$, we obtain

$$x_0 = \frac{eE_0/m}{\omega_r^2 - \omega^2 - i\beta\omega} \qquad (7.70)$$

and

$$\alpha(\omega) = \frac{e^2/m}{\omega_r^2 - \omega^2 - i\beta\omega} \qquad (7.71)$$

Again, α is a complex function of the frequency. To separate it into real and imaginary parts, we multiply numerator and denominator by the complex conjugate of the latter:

$$\alpha(\omega) = \frac{(e^2/m)(\omega_r^2 - \omega^2 + i\beta\omega)}{(\omega_r^2 - \omega^2 - i\beta\omega)(\omega_r^2 - \omega^2 + i\beta\omega)} = \frac{(e^2/m)(\omega_r^2 - \omega^2 + i\beta\omega)}{(\omega_r^2 - \omega^2)^2 + \beta^2\omega^2}$$

from which we obtain

$$\alpha_{re}(\omega) = \frac{(e^2/m)(\omega_r^2 - \omega^2)}{(\omega_r^2 - \omega^2)^2 + \beta^2\omega^2} \qquad (7.72)$$

$$\alpha_{im}(\omega) = \frac{(e^2/m)\beta\omega}{(\omega_r^2 - \omega^2)^2 + \beta^2\omega^2} \qquad (7.73)$$

We can write a similar expression for the dielectric constant K. Under normal conditions, the polarization in a gas is small compared with the electric field, so that

$$\vec{E}_{loc} = \vec{E} + \frac{1}{3\varepsilon_0}\vec{P} \approx \vec{E}$$

then from equations (7.42) and (7.55) we find

$$\chi = \frac{N}{\varepsilon_0}(\alpha_0 + \frac{p^2}{3K_B T})$$

and so

$$K = 1 + \chi = 1 + \frac{N}{\varepsilon_0}(\alpha_0 + \frac{p^2}{3K_B T})$$

For simplicity, we assume N to be small so that the local field correction is negligible, and then the above expression for K reduces to

$$K(\omega) = 1 + \frac{N\alpha_0(\omega)}{\varepsilon_0} = 1 + \frac{N}{\varepsilon_0}[\alpha_{re}(\omega) + i\alpha_{im}(\omega)]$$

from which we find

$$K_{re}(\omega) = 1 + \frac{Nq^2}{m\varepsilon_0}\frac{\omega_r^2 - \omega^2}{(\omega_r^2 - \omega^2)^2 + \beta^2\omega^2} \tag{7.74}$$

$$K_{im}(\omega) = \frac{Nq^2}{m\varepsilon_0}\frac{\omega_r^2 - \omega^2}{(\omega_r^2 - \omega^2)^2 + \beta^2\omega^2} \tag{7.75}$$

The form of these functions has the typical resonance shape. The peaks get narrow and higher as β is made smaller, approaching equation (7.66) as $\beta \rightarrow 0$. Thus, K_{im} is very small except in the vicinity of the resonance. It can be shown that K_{im} is associated with absorption of energy from the field.

In a real material there are a variety of resonance and relaxation processes. The electronic resonances are the fastest and occur in the visible, ultraviolet, and x-ray frequency ranges. Far beyond the highest of them, the material behaves essentially like empty space. Processes involving atomic and molecular motions typically occur in the infrared, and those involving large groups of molecules are very much slower.

a. Kramers-Kronig Relations

Even though the expressions for K_{re} and K_{im} resulting from the damped oscillator model look rather complicated, the real-life situation is far worse. There are always so many resonances over such a wide range of frequencies that a single expression cannot come close to an accurate description. Only a direct measurement will suffice. But because of the so-called *Kramers-Kronig relations*, it is not necessary to measure both functions. If either one is known at all frequencies, the other can be obtained at any desired frequency.

One of the Kramers-Kronig relations is

$$K_{re} = 1 + \frac{2}{\pi} P \int_0^\infty \frac{\omega' K_{im}(\omega')}{\omega'^2 - \omega^2} \, d\omega' \qquad (7.76)$$

where P designates the principal part of the integral: The integral is evaluated in two pieces with a gap around the singularity at $\omega' = \omega$, and the limit is taken as the gap shrinks to zero. Usually, the integrals have to be evaluated numerically.

The Kramers-Kronig relations are independent of any specific model and are based on the linearity and causality (a cause precedes its effect) of the dielectric response. These relations are of great practical and fundamental importance.

Problems

1. Suppose that in solid hydrogen in an electric field the average electron positions were displaced 0.01 Å from the proton positions, all in the same direction. Find the polarization. The interatomic spacing is about 3 Å.

2. Consider a cylinder extending from $z = 0$ to $z = l$. It is polarized non-uniformly: $\vec{P} = Cz^2 \hat{k}$ (**Figure 7.26**). Show that the total charge is zero.

3. A sphere of radius a is polarized along the radius vector so that $P = cr$. Find the polarization charge densities and show that the total charge is zero.

4. A thin rod of cross-section A and length l lies on the x-axis with its nearer end at a distance b from the origin. The polarization is cx parallel to the x-axis. Find the charge at each end and the volume density of charge in the interior of the rod and show that the total charge is zero. Also, find the potential at the origin due to the rod.

5. Given a uniform electric field $\vec{E}$,
   (a) A thin disk of dielectric constant K is placed in it. If the disk is placed perpendicular to $\vec{E}$, find the electric field outside and inside the disk.
   (b) A thin cylinder of dielectric constant K is placed in it, with its axis parallel to the field. Find the electric field inside and outside the cylinder.

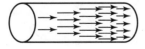

I FIGURE 7.26

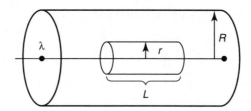

I FIGURE 7.27

6. The following cavities are cut in a dielectric in which the polarization P is uniform and fixed: (a) a narrow slit at right angles to the polarization, (b) a sphere, and (c) a long needle-like cavity with its axis parallel to the polarization. Find the electric intensity at the center of each due to the charge produced on the surface of the cavity.

7. A long cylindrical cavity at right angles to the polarization is cut in the dielectric of the preceding problem. Find the electric field intensity at its center due to the charges on the surface of the cavity.

8. An artificial dielectric consists of a large number of metal spheres, all macroscopic size, arranged in a three-dimensional lattice structure. Find the permittivity of the dielectric.

9. A long straight wire, carrying uniform line charge λ, is surrounded by rubber insulation out to a radius R (**Figure 7.27**). Find the electric displacement.

10. Given a dielectric sphere of radius R with a point charge Q embedded at the center. Find
    (a) The electric displacement vector $\vec{D}$, the electric field $\vec{E}$, and the polarization vector $\vec{P}$ at a point r, where $r < R$;
    (b) The polarization charge density σ_P at the outer surface of the sphere;
    (c) The volume polarization charge density ρ_P.

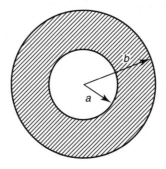

I FIGURE 7.28

11. A metal sphere of radius a carries a charge Q (**Figure 7.28**). It is surrounded, out to radius b, by linear dielectric material of permittivity ε. Find the potential at the center (relative to infinity).

12. Two oppositely charged parallel metallic plates are separated by two sheets of dielectric material of permittivity ε_1 and ε_2, respectively (**Figure 7.29**). The thicknesses of the dielectric sheets are t_1 and t_2. The electric field at any point in dielectric 1 is E_1 and in dielectric 2 is E_2. Find
    (a) The displacement in each dielectric,
    (b) The surface density of free charge,
    (c) The surface polarization charge on the surface of each dielectric,
    (d) The component of electric field due to induced charges,
    (e) The potential difference between the plates, and
    (f) The electric susceptibility of each dielectric.

13. A cable of radius R_1 is insulated with a dielectric of outer radius R_2 that is itself enclosed in a grounded conducting sheath. Find the force per unit volume exerted the dielectric.

14. In linear dielectrics both the polarization vector $\vec{P}$ and the displacement vector $\vec{D}$ are proportional to $\vec{E}$. Show that even though
    $$\nabla \times \vec{E} = 0, \quad \nabla \times \vec{D} \neq 0, \quad \nabla \times \vec{P} \neq 0.$$

15. Determine the work done in moving apart the plates of a plane-parallel capacitor by a distance d. The area of each plate is A and the charge on it is Q. Consider the numerical example: $Q = 10^{-19}$ coulomb, $A = 100$ cm^2, $d = 1$ cm.

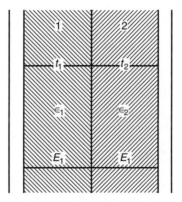

I FIGURE 7.29

16. Consider a parallel plate capacitor. We first charge it up and then insert a dielectric slab while maintaining the potential difference across the plates. What is the total amount of external work done?

17. Assuming a simple classical model for an atom, derive an expression for the induced dipole moment and hence for its polarizability.

18. For molecules at room temperature, pE_{loc}/K_BT is a small quantity even if the electric field is as large as 10^7 coulomb-meter. Show that, for $pE_{loc}/K_BT \ll 1$, equation (7.47) can be approximated by equation (7.48).

19. Using the Clausius-Mossotti equation, estimate the dielectric constant of a typical nonpolar solid dielectric. N is typically $\cong 5 \times 10^{28}$ atoms/meter^3 and take $\alpha \approx 10^{-40}$ C-m^2/V.

20. Given equation (7.65), the equation of motion of a forced damped oscillator, show that a solution of the form $x = x_0 e^{-i\omega t}$ leads to equations (7.66) and (7.67).

21. Show that in a linear dielectric medium, Poisson equation has the form

$$\nabla^2 V = -\rho/K\varepsilon_0$$

8 The Physics of Electrical Conductivity

In this chapter we discuss the interaction between electric fields and electric conducting media. Conducting media are distinguished by the presence of free, or nearly free, charge carriers. The application of a static electric field can produce steady current. This is in contrast with what we learned in Chapter 7 on dielectrics, where the application of a static electric field produced a polarization and no steady current.

We first consider transfer of electric charges in solids, and then we take a brief look at electrical conductivity of gases and liquids. Finally, we consider superconducting solids, a special class of materials of very great theoretical interest and growing practical importance.

8.1 Electrical Conductivity of Solids

It is found experimentally that in many solid conductors at constant temperature the current density $\vec{j}$ in the steady state is linearly proportional to the applied static electric field $\vec{E}$, which can be written as

$$\vec{j} = \sigma \vec{E} \tag{8.1}$$

where σ is called electrical conductivity of the material. Equation (8.1) is known as Ohm's law, and the materials that obey Ohm's law are said to be Ohmic.

For a homogeneous isotropic conductor σ is constant independent of position, but it is temperature dependent. In nonisotropic conductors such as certain crystals or plasma, the current may flow more easily in some directions than in others. In these situations the simple relationship (8.1) is replaced by a tensor equation.

Consider a linear conductor of cross-sectional area A. A steady current I in it can be expressed in terms of conductivity σ of the conductor

$$I = \int_S \vec{j} \cdot \hat{n} dS = \sigma \int_S \vec{E} \cdot \hat{n} dS = \sigma A E$$

If the potential difference across the ends of the conductor is V and the length of the conductor is l, then the uniform electric field within the conductor is given by

$$E = V/l$$

Combining this with the previous equation, we obtain

$$I = \sigma A V/l$$

or

$$V/I = R \qquad\qquad (8.2)$$

where

$$R = l/\sigma A \qquad\qquad (8.3)$$

and it is called the resistance of the conductor. Equation (8.2) is the elementary form of Ohm's law found in general physics textbooks.

The MKS unit of resistance is the ohm (denoted by Ω, 1 ohm = 1 volt per ampere) and its reciprocal is called the mho (1 mho = 1 ampere per volt). Thus, conductivity is measured in mho per meter. The reciprocal of conductivity is called the resistivity, measured in ohm-meters.

Ohm's law is an approximation to the observed behavior of many conducting materials. It is in that sense an empirical and not a fundamental law like those we met in electrostatics. Nonetheless, it is a very good approximation not only for a wide class of materials but also for a range of field strengths. There are many instances when Ohm's law fails. A common example of non-ohmic behavior occurs at semiconductor junctions. Another non-ohmic behavior is the flow of electrons in the space between the electrodes of a vacuum tube.

We now consider in some detail the concept of conductivity. Specifically, we would like to establish the linear behavior expressed by equation (8.1). Although the details of the argument are oversimplified, the classical crude model to be offered here gives an essentially correct picture. Actually, the problem requires quantum mechanical treatment.

To explain the transfer of charge in a conductor, we must account for the origin of the charge carriers and their motions under the action of forces.

In the electronic configuration of an isolated atom, certain electrons form closed shells, whereas others belong to incomplete outer shells and are collectively referred to as valence electrons. For example, a neutral atom of copper has 29 electrons, 28 of which are distributed in the K, L, and M shells whereas the 29th is the valence electron. When free atoms are brought together to form the crystalline state, the inner shell electrons remain more or less tightly bound to their cores, but the valence electrons behave as though they were nearly free, in the sense that their potential or binding energy is small relative to their kinetic energy.

Let us picture the interior of a solid conductor as a three-dimensional lattice of atoms with free electrons moving about between the atoms and colliding frequently with them. As the effective radius of an atom is of the order of 10^{-8} cm, whereas that of an electron is of the order of 10^{-13} cm, we may neglect the collisions between electrons in comparison with the collisions between atoms and electrons. The atoms vibrate about their equilibrium positions and the electrons move, with a mean energy determined by the temperature. Atoms have the usual thermal agitation of any solid, whereas the electrons collectively behave like a gas at the temperature of the conductor, or approximately so, at least.

In the absence of an externally applied field, the average velocity of an assemblage of electrons is zero, because of the random thermal motion. When an external field is applied, the electron velocities are modified. Superimposed upon their thermal motions, the electrons acquire a component of velocity referred to as the drift velocity. It is the drift component of velocity that constitutes the electric current. The drift velocity is small compared with the thermal velocities, so that the two motions may be treated as independent. The *mean free time* τ between successive collisions of a free charge with some atom is thus determined by the structure of the conductor and the temperature, but not by $\vec{E}$.

Under the action of an applied electric field E, each electron acquires an acceleration in a direction opposite to that of the field. If there were no other forces, the electron would be accelerated indefinitely under the action of the field. Because of the collisions that occur, we must introduce a resistive or a frictional type of damping force proportional to the drift velocity. The equation of motion is then written as

$$m \frac{dv}{dt} + \frac{mv}{\tau} = eE \tag{8.4}$$

where v is the drift velocity superimposed upon the thermal velocities. The coefficient of v in the resistive term is the damping constant (force per unit velocity). It contains a characteristic time τ, the meaning of which will be clarified as we discuss the solution of (8.4). The differential equation in (8.4) can be solved by separating variables. Upon rearrangement we obtain

$$\frac{dv}{(v-c)} = -\frac{dt}{\tau} \tag{8.5}$$

where $c = eE\tau/m$. At the time $t = 0$, the drift velocity $v = 0$. With this initial condition the integration yields

$$v = c(1 - e^{-t/\tau}) \tag{8.6}$$

For $t \gg \tau$, v approaches a terminal value v_f given by the constant c:

$$v_f = c = (e\tau/m)\,E \tag{8.7}$$

Equation (8.6) shows that the parameter τ, called the relaxation time, is a characteristic time during which the drift velocity v reaches the value of $(1 - 1/e)v_f$, or in time τ, v attains the value $v = 0.62v_f$. A long relaxation time means a weak frictional force. From equation (8.5) it appears that the drift velocity of the conduction electron increases with a constant acceleration eE/m. If τ is very short, the frictional force is large, but v attains 62% of v_f in a correspondingly short time. For the purposes of this calculation, τ may be interpreted as the time during which the electron does not suffer a collision and the applied field is free to accelerate it. So τ is often called the mean free time.

Let us return to the current density j, which can be written as

$$j = nev_f$$

When the value of v_f in (8.7) is used as the drift velocity, the expression for j becomes

$$j = (ne^2\tau/m)\,E$$

or

$$j = \sigma E \tag{8.8}$$

where

$$\sigma = ne^2\tau/m \tag{8.9}$$

The result in (8.8) indicates that the current density is proportional to the field, as Ohm's law stated. The proportionality constant a in (8.9) is the conductivity of the conductor.

The conductivity σ is a readily measurable quantity, and its experimental value makes it possible to evaluate τ for a particular conductor. Because most conducting wire is made of copper, let us use copper as an example. From Avogadro's number, the gram atomic weight, and the density we can find the concentration of Cu atom. This is also the concentration of electrons, because there is one valence electron per Cu atom. Therefore

$$n = \frac{6.03 \times 10^{23} \text{atoms/mole}}{63.5\text{g/mole}} \times 8.90 \times 10^6 \text{g/m}^3$$

$$= 8.46 \times 10^{28} \text{atoms/m}^3$$

The electron constants are

$$m = 9.10 \times 10^{-31} \text{ kg}, \quad e = -1.6 \times 10^{-19} \text{ coulomb}$$

and the conductivity of Cu at room temperature is

$$\sigma = 5.88 \times 10^7 \ (\Omega\text{-m})^{-1}$$

From these data we find

$$\tau = \frac{m\sigma}{ne^2} = \frac{9.1 \times 10^{-31} \times 5.88 \times 10^7}{8.46 \times 10^{28} \times (1.6 \times 10^{-19})^2} = 2.48 \times 10^{-14} \text{sec}$$

This means that in a time interval of about 3×10^{-14} sec, the steady-state velocity would be attained in the presence of a static electric field in Cu. This observation gives a support to our selection of v_f as the appropriate drift velocity in the discussion of the transport of charge in the steady state.

The drift velocity in conductors is quite small. Let the steady current in a copper conductor 1 cm in diameter be 100 amp distributed uniformly over the cross-section. The current density j is 1.28×10^6 amp/m^2. Using the values of n and e, the drift velocity is

$$v_f = j/ne = 9.55 \times 10^{-5} \text{ m/sec} \approx 0.1 \text{ cm/sec}$$

which is indeed quite small.

8.2 Energy Loss and Joule Heating

In the collision processes, conduction electrons transfer kinetic energy acquired between collisions to the material as a whole; this energy appears as the joule heat. The average rate at which work is done on an electron by the applied electric field $\vec{E}$ is $e\vec{E} \cdot \vec{v}$. Averaged over all the electrons, the mean value of $\vec{v}$ is related to $\vec{j}$ by the equation $\vec{j} = ne\vec{v}$. Hence, the mean rate at which the field does work on the n electrons per unit volume of the conductor is

$$N e\vec{E} \cdot \vec{v} = \vec{E} \cdot \vec{j} = \sigma E^2 = \frac{j^2}{\sigma}$$

and the total rate of energy loss is

$$\int \vec{j} \cdot \vec{E} d\vec{r} = \int \sigma E^2 d\vec{r} \tag{8.10}$$

This energy loss is sometimes referred to as the Joule heat loss. So steady current flow is possible only if sources of electric field are provided.

We have derived the above result using an oversimplified classical model, but the expression $\vec{E} \cdot \vec{j}$ for the Joule heating per unit volume in a conductor is valid generally.

8.3 Equations of the Static Field and Flow

For electrostatic fields, whether or not steady current flow, $\nabla \times \vec{E} = 0$ and $\nabla \cdot \vec{E} = \rho/\varepsilon_0$ are valid. The electric field is still conservative, and Gauss' law is still valid. For ohmic media, we have Ohm's law ($\vec{j} = \sigma \vec{E}$). From conservation of charge, we also have the equation of continuity $\nabla \cdot \vec{j} + \partial \rho/\partial t = 0$.

For ohmic media, the two equations for $\vec{E}$ may be transcribed into equations for $\vec{j}$ with the help of Ohm's law ($\vec{E} = \vec{j}/\sigma$):

$$\nabla \times \vec{E} = 0 \rightarrow \nabla \times (\vec{j}/\sigma) = 0 \tag{8.11}$$

$$\nabla \cdot \vec{E} = \rho/\varepsilon_0 \rightarrow \nabla \cdot (\vec{j}/\sigma) = \rho/\varepsilon_0 \tag{8.12}$$

The integral equations corresponding to equations (8.11) and (8.12) are

$$\oint \vec{E} \cdot d\vec{r} = 0 \rightarrow \oint (\vec{j}/\sigma) \cdot d\vec{r} = 0 \tag{8.11a}$$

and

$$\oint \vec{E} \cdot d\vec{a} = Q/\varepsilon_0 \rightarrow \oint_S (\vec{j}/\sigma) \cdot d\vec{a} = Q/\varepsilon \tag{8.12a}$$

where Q is the net charge inside a volume τ whose (closed) surface is S.

If σ is constant, then equation (8.11) becomes

$$\sigma^{-1}\nabla \cdot \vec{j} = \rho/\varepsilon_0$$

On the other hand, for steady condition, $\partial\rho/\partial t = 0$ and the continuity equation reduces to

$$\nabla \cdot \vec{j} = 0$$

Combining these two results, we see that $\rho = 0$ inside the medium for steady flow.

If we consider only media with constant conductivities, i.e., homogeneous materials, equations (8.11), (8.11a), (8.12), and (8.12a) become

$$\left.\begin{array}{c} \nabla \times \vec{j} = 0 \\ \oint \vec{j} \cdot d\vec{r} = 0 \end{array}\right\} \quad \text{and} \quad \left.\begin{array}{c} \nabla \cdot \vec{j} = 0 \\ \oint \vec{j} \cdot d\vec{a} = 0 \end{array}\right\} \tag{8.13}$$

We also have a differential equation for the potential difference along a conductor, the Laplace equation:

$$\nabla \cdot \vec{j} = \nabla \cdot (\sigma\vec{E}) = \sigma\nabla \cdot (-\nabla V) = 0$$

or

$$\nabla^2 V = 0 \tag{8.14}$$

The boundary conditions that V must satisfy are different from the boundary conditions that apply to this equation in electrostatics. We may suppose that the potentials of the terminals in contact with the conductor that drive the current flow are specified. There can be no current flow normal to the surface of the conductor. Because $\vec{j} = \sigma\vec{E}$, it follows that the normal component of the electric field, $\vec{E} \cdot \hat{n} = -\partial V/\partial n$, is zero at the surface of the conductor. These conditions suffice to make the solution to the potential problem unique; if the potential problem is solved, the current flow may be calculated using $\vec{j} = \sigma\vec{E} = -\sigma\nabla V$.

8.4 Insufficiency of the Classical Theory of Electrical Conductivity

Under all ordinary conditions the conductivity is a characteristic of the conducting substance independent of j and E. It does, however, vary with the temperature. Experimentally, it is found that σ is more nearly proportional to the inverse first power of the temperature. It is impossible to explain this relationship on the basis of equation (8.9), because, according to the kinetic theory of gases, τ is inversely proportional to the square root of the temperature. This contradiction cannot be overcome merely by improvements of the classical model.

Quantum theory gives a completely different treatment of the electrical conductivity from the point of view of the band theory of solids. This subject is beyond our scope, but we can give a rough idea of the silent features so as to glean some understanding of conductive processes.

According to quantum ideas, conduction electron should be treated as waves, with wavelength λ related to momentum p, as suggested by de Broglie:

$\lambda = h/p$

where h is Planck constant, 6.7×10^{-34} J·s. Thus, quantum theory gives electrons (and other atomic and subatomic particles) a combined wave-particle character and describes their motion in terms of energy levels. In isolated atoms the electrons can have only certain states of motion (roughly speaking, orbits) each of which has definite discrete energy. The electrons seek the lowest possible energy levels, but the Pauli exclusion principle (proposed by W. Pauli in 1925) prevents more than one from occupying any single state. Thus, the levels are filled from the lowest energy up, with vacant levels generally present above the occupied ones. As many atoms are brought close together in a crystal, two important effects take place: (1) the states no longer are confined to single atoms but are spread over the entire crystal, and (2) the energies are perturbed to various degrees by the neighboring atoms so that they are no longer discrete but spread out into an energy band. **Figure 8.1** is a typical energy-level diagram for a metal. A level in such a diagram is an explicit or implicit horizontal line whose height above a reference level indicates a particular quantized energy value that an electron can take. Figure 8.1 shows two bands that each contain many possible levels for conduction-electron motion. The spacing of levels is so close that it can be considered continuous; for instance, the lower band in Figure 8.1 might consist of 10^{24} distinct levels!

The lowest band is only partially filled. Except for the fluctuations produced by thermal agitation, there is a definite level called the "Fermi level" representing the top of the filled group. Just above the Fermi level there are levels into which electrons can be lifted either by thermal agitation or by a field E, no matter how small in magnitude. But excitation by thermal agitation makes no change in the completely random directionality of electron motions, although of course it leaves some levels vacant below the Fermi

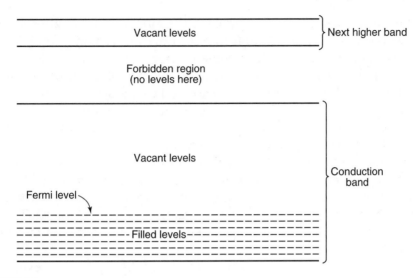

I FIGURE 8.1

level and fills some levels just above. The function that describes the relative occupation of a level at any temperature T above absolute zero is the so-called Fermi distribution function, $f(U) = \left[1 + e^{(U - U_F)/kT}\right]^{-1}$ where U is the energy of the level in question, U_F is the Fermi level energy, and k is Boltzmann's constant.

The effect of an electric field E, however, is to produce a net electron drift v proportional to $-E$. More specifically, a field E sets up a continuous process of electrons jumping into empty levels, corresponding to flow in the $-E$ direction and away from levels of the opposite kind. At the same time, collisions with crystal imperfections, impurity atoms, and thermally agitated atomic nuclei produce backward jumps that counteract the effect of E. Thus, a steady state of flow is reached in a very short time for each value of E. The proportionality of v and $-E$ is the basis of Ohm's law.

We refer the reader who is interested in pursuing this subject further to the book by Arthur F. Kip, who gave an excellent and relatively elementary presentation of the quantum mechanical model of the electrical conductivity (*Fundamentals of Electricity and Magnetism*, second edition, McGraw-Hill Book Company).

8.5 Magnetic Field Effects

In this section we discuss two magnetic field effects, magnetoresistance and Hall effects.

When a magnetic field $\vec{B}$ is applied, the moving carriers are subject to Lorentz force, $e\vec{v} \times \vec{B}$, in addition to electrical and frictional force (due to collisions). Lorentz force deflects some moving carriers in one direction or the other and thus makes less headway along the direction of the electric field between collisions. Thus, a decrease in the conductivity of the conductor or an increase in the resistivity of the conductor results.

The magnetoresistance can be reduced by using the Hall field. In 1879, E. H. Hall discovered that when a conductor is placed in a magnetic field perpendicular to the direction of the current flow, a voltage is developed in a direction perpendicular to both the current and the magnetic field.

To illustrate the Hall effect, we consider a thin conducting rectangular bar with a current I flowing from left to right, as shown in **Figure 8.2**. The current is carried by electrons of a charge $-e$ from right to left with mean drift velocity $\vec{v}$.

Under the influence of the magnetic field, the electrons experience Lorentz force $\vec{F} = -e\vec{v} \times \vec{B}$ that tends to deflect them toward the side P of the bar. Electrons are forced toward side P until a charge distribution is set up that generates an electric field $\vec{E}_{H}$, the Hall field, just big enough to prevent more electrons from accumulating on the side P. The electric force $-e\vec{E}_{H}$ on the electrons is then equal and opposite to that due the magnetic field,

$$e\vec{E}_{H} = e\vec{v} \times \vec{B}$$

The Hall field tends to minimize the deflection, and the magnetoresistance can be reduced by maximizing this field.

Returning to the discussion of Hall effect, the electric field in the bar has constant magnitude $v_{x}B$ and is in the direction from side Q to P. If the width

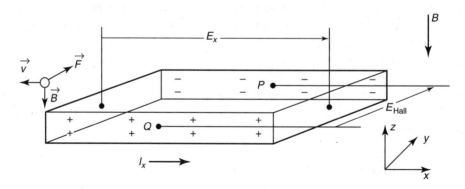

I FIGURE 8.2

of the bar is w, i.e., the distance between sides P and Q, the Hall voltage V_H will be

$$V_H = wE_H = Bv_x w$$

The current I_x is given by

$$I_x = nev_x wl$$

where n is the number of conduction electrons per unit volume and $A = wl$ is the cross-sectional area of the bar, l being the bar's thickness.

The Hall voltage V_H is thus

$$V_H = \frac{I_x}{nel} B \tag{8.15}$$

from which we have

$$\frac{V_H l}{BI_x} = \frac{1}{ne} = R_H$$

where

$$R_H = 1/ne \tag{8.16}$$

is defined as the Hall coefficient, which is determined essentially by the sign and density of the charge carriers. Moreover, $\sigma = Ne\mu$, so there follows the simple relationship

$$\mu = R_H \sigma \tag{8.17}$$

The Hall effect provides an indirect method for determining magnetic field. In a given experiment, if the Hall coefficient R_H and the conductivity σ are measured, the mobility μ and the carrier concentration e can be found. Then a measurement of the Hall voltage across the bar gives the magnitude of the magnetic field $\vec{B}$.

8.6 Frequency-Dependent Conductivity

When the electric field is not static but a sinusoidal function of the time, the charge carriers are not subject to a constant acceleration between collisions as we assumed in the above discussion. Let us now solve equation (8.4), the equation of motion for the charge carriers, with both $\vec{E}$ and $\vec{v}$ being sinusoidal functions of time:

$$\vec{E}(t) = E_0 e^{-i\omega t}, \quad \vec{v}(t) = \vec{v}_0 e^{-i(\omega t - \delta)}$$

We work in the complex exponential notation, $\vec{v}$ has the same dependence, but not necessary in phase. Then equation (8.4) gives

$$-im\omega\vec{v}_0 e^{i\delta} + (m/\tau)\,\vec{v}_0 e^{i\delta} = e\vec{E}_0$$

or

$$\vec{v}_0 = \frac{e\tau/m}{1 - i\omega\tau}\vec{E}_0 e^{-i\delta}$$

and

$$\vec{j} = ne\vec{v} = ne\left(\frac{e\tau/m}{1 - i\omega\tau}\vec{E}_0 e^{-i\delta}\right)e^{-i(\omega t - \delta)}$$

$$= ne\frac{e\tau/m}{1 - i\omega\tau}\vec{E} = \Sigma\vec{E}$$

where the conductivity Σ is

$$\Sigma = \frac{\vec{j}}{\vec{E}} = \frac{ne^2\tau}{m}\frac{1}{1 - i\omega\tau} = \frac{ne^2\tau/m}{1 + \omega^2\tau^2}(1 + i\omega\tau) = \sigma' + i\sigma'' \tag{8.18}$$

where

$$\sigma' = \frac{ne^2\tau/m}{1 + \omega^2\tau^2} = \frac{1}{1 + \omega^2\tau^2}\sigma, \quad \sigma'' = \frac{\omega\tau}{1 + \omega^2\tau^2}\sigma \tag{8.19}$$

and $\sigma = ne^2\tau/m$, as given by equation (8.9). σ' and σ'' are plotted in **Figure 8.3**.

Let us now write $\vec{j}$ in terms of σ' and σ'':

$$\vec{j}(t) = \Sigma\vec{E} = (\sigma' + i\sigma'')\vec{E}_0 e^{-i\omega t}$$

from this we may write the real current

$$j(t) = \sigma'E_0\cos\omega t + \sigma''E_0\sin\omega t$$

We see that the two components of the current corresponding to σ' and σ'' are $\pi/2$ out of phase.

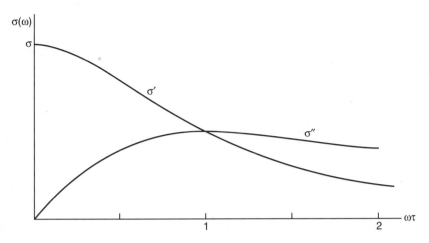

I FIGURE 8.3

It is also interesting to examine the rate at which work is done by the electric field on the medium. We leave this as homework.

8.7 Skin Effect

In a conducting medium, if $\vec{j}$ varies in time, the associated magnetic field varies also. As a consequence, an induced electric field $\vec{E}_{in}$ is set up, which acts on the charge carriers so as to alter the current density. How do we find the current density? We may use the electromagnetic field equations that are developed in Chapter 11. However, a satisfactory approximate solution of the problem can be found by considering a quasistatic case, in which the variation of the current density is very small, and so we can ignore the displacement current. Then in a nonmagnetic conductor we have

$$\nabla \times \vec{B} = \mu_0 \vec{j}$$

If the magnetic properties of the conductor are isotropic and homogeneous, we have

$$\nabla \times \vec{B} = \mu \vec{j} \tag{8.20}$$

where μ is the magnetic permeability of the medium. If the conductor is stationary and it is ohmic, then we have

$$\nabla \times \vec{E} = -\partial \vec{B}/\partial t \qquad (8.21)$$

and

$$\vec{j} = \sigma \vec{E} \qquad (8.22)$$

Taking the curl of Faraday's law, equation (8.22), we obtain

$$\nabla \times (\nabla \times \vec{E}) = -\partial (\nabla \times B)/\partial t$$

It becomes, after substituting equations (8.22) and (8.20),

$$\nabla \times (\nabla \times \vec{j}/\sigma) = -\partial (\mu\vec{j})/\partial t$$

Expanding the curl and assuming that $\vec{j}$ is solenoidal ($\nabla \cdot \vec{j} = 0$), we get

$$\nabla^2 \vec{j} = \sigma\mu \frac{\partial \vec{j}}{\partial t} \qquad (8.23)$$

This is the equation governing current distribution in conductor, and this current is called the eddy current.

We can solve equation (8.23) by the method of separation of variables. But we consider an example of an ideal problem. Assume that a current varying as $e^{i\omega t}$ is present in a conductor σ that fills the region above $z = 0$ (**Figure 8.4**) and that near the origin (near the surface of the conductor) $\vec{j}$ is unidirectional along the y-axis. At the surface the current density is $j_0\hat{y}$. The region $z < 0$ is a vacuum and $\vec{j}\,(z < 0)$ is zero. We take

$$\vec{j}(z,t) = j_0 Z(z) e^{i\omega t} \hat{y} \qquad (8.24)$$

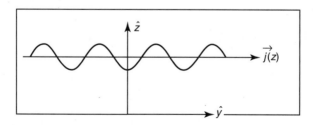

I FIGURE 8.4

Substituting this into equation (8.23) we obtain

$$\frac{d^2Z}{dz^2} = i\omega\sigma\mu Z(z) = \alpha^2 Z(z) \tag{8.25}$$

where

$$\alpha^2 = i\omega\sigma\mu \quad \text{or} \quad \alpha = (1+i)\left(\frac{\sigma\omega\mu}{2}\right)^{1/2}$$

In the last step we have made use of the identity $\sqrt{i} = (1+i)/\sqrt{2}$.
Equation (8.25) has the solution

$$Z(z) = Ae^{\alpha z} + Be^{-\alpha z}$$

To maintain finite $\vec{j}$ we assume that as $z \to \infty$, $j \to 0$. Then the coefficient A must be zero, and we finally find

$$\vec{j}(z,t) = j_0 e^{-\alpha z} e^{i\omega t} \hat{y} \tag{8.26}$$

Let us rewrite α as

$$\alpha = (1+i)\beta, \quad \text{with } \beta = \sqrt{\sigma\omega\mu/2}$$

then equation (8.26) becomes

$$\vec{j}(z,t) = j_0 e^{-\beta z} e^{i(\omega t + \beta z)} \hat{y}$$

We see that $\vec{j}$ decreases rapidly as z increases. The depth at which the current is decreased by a factor e^{-1} is $d = 1/\beta$. As the frequency $\omega/2\pi$ increases, the penetration depth d decreases as $1/\omega^{1/4}$; thus, at high frequencies the current flows mainly on the surface, reducing the effective conductivity of the medium.

8.8 Electrical Conductivity of Gases and Liquids

Under normal conditions, all gases are poor conductors of electricity. A gas becomes a conductor only in the ionized state, that is, when electrons and ions appear. Positive ions are atoms, molecules, or group of molecules that have lost one or more electrons, and negative ions are atoms, molecules, or group of molecules that have gained electrons. Let us consider an ionized gas of positive ions of mass M and charge e and electrons of mass m and charge $-e$.

The number of electrons and ions per unit volume are taken to be equal to n.

Under the action of a (static) electric field, the electron experiences a force $-eE$ and the ion a force $+eE$, and they accelerate according to the equations, respectively,

$$M\frac{d\vec{V}}{dt} = e\vec{E}, \quad m\frac{d\vec{v}}{dt} = -e\vec{E} \tag{8.27}$$

from which we get, by integration

$$\vec{V} = \vec{V}_0 + \frac{e}{M}\vec{E}t, \quad \vec{v} = \vec{v}_0 - \frac{e}{m}\vec{E}t \tag{8.28}$$

The charge flow per particle due to electron and ion are given by, respectively,

$$\vec{j}_- = -e\vec{v} = = e\vec{v}_0 + \frac{e^2}{m}\vec{E}t, \quad \vec{j}_+ = e\vec{V} = e\vec{V}_0 + \frac{e^2}{M}\vec{E}t \tag{8.29}$$

Note that the component of the current induced by the external electric static field has the same sign for the two charge carriers, the negative electron and the positive ion. The total charge flow due to electron and ion is then given by

$$\vec{j}_c = \vec{j}_- + \vec{j}_+ = e^2\left(\frac{1}{m} + \frac{1}{M}\right)\vec{E}t \tag{8.30}$$

So far we have ignored the collisions between the two charge carriers. The time t in equation (8.30) should be replaced by the mean free time τ between collisions. The mean free time τ is equal to the mean free path between collisions divided by the mean drift velocity of the charge carrier. The drift velocity can read from equation (8.30) with t replaced by τ:

$$\langle\vec{v}\rangle = -\frac{\vec{j}_-}{e} = \frac{e\tau}{m}\vec{E}, \quad \langle\vec{V}\rangle = \frac{\vec{j}_+}{e} = \frac{e\tau}{M}\vec{E}$$

We see that the drift velocity is proportional to the electric field. This relation is often expressed in terms of "mobility" μ:

$$\langle v\rangle = \mu_-\left|\vec{E}\right|, \quad \langle V\rangle = \mu_+\left|\vec{E}\right|$$

Comparing this with the last equation, we obtain for the mobilities

$$\mu_- = e\tau/m, \quad \mu_+ = e\tau/M \tag{8.31}$$

It is clear that the mobility of the electron is higher and the mobility should be inversely proportional to the density of the gas, as the mean time τ is inversely proportional to the density. It should be noted that experiments give mobility values smaller than the theoretical values. Thus, equation (8.31) does not explain the experimental facts completely. For example, the ions may acquire dipole moments as they approach each other during a collision, and this alters the nature of collisions. It is beyond this book to discuss this kind of advanced topic.

Now let us go back to electric current. The current density is given by

$$\vec{j} = n\vec{j}_c = ne^2\left(\frac{1}{m} + \frac{1}{M}\right)\vec{E}\tau \tag{8.32}$$

which becomes, in terms of mobility,

$$\vec{j} = ne(\mu_- + \mu_+)\vec{E} \tag{8.32a}$$

The current density can also be written in terms of electrical conductivity σ:

$$\vec{j} = \sigma\vec{E} \tag{8.33}$$

where

$$\sigma = ne^2\left(\frac{1}{m} + \frac{1}{M}\right)\tau \tag{8.34}$$

It is clear that because $M \gg m$, most of the current is carried by the electrons.

As for liquids, most pure liquids contain very few charge carriers and so are poor conductors. Solutions of salts, acids, and impurities in water and some other liquids are good conductors of electricity, because the molecules of the solute dissociate into positive and negative ions. It is these ions that are responsible for the conductivity of the solution. The current density and the mobilities of the positive and negative ions are defined in exactly the same way as for the ions in gases [i.e., given by equations (8.31) and (8.32) or (8.32a)], but the mobility of ions in liquids is much lower than that in gases.

We now show that at low concentration, the conductivity is proportional to the concentration of ions. The latter depends on the degree of dissociation. The dissociation is described by a dissociation coefficient α, which is defined as the ratio of the concentration of the dissociated molecules n to the concentration of the solvent molecules n_0

$$\alpha = n/n_0 \quad \text{or} \quad n = \alpha n_0 \tag{8.35}$$

The concentration of nondissociated molecules is then equal to

$$n' = (1 - \alpha)n_0 \tag{8.36}$$

In a solution the dissociation of molecules and the recombination of ions to form neutral molecules is a continuous process in a solution. In equilibrium the two processes balance each other. Now, the number of dissociating molecules is proportional to the number of undissociated molecules, i.e.,

$$\Delta n = \beta(1 - \alpha)n_0$$

where β is a coefficient of proportionality. The number of recombining molecules $\Delta n'$ is proportional to the product of the concentrations of the positive and negative ions, i.e.,

$$\Delta n' = \gamma \alpha^2 n_0^2$$

where γ is coefficient of proportionality. In equilibrium, $\Delta n = \Delta n'$

$$\beta(1 - \alpha)n_0 = \gamma \alpha^2 n_0^2$$

From this we find

$$\frac{1 - \alpha}{\alpha^2} = \frac{\gamma}{\beta} n_0 \tag{8.37}$$

which shows that the dissociation coefficient α depends on the concentration of the solute. At very weak concentrations, when $n_0 \approx 0$, we see that, from equation (8.37)

$$\alpha \approx 1 \tag{8.38}$$

On the other hand, when there is little dissociation, $\alpha \ll 1$, then equation (8.15) gives

$$\alpha = \sqrt{\frac{\beta}{\gamma}} \frac{1}{\sqrt{n_0}} \tag{8.39}$$

which indicates that the dissociation coefficient decreases as the concentration of the solute increases.

Next, we write the current density in terms of α:

$$\vec{j} = e\,(\mu_- + \mu_+)\,n\vec{E} = e\,(\mu_- + \mu_+)\,\alpha n_0 \vec{E} \qquad (8.40)$$

from which we obtain the conductivity σ

$$\sigma = e\,(\mu_- + \mu_+)\,\alpha n_0 \qquad (8.41)$$

At low concentration, α is constant [given by equation (8.38)] and the sum of the mobilities of the ions is also approximately constant. Thus, at low concentration, the conductivity is proportional to the concentration n_0. At high concentrations, α depends on β as given by equation (8.37), and the mobilities of the ions begins to depend markedly on n_0. No direct proportionality between the conductivity and the concentration is observed.

8.9 Superconductivity

Superconductivity was first discovered as an anomaly in conductivity. In 1911 the Danish physicist Kamerlingh Onnes observed that the electrical resistance of mercury went abruptly to zero when the temperature was reduced below 4.15 K: A steady current flowed in a ring without a sustaining electric field. This discovery followed from his success in 1908 of liquefying helium, thus making temperatures down to about 1 K accessible to experiment. This phenomenon of superconductivity was subsequently found in many other metals and alloys; until recently the critical temperatures, T_c, at which the transition to superconducting state occurs had not exceeded 24 K. All superconducting technology depended on the availability of expensive liquid helium. But in 1986 new classes of "high-T_c" superconductors were discovered having critical temperatures that exceeded the boiling point at atmospheric pressure of liquid nitrogen, 77.4 K. These new superconductors are complex compounds, such as $Yba_2Cu_3O_7$; their properties and possible technological applications are being intensively studied. Today, we can study these superconductors with much cheaper liquid nitrogen.

The critical temperature T_c depends to some extent on the chemical purity and the crystalline perfection of the sample being studied. If a large magnetic field, H, is applied parallel to a superconducting wire, it is found that the sample becomes normal. This critical field depends on both the material and the temperature given by the following expression and as shown in **Figure 8.5**.

$$H_c = H_0\left[1 - \left(T/T_c\right)^2\right] \qquad (8.42)$$

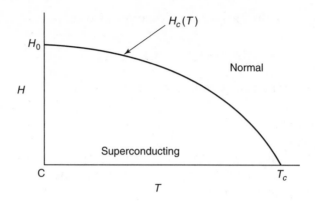

I FIGURE 8.5

8.10 The Meissner Effect

The application of Maxwell equations to a superconductor led to the conclusion that the time rate of change of the magnetic field in the interior of superconductor should be zero. Thus, as the sample was cooled below to T_c in the presence of a magnetic field, magnetic flux was trapped. But in 1933 W. Meissner and R. Ochsefield disproved this experimentally. They found that on cooling a sphere through the transition temperature in a weak field, the magnetic field was repelled out of the sample, as shown in **Figure 8.6**. This effect is called flux exclusion or, more commonly, the Meissner effect. Above some critical value H_c of the applied field, the Meissner effect, and with it the superconductivity, is destroyed as the magnetic field again penetrates the

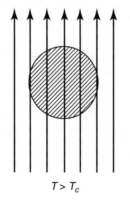

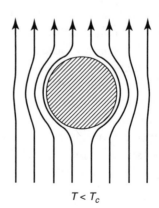

I FIGURE 8.6

material. Such materials are known as type I superconductors, and they are mainly pure elements such as mercury, lead, and aluminum. In some superconductors there is only a partial penetration, so that superconductivity persists. Such materials are called type II superconductors, and they are mainly alloys.

The Meissner effect demonstrates that a superconductor is characterized by more complex electromagnetic properties than simple infinite conductivity. A superconductor behaves as though it has zero permeability or a perfect diamagnetic susceptibility.

8.11 Electromagnetic Properties of Superconductors

The two unique properties of infinite conductivity and the Meissner effect of a superconductor are independent, in the sense that neither implies the other. The Meissner effect is always accompanied by a vanishing electrical resistivity of the metal, superconductivity. The reverse is not true, however; certain superconducting alloys do not exhibit the Meissner effect. Any satisfactory theory of superconductivity must explain these two unique properties: the infinite superconductivity and the flux exclusion (the Meissner effect).

Regarding electromagnetic properties of superconductors, there are two approaches; they are equivalent if interpreted properly. The first approach is to say that inside a superconductor $\vec{B} = \mu_0(\vec{H} + \vec{M}) = 0$ and at boundaries between superconductors and other media, the tangential component of $\vec{H}$ and the normal component of $\vec{B}$ are continuous. This approach views the superconductor as a magnetic material with susceptibility $\chi_m = -1$, i.e., a medium that exhibits perfect diamagnetism. At the surface of the superconductor, magnetization current flow with surface density $\vec{j}_M = \hat{n} \times (\vec{M}_{out} - \vec{M}_{in})$ with $\hat{n}$ the outward drew normal to the surface. $\vec{M}_{out}$ is usually zero; inside the superconductor, volume magnetization currents flow with density $\vec{j}_M = \nabla \times \vec{M}$.

The second approach puts $\vec{B} = \vec{H} = \vec{M} = 0$ inside the superconductor and invokes a real surface current $\vec{j}_S = \hat{n} \times \vec{H}_{out}$ (because $H_{in} = 0$). In this approach there are no currents of any kind flowing in the interior of the superconductor (flux exclusion by surface transport current).

These two descriptions of a superconductor are so strikingly different. How are they related? They are equivalent when properly interpreted. Let us illustrate this with a simple example.

Example 1

Consider a superconducting sphere of radius a placed in a uniform external field $B_0\hat{k}$. Find the magnetic field $\vec{B}$ both inside and outside the sphere.

Solution

The first approach treats the superconductor as a magnetic material, and the boundary value problem takes the form

Outside: $\vec{B} \to B_0 \hat{k}$ as $r \to \infty$

$$\nabla \cdot \vec{B} = 0; \quad \nabla \times \vec{H} = 0; \quad \vec{B} = \mu_0 \vec{H}$$

Inside: $\vec{B} = \mu_0(\vec{H} + \vec{M}) = 0, \quad or \quad \vec{B} = 0, \quad \vec{H} = -\vec{M}$

$\nabla \times \vec{H} = 0; \quad \nabla \cdot \vec{M} = 0$ (there are no magnetic poles inside the superconducting sphere)
At $r = a B_r$ continuous; H_θ continuous.

Two magnetic scalar potentials, $\phi_1{}^*$ outside and $\phi_2{}^*$ outside, may be introduced. Utilizing spherical coordinates and taking account of the boundary condition, $\vec{B} \to B_0 \hat{k}$ as $r \to \infty$, we have

$$\phi_1{}^* = \frac{B_0}{\mu_9} r \cos\theta + \sum_{l=0}^{\infty} C_l r^{-(l+1)} P_l(\cos\theta)$$

then

$$\vec{B} = -\mu_0 \nabla \phi_1{}^*$$

Note in spherical coordinates

$$\nabla f = \frac{\partial f}{\partial r} \hat{r} + \frac{\partial f}{r \partial \theta} \hat{\theta} + \frac{1}{r \sin\theta} \frac{\partial f}{\partial \phi} \hat{\phi}$$

And we have

$$B_r = B_0 \cos\theta + \mu_0 \sum_{l-0}^{\infty} (l+1) r^{-(l+1)} P_l(\cos\theta)$$

Similarly, we can find $B_\theta = \dfrac{\partial \phi_1{}^*}{r \partial \theta}$. Evidently, $B_\varphi = 0$.

Because $\vec{B}$ is zero inside and B_r is continuous across $r = a$, each C_l except C_1 must be zero and $C_1 = -B_0 a^3 / 2\mu_0$

$$\phi_1{}^* = -\frac{B_0}{\mu_0} \cos\theta + C_1 r^{-2} P_1(\cos\theta) \quad \text{(outside)}$$
$$B_r = B_0 \cos\theta + 2\mu_0 C_1 r^{-2} P_1(\cos\theta)$$

Inside the sphere, $\phi_2{}^*$ must be regular at $r = 0$ and match the boundary conditions at $r = a$. These limit $\phi_2{}^*$ to be of the form

$$\phi_2{}^* = d_2 r \cos\theta$$

with d_2 a constant to be determined later. Then, $\vec{H} = -\nabla \phi_2^*$, we have

$$H_r = -d_2 \cos\theta$$
$$H_\theta = d_2 \sin\theta \quad \text{(inside)}$$

But outside $H_\theta = -\dfrac{3}{2} \dfrac{B_0}{\mu_0} \sin\theta$, it follows that $d_2 = -\dfrac{3}{2} \dfrac{B_0}{\mu_0}$. There is no surface transport current. But $\vec{M}_{in} = -\vec{H}$, and $\vec{j} = \hat{n} \times (-\vec{M}_{in})$, a surface magnetization current $\vec{j}_M = -\dfrac{3}{2} \dfrac{B_0}{\mu_0} \sin\theta \hat{\phi}$ exists. We now summarize these results as follows:

Outside: $\vec{B} = \mu_0 \vec{H} = B_0 \hat{k} - B_0 \dfrac{a^3}{r^3} \cos\theta \hat{r} - \dfrac{1}{2} B_0 \dfrac{a^3}{r^3} \sin\theta \hat{\theta}$

Inside: $\vec{B} = 0, \quad \vec{H} = \dfrac{3}{2} \dfrac{B_0}{\mu_0} \hat{k}, \quad \vec{M} = -\dfrac{3}{2} \dfrac{B_0}{\mu_0} \hat{k}$

At $r = a$: $\vec{j}_M = -\dfrac{3}{2} \dfrac{B_0}{\mu_0} \sin\theta \hat{\phi}$

The second approach is identical for the outside region but takes the form $\vec{B} = \vec{H} = \vec{M} = 0$ for the inside. There is also a real transport surface current $\vec{j}_S = \hat{n} \times (\vec{H}_{out}) = -(3B_0/2\mu_0) \sin\theta \hat{\phi}$. This description can be summarized as follows:

Outside: $\vec{B} = \mu_0 \vec{H} = B_0 \hat{k} - B_0 \dfrac{a^3}{r^3} \cos\theta \hat{r} - \dfrac{1}{2} B_0 \dfrac{a^3}{r^3} \sin\theta \hat{\theta}$

Inside: $\vec{B} = 0, \quad \vec{H} = 0, \quad \vec{M} = 0.$

At $r = a$: $\vec{j}_S = -\dfrac{3}{2} \dfrac{B_0}{\mu_0} \sin\theta \hat{\phi}$

8.12 The London Equations

Ginzberg and Landau set up a very successful phenomenological theory of superconductivity in 1950; in 1967 Barden, Cooper, and Schreiffer provided a quantum mechanical model successfully explaining the mechanism of superconductivity, and their theory is popularly called the BCS theory. We cannot present the BCS theory here because it is beyond the scope of this book, but we can quickly summarize the essential points of the theory.

Electrical conduction occurs when some of the electrons in a material are able to move about freely; in the case of superconductivity, however, pairs of spin-up and spin-down electrons (called Cooper pairs) condense out in the material. Superconductivity can be interpreted by saying that it occurs because the number of pairs is not rigidly determined, and thus the state does not change even if the pairs float around from place to place. The strong diamagnetism of a superconductor is also due to the existence of a large number of Cooper pairs. As we know when a uniform static magnetic field is applied to a freely moving charged particle, the particle is caused to move in such a way that the magnetic field it produces is opposed to the applied field. A freely moving charged particle is thus diamagnetic and attempts to cancel out the applied field. A single particle is not very effective in doing this. In the normal state of a metal we have a large number of free charged particles, the electrons, and it would seem that they should be able to reinforce each other and give rise to a strong magnetic field opposing the applied field. However, because of the Pauli principle, the electrons all have different momenta, and this weakens the cumulative effect of the electrons. Indeed, the magnetic fields produced by the electrons tend to cancel each other out, and the resultant diamagnetism (known as the Landau diamagnetism) is very weak. In the superconductive state, however, we have a large number of freely moving Cooper pairs, all of which have the same momentum. In the absence of any electric currents, this common momentum is zero. Thus, when a magnetic field is applied, the Cooper pairs are able to reinforce in an effective manner each other's attempt to cancel out the applied field, and a very strong diamagnetism results.

In the following we introduce a simple but useful electromagnetic description of superconductivity offered by the brothers Fritz and Henry London.

From the above discussion we see that supercurrents are different from ordinary conduction current in that they are caused primarily by a magnetic field. In effect, $\vec{B}$ acts as a source density for $\vec{J}(\vec{r})$, which flows in loop around it. To describe these relationships, in 1934 Fritz and Henry London suggested that in a superconductor a current $\vec{J}(\vec{r})$ is spontaneously generated in response to any magnetic field $\vec{B}(\vec{r})$, related by the equation

$$\nabla \times (\mu_0 \Lambda^2 \vec{J}) = -\vec{B} \tag{8.43}$$

where Λ is a material and temperature dependent parameter.

This *London equation* supplements the Maxwell equations in a superconductor, in the same way as the equation $\vec{J} = \sigma \vec{E}$ relates current density and electric field and supplements the Maxwell equations in a normal conductor. We see that the Meissner effect follows from the London equation, and the equation also provides a basis for understanding the distinction between type I and type II superconductors (to be defined later).

To derive equation (8.43), we consider moving a magnet close to a sample of superconductor. The magnetic field in the sample changes with time and, from Maxwell equations, there is an electric field. The mean Lorentz force acting on a Cooper pair accelerates it. If $\vec{v}(\vec{r},t)$ is the velocity of a Cooper pair centered on $\vec{r}$ at time t,

$$m_C \frac{d\vec{v}}{dt} = -2e\left(\vec{E} + \vec{v} \times \vec{B}\right) \tag{8.44}$$

The pair carries charge $-2e$. The mass of the pair, m_C, is an effective mass of order of magnitude $2m_e$. Because the resistance is zero, there is no damping term. The acceleration of the pair is $(d\vec{v}/dt)$ and is the *total* time derivative. Of course, the London brothers did not use the Cooper pair, which was not known to them.

Because the velocity field $\vec{v}(\vec{r},t)$ depends on position,

$$\frac{d\vec{v}}{dt} = \frac{\partial \vec{v}}{\partial t} + \frac{\partial \vec{v}}{\partial x}\frac{dx}{dt} + \frac{\partial \vec{v}}{\partial y}\frac{dy}{dt} + \frac{\partial \vec{v}}{\partial z}\frac{dz}{dt} = \frac{\partial \vec{v}}{\partial t} + (\vec{v} \cdot \nabla)\vec{v} \tag{8.45}$$

Using equation (8.45) and the identity

$$\nabla(\vec{v} \cdot \vec{v}) = 2(\vec{v} \cdot \nabla)\vec{v} + 2\vec{v} \times (\nabla \times \vec{v})$$

we can rewrite equation (8.44) as

$$\frac{\partial \vec{v}}{\partial t} + \frac{1}{2}\nabla(v^2) + \frac{2e}{m_C}\vec{E} = \vec{v} \times \left(\nabla \times \vec{v} - \frac{2e}{m_C}\vec{B}\right) \tag{8.46}$$

Taking curl of this equation gives

$$\frac{\partial}{\partial t}\left(\nabla \times \vec{v} - \frac{2e}{m_C}\vec{B}\right) = \nabla \times \left[\vec{v} \times \left(\nabla \times \vec{v} - \frac{2e}{m_C}\vec{B}\right)\right] \tag{8.47}$$

because $\nabla \times \nabla(v^2) = 0$ and $\nabla \times \vec{E} + \partial\vec{B}/\partial t = 0$.

It is clear that one solution of equation (8.47) is

$$\nabla \times \vec{v} - \frac{2e}{m_C}\vec{B} = 0 \tag{8.48}$$

Using equation (8.48), equation (8.46) reduces to

$$\frac{\partial \vec{v}}{\partial t} + \frac{1}{2} \nabla (v^2) + \frac{2e}{m_C} \vec{E} = 0 \qquad (8.49)$$

It is this pair of equations, (8.48) and (8.49), that form the basis of the electromagnetic theory of superconductivity.

If we start with the magnet far removed from the superconductor and with no supercurrent, so that $\vec{B} = 0$ and $\vec{v} = 0$, then initially $\nabla \times \vec{v} - \frac{2e}{m_C} \vec{B} = 0$ everywhere in the sample, and equation (8.46) shows that this remains true at all subsequent times.

In terms of the number density of Cooper pairs N_C, the supercurrent density is

$$\vec{J}(\vec{r}, t) = -2e N_C \vec{v}(\vec{r}, t)$$

so that equation (8.48) may be written as

$$\nabla \times \left(\frac{m_C}{4e^2 N_C} \right) \vec{J} + \vec{B} = 0 \qquad (8.50)$$

This is the London equation (8.43) if we identify

$$\Lambda = \left(\frac{m_C}{4 \mu_0 e^2 N_C} \right)^{1/2} \qquad (8.51)$$

Equation (8.47) has other solutions. The London equation is the solution corresponding to a superconducting state. Equation (8.50) gives

$$\nabla \times (\Lambda^2 \partial \vec{J} / \partial t) + \partial \vec{B} / \partial t = 0$$

from which it follows that

$$\Lambda^2 \frac{\partial \vec{J}}{\partial t} = E \qquad (8.52)$$

Equation (8.52) shows that any steady current can persist indefinitely in the absence of $\vec{E}$ field. This equation and equation (8.43) are often called the London equations.

8.13 The London Equations and the Meissner Effect

We see that the Meissner effect follows from equation (8.43). With the help of the steady-state Maxwell equation $\nabla \times \vec{B} = \mu_0 \vec{J}$, we can eliminate $\vec{J}$ from equation (8.43) and obtain

$$\nabla \times (\Lambda^2 \nabla \times \vec{B}) = -\vec{B} \qquad (8.53)$$

In fields that are weak compared with B_C, we can neglect any spatial dependence of Λ (or, equivalently, N_c). Then, using the vector identity $\nabla \times (\nabla \times \vec{B}) = \nabla(\nabla \cdot \vec{B}) - \nabla^2 \vec{B}$ and noting that $\nabla \cdot \vec{B} = 0$, equation (8.53) becomes

$$\nabla^2 \vec{B} = \Lambda^{-2} \vec{B} \qquad (8.54)$$

The significance of this equation may be seen best by considering a large slab of superconductor with a free surface at $z = 0$. Suppose the externally applied field is uniform and parallel to the surface, in the x-direction. Inside we can expect the field to be of the form $(B(z), 0, 0)$ so that equation (8.54) reduces to the ordinary differential equation

$$\frac{d^2 B}{dz^2} = \frac{1}{\Lambda^2} B \qquad (8.55)$$

The solution of this equation that remains finite at all $z > 0$ is

$$B(z) = B_0 e^{-z/\Lambda} \qquad (8.56)$$

where B_0 is the field at the surface. Equation (8.56) indicates that the field decays exponentially with distance from the surface. Λ is called the London penetration depth, and a type I superconductor well below its critical temperature is typically 10^{-8} to 10^{-7} m. This is small, but can be measured experimentally. The field is negligible a few penetration depths from the surface. Representative values of Λ are given in the following table:

Elements	Λ, 10^{-10} m
Al	80
Sn	170
Cd	550
Pb	185

The current density is similarly confined to the surface layer. Using $\nabla \times \vec{B} = \mu_0 \vec{J}$

$$\vec{J} = \left(0, - (B_0/\mu_0 \Lambda) e^{-z/\Lambda}\right)$$

The "surface current" density is of magnitude

$$\int_0^\infty J_y\,dz = -B_0/\mu_0$$

The London equation is an approximation of the results of more exact theories. In strong fields, the more sophisticated Ginzberg and Landau equations are needed. Nevertheless, even in strong fields the London equation gives a qualitative understanding of the phenomena.

8.14 Flux Quantization

The phenomenon of flux exclusion by a superconductor described earlier refers only to simply connected objects, i.e., one with no holes or their topological equivalent. Now let us consider the production of persistent current in a superconducting ring. As shown in **Figure 8.7**, first a ring in normal state is placed in a magnetic field (a) and then the ring is cooled into superconducting state and persistent current is established (b). The field is forced out of the ring but remains in the hole even when the applied field is removed. The trapped magnetic flux generated by the current remains constant because the current in the ring does not attenuate. Realizing that superconductivity is fundamentally a quantum phenomenon, F. London suggested that the trapped magnetic flux should be quantized in units of h/e. The electric charge e in the denominator arises because London assumed the supercurrent was carried by single electrons. Subsequent delicate measurements on very small superconducting hollow cylinders showed that the flux quan-

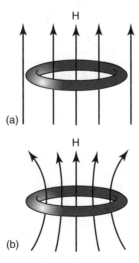

I FIGURE 8.7

tum is one-half the value postulated by London. That is, the magnetic flux Φ is quantized in units of $h/2e$, or

$$\Phi = nh/2e = n\Phi_0$$

where n is an integer and

$$\Phi_0 = h/2e = 2.0679 \times 10^{-15} \text{ T-m}^2$$

is the magnetic flux quantum. This is in agreement with the BCS theory and the Cooper pair concept.

8.15 Josephson Junctions

The remarkable properties of zero resistance and the Meissner effect make superconductors ideal for many applications. Some of these have been in use for years, whereas others may become feasible only if higher temperature superconductors become available.

One of the applications of superconductors was in a device known as a Josephson junction. In 1962 Brian Josephson predicted that electron pairs could tunnel from one superconductor through a thin layer of insulator into another superconductor. The superconductor–insulator–superconductor layer constitutes the Josephson junction. In the absence of any applied magnetic or electric field, a DC current flows across the junction (the DC Josephson effect). For an applied voltage across the insulator of

$$V(t) = V_0 + V_1 \sin \omega t$$

A current will cross the insulator if and only if the angular frequency ω of the oscillating voltage is related to the constant voltage V_0 by

$$\omega = \frac{e}{\pi h} V_0$$

This is the AC Josephson effect. At a potential of 1 mV the angular frequency ω is 7.7×10^{10} rad/s, which is in the microwave range. Because frequencies can be very accurately measured, the Josephson junction now constitutes the most accurate method of measuring voltage. The Josephson junction is used for just this purpose at the National Institute for Standards and Technology (NIST, formerly the National Bureau of Standards). With this device, the accuracy of the voltage standard is approximately one part in 10^{10}.

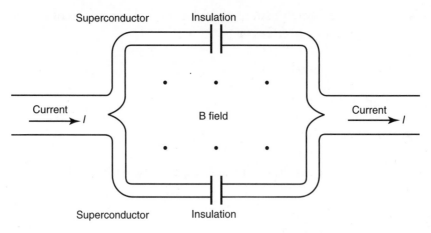

Superconductor   Insulation

Current   *I*

B field

Current   *I*

Superconductor   Insulation

I FIGURE 8.8

The most accurate method of measuring small magnetic fields is also based on the Josephson effect. If two Josephson junctions are put in parallel, as in **Figure 8.8**, the current depends on the magnetic flux Φ from an external field that threads the loop of the circuit containing the junctions. The current is given by

$$I = I_0 \cos\left(\frac{2\pi e}{h}\,\Phi\right)$$

By observation of changes in the current, changes in magnetic flux on the order of h/e can be observed.

A commercially available apparatus for magnetic field measurement based on this principle is the SQUID (acronym for superconducting quantum interference device). SQUIDs are used in medical physics to measure magnetic fields as small as 10^{-11} T in human organs. These fields are produced by nerve and muscle electric currents and by particles of magnetized iron. The magnetic study of heart action, the magnetocardiogram (the analog of the electrocardiogram), has the advantage that the magnetic effects come directly from the area of interest and not from changes in skin properties.

Problems

1. Given a cylinder of a linear conductor of length l, area A, and conductivity σ, find the relation between total current and potential difference along the cylinder. Also show that its resistance is given by $l/\sigma A$.

2. The conductivity of copper at room temperature is 6×10^7 mho/m. Find the resistance of a copper wire of 1 km in length and of 1 mm in diameter.

3. In any circuit segment ab, the current is driven by the algebraic sum of potential differences and *emf*'s and is limited by the resistance, so that Ohm's law is generalized to

$$I_{a \to b} R_{ab} = U_a - U_b + \mathcal{E}_{ab}$$

Find the potential difference between points a and b in **Figure 8.9**.

4. Breakdown of Ohm's law.
Consider a parallel plate diode, shown in **Figure 8.10**. The surface area of the electrodes is assumed to be large enough so that the edge effects may be neglected. Under operating conditions electrons leave the cathode ($V = 0$ at $x = 0$) and are accelerated toward the anode ($V = V_0$ at $x = d$). During the process the space charge builds up in the region between the electrodes and limits the flow of the current. The region between the plates is vacuum. Solve Poisson equation

$$\frac{d^2 V}{dx^2} = -\rho/\varepsilon_0$$

where ρ is the charge density, and express your result as a current–voltage relation. You will see that it does not obey Ohm's law.

5. In a conductor of uniform cross-section, such as a long uniform wire, the current density is the same at all points within the conductor. Now given a conductor of arbitrary shape, with a nonuniform cross-section, show that the distribution of current throughout the volume of the conductor is given by

$$\vec{j} = -\sigma \nabla V$$

where the potential difference V satisfies Laplace equation $\nabla^2 V = 0$.

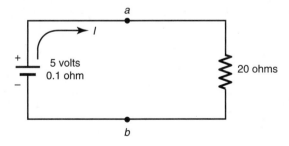

a

5 volts
0.1 ohm

20 ohms

b

I FIGURE 8.9

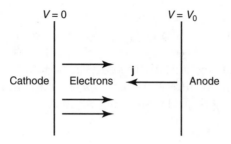

I FIGURE 8.10

6. According to quantum mechanics, conduction electrons must treated as waves, with wavelength λ related to momentum p, as suggested by de Broglie $\lambda = h/p$. As a result, we cannot determine the position and momentum of an electron accurately simultaneous. The uncertainties in position and momentum, Δx and Δp, are given by Heisenberg's uncertainty principle

$$\Delta x \Delta p \geq \hbar \quad (\hbar = h/2\pi)$$

For an electron in a cubical box with side L, the uncertainty in the coordinate is $\Delta x = L$, and the uncertainty in momentum is $\Delta p = \hbar/L$. Now let us plot the components of the momentum p_x, p_y, and p_z on the axes of a spatial system of coordinate to specify the momentum of the electron. The space obtained is called the momentum space or phase space. If all momentum space is divided into cell of cube with the side $\hbar/L$, each cell can take only two electrons (one spin up and one spin down), as dictated by Pauli's principle (**Figure 8.11**).

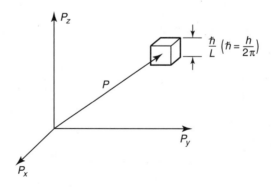

I FIGURE 8.11

Let us now discuss the behavior of the electrons in a metal at a temperature 0 K. All the electrons tend as close as possible to the origin of coordinate in the momentum space where they have the minimum kinetic energy $E_{kin} = mv^2/2 = p^2/2m$. Each cell (state) can take no more than two electrons, so the cells are filled up from the lowest momentum up. The top filled level is called the Fermi level with momentum p_F. Show that the Fermi momentum and the corresponding Fermi energy are given by

$$p_F = \hbar \, (3n/8\pi)^{1/3}, \quad E_F = (\hbar^2/2m)(3n/8\pi)^{2/3}$$

7. A long solenoid of radius b is wound coaxially around a type I superconducting cylinder of radius a. Find the magnetic field at all points both (a) above and (b) below the transition temperature T_c.

9 Magnetic Properties of Matter

So far, we have mainly dealt with the electric properties of materials and the magnetic effects of electric current in vacuum. In this chapter we outline macroscopic and microscopic descriptions of the magnetic properties of matter.

9.1 Magnetic Materials

All matter is composed of atomic systems that contain moving charges. The motion of an electron in an atom consists of a revolution in an orbit about the nucleus and a rotation or spin about an axis. Nuclei also have a spin about an axis, but this motion is not significant for our present purpose. The orbital motion of an electron is equivalent to a circulation of charge, or a current loop. These atomic current loops, as in the case of the current in a loop of wire, constitute magnetic dipole. They are affected when an external field is applied to matter. The magnitude of the angular momentum associated with the spin is an intrinsic property of the electron and is not altered by the applied field. The orientation of the spin axis in the direction of the field, however, appears to be the main reason for the unusual magnetic properties of special materials such as iron, cobalt, and nickel.

The response of the materials to an external applied magnetic field depends on the properties of the individual atoms and molecules and on their interactions. In many materials the small electric currents associated with the orbital motion and spin of the electrons average to zero. When such atoms are placed in a magnetic field, minute electron currents are generated by induction in the clouds of electrons surrounding the nucleus. By Lenz's law, the direction of these currents is such that the magnetic field associated with them opposes the inducing magnetic field. These are known as diamagnetic substances.

There are other materials in which the atoms do have an intrinsic magnetic moment because currents from the orbital motions and spins of the electrons do not average to zero. Although electron spins tend to pair and cancel one another, there are atoms in which pairing is incomplete. When such materials are placed in a magnetic field, the magnetic moments of the atoms and the induced magnetism enhances the external field. This effect is more pronounced at low temperatures (so thermal motion is small). Such substances are called paramagnetic substances.

Our picture on diamagnetism and paramagnetism is based on the vanishing or not of the net magnetic moment. When we try to calculate classically an average of the magnetic moment for an assembly of atoms in thermal equilibrium, we find it is always zero in all cases. This indicates that any respectable theory of magnetization in materials cannot be a classical formulation but must be based on a quantum mechanical treatment. But as far as diamagnetic and paramagnetic behavior is concerned, classical arguments give us some idea of what is going on, and so we use them nonetheless.

There is another class of materials known as ferromagnetic substances that can be understood only in terms of quantum mechanics. Ferromagnetic materials—iron, cobalt, nickel, and certain alloys—when brought near a magnet are attracted very strongly. Unlike paramagnetic materials, they do not readily shed their magnetic properties when the temperature is increased. For example, nickel has to be heated to over 630 K and iron to about 1040 K before their ferromagnetic properties are reduced to weakly paramagnetic. Curie discovered this property, and the critical temperature at which this change of behavior occurs is known as the Curie temperature.

This strange behavior has its origin in the interactions between unpaired electron spins and so is inherently quantum mechanical. Roughly speaking, what happens in, for example, iron is that there are five unpaired electrons. These electrons interact in such a way that their spins align in the same direction. Moreover, this alignment within one atom of iron is strong enough to influence its neighbors so that the atoms behave collectively rather than as individuals. Prof. Bitter of MIT first observed this collective behavior of ferromagnetic crystals. By allowing very fine iron powder to settle on the surface of a crystal, he found that irregular *domains* were formed. Domains are typically 10^{-12} to 10^{-16} m^3 in volume, containing on average 10^{15} atoms. Within the domains there is almost complete alignment of the electron spins of these atoms, but the domains themselves are randomly oriented so that there is no overall magnetization of the specimen. When, however, a ferromagnetic substance is placed in an external magnetic field, the domains rotate collectively and line up with the field.

9.2 Magnetization and Magnetization Current

We saw in Chapter 4 that a small current loop is equivalent to a magnetic dipole of moment $\vec{m}$. Likewise, the atomic current loop contributes a certain amount of magnetic dipole moment. These contributions add up vectorially to make up the atomic magnetic dipole moment $\vec{m}$. Thus, in analog of the polarization, we may define the magnetization $\vec{M}$ over a volume element $d\vec{r}'\,(=\Delta V)$

$$\vec{M} = \lim_{\Delta V \to 0} \left(\sum \vec{m}_i / \Delta V \right) \qquad (9.1)$$

where $\vec{m}_i$ is the magnetic moment of the ith atom, and we sum up vectorially all of the dipole moments in ΔV, which is small from a macroscopic point of view but it still contains a statistically large number of atoms.

We now rewrite equation (9.1) as

$$d\vec{m} = \vec{M} d\vec{r}'$$

Then from equation (4.29) of Chapter 4,

$$\vec{A}(\vec{r}) = \frac{\mu_0}{4\pi} \int \vec{M}(\vec{r}') \times \nabla' \left(\frac{1}{|\vec{r} - \vec{r}'|} \right) d\vec{r}' \qquad (9.2)$$

$$= \frac{\mu_0}{4\pi} \left[\int \frac{\nabla' \times \vec{M}}{|\vec{r} - r'|} d\vec{r} - \int \nabla' \times \left(\frac{\vec{M}}{|\vec{r} - r'|} \right) d\vec{r} \right]$$

We now use the fact that for any vector $\vec{\alpha}$

$$\int \nabla \times \vec{\alpha}\, d\vec{r} = -\int \vec{\alpha} \times d\vec{S}' \qquad (9.3)$$

If you are not familiar with this, it is easy to show it. Consider any constant vector $\vec{\beta}$, then

$$\vec{\beta} \cdot \int \vec{\alpha} \times d\vec{S}' = \int (\vec{\beta} \cdot \vec{\alpha}) \cdot d\vec{S}' = \int \nabla' \cdot (\vec{\beta} \times \vec{\alpha})\, d\vec{r}' = -\vec{\beta} \cdot \int \nabla \times \vec{\alpha}\, d\vec{r}'$$

which, because $\vec{\beta}$ is an arbitrary vector, gives equation (9.3). Using equation (9.3) in the second integral in equation (9.2) gives

$$\vec{A}(\vec{r}) = \frac{\mu_0}{4\pi} \left[\int \frac{\nabla' \times \vec{M}}{|\vec{r} - \vec{r}'|} d\vec{r}' + \int \frac{\vec{M} \times d\vec{S}'}{|\vec{r} - \vec{r}'|} \right] \qquad (9.4)$$

$$= \frac{\mu_0}{4\pi} \left[\int \frac{\nabla' \times \vec{M}}{|\vec{r} - \vec{r}'|} \, d\vec{r}' + \int \frac{\vec{M} \times \hat{n}' dS'}{|\vec{r} - \vec{r}'|} \right]$$

The integrals are taken over the volume and surface of the magnetized matter, as shown in **Figure 9.1**.

From equation (4.26) of Chapter 4, the vector potential produced by volume current $\vec{j}(\vec{r}')$ is

$$\vec{A}(\vec{r}) = \frac{\mu_0}{4\pi} \int \frac{\vec{j}(\vec{r}')}{|\vec{r} - \vec{r}'|} \, d\vec{r}'$$

It is easy to show that the vector potential produced by a surface current $\vec{K}(\vec{r}')$ is given by a similar expression:

$$\vec{A}(\vec{r}) = \frac{\mu_0}{4\pi} \int \frac{\vec{K}(\vec{r}')}{|\vec{r} - \vec{r}'|} \, d\vec{r}'$$

Upon comparing these with equation (9.4), we see that the vector potential given by equation (9.4) would be produced by two magnetization currents: a volume current density $\vec{j}_M$ and a surface current density $\vec{K}_M$:

$$\vec{j}_M(\vec{r}') = \nabla' \times \vec{M}, \quad \text{and} \quad \vec{K}_M = \vec{M} \times \hat{n}'$$

$$(9.5)$$

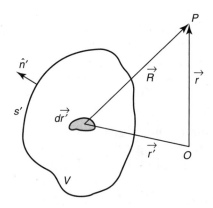

I FIGURE 9.1

Thus, as far as the effects outside a magnetic material are concerned, the material can be replaced by a distribution of volume and surface current densities that are related to the magnetization $\vec{M}$ by means of equation (9.5). These magnetization current densities are also known as Amperian current densities or bound current densities.

We can omit the primes on equation (9.5) and simply write

$$\vec{j}_M = \nabla \times \vec{M}, \quad \text{and} \quad \vec{K}_M = \vec{M} \times \hat{n} \tag{9.6}$$

but we have to remember that the differentiation is with respect to source point coordinates and $\hat{n}$ is the outward normal. Note that $\vec{M} \times \hat{n}$ is tangent to the surface because it is perpendicular to $\hat{n}$; this shows that it is the tangential component of $\vec{M}$ that leads to the surface current.

The subscript M is added to distinguish $\vec{j}_M$ from $\vec{j}$, the current density due to free or conduction electrons, that due to $\vec{M}$, the resultant magnetic moment per unit volume or the magnetization, originates from the Amperian currents. Unlike the conduction current, these neither transport charge nor produce ohmic heating.

Example 1

A cylinder has uniform magnetization $\vec{M}$ in the direction of its length. Calculate the magnetization current distribution.

Solution

$\vec{M} = M\hat{k}$; $\vec{j}_M = 0$; $\vec{K}_M = 0$ on end faces and $\vec{K}_M = M\hat{e}_\phi$ on cylindrical surfaces. Can you show that the magnetization current distribution is the same as an ideal solenoid, with M replacing NI/L?

Physically we may understand equation (9.6) by considering a simplified model of magnetized matter as though it consisted of atomic loop currents circulating in the direction, side by side. If the magnetization is uniform, the current in the various loops tend to cancel each other out, and there is not net effect current in the interior of the material. But at the surface there are no adjacent currents to produce a cancellation, and because the currents in the whirls are all circulating in the same sense, the result in effect is that of a surface current circulating on the surface.

If the magnetization is nonuniform, the cancellation is not complete. As an example of nonuniform magnetization, consider the abrupt change of magnetization shown in **Figure 9.2**. It is evident that between the two broken lines there is more current moving down than that moving up; and there is a resultant current in the interior. To find the relationship between $\vec{j}_M$ and $\vec{M}$, let us consider two small volume elements located next to each other in

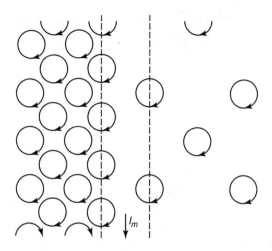

I FIGURE 9.2

the direction of the y-axis (**Figure 9.3**), each element of volume $\Delta x \Delta y \Delta z$. If the magnetization in the first volume element is $\vec{M}(x, y, z)$, then the magnetization in the second volume is

$$\vec{M}(x, y, z) + \frac{\partial \vec{M}}{\partial y} \Delta y + higher-order\ terms$$

The x-component of magnetic moment of the first element, $M_x \Delta x \Delta y \Delta z$, may be written in terms of a circulating current, I'_c:

$$M_x \Delta x \Delta y \Delta z = I'_c \Delta y \Delta z$$

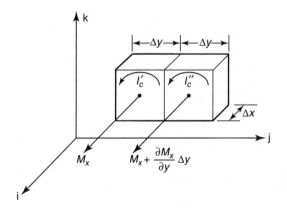

I FIGURE 9.3

Similarly, the x-component of magnetic moment of the second element, neglecting higher order terms, is

$$\left(M_x + \frac{\partial M_x}{\partial y}\, \Delta y\right) \Delta x \Delta y \Delta z = I''_c\, \Delta y \Delta z$$

The net upward current in the middle region of the two volume elements is

$$I'_c - I''_c \;=\; -\frac{\partial M_x}{\partial y}\, \Delta x \Delta y \tag{9.7}$$

We next consider two adjacent volume elements along the x-axis and focus our attention on the y-component of the magnetization in each volume element. In the middle region of the two elements, the net upward current due to the circulating currents that define the magnetic moments is

$$(\vec{I}_c)_{up} = \frac{\partial M_y}{\partial x}\, \Delta x \Delta y \tag{9.8}$$

These are the only circulating currents of a particular element that give rise to a net current in the z-direction. This net current, which comes about from nonuniform magnetization, is the magnetization current. This current is not a transport current but derives, as we have seen, from circulatory currents, that is, from atomic currents in the material. The effective area for each of the currents in equations (9.8) and (9.9) is $\Delta x \Delta y$. Thus,

$$(j_M)_z = \frac{\partial M_y}{\partial x} - \frac{\partial M_x}{\partial y}$$

or

$$\vec{j}_M = \nabla \times \vec{M}$$

as given by equation (9.6).

We now return to equation (9.2) and rewrite it in terms of $\vec{j}_M$ and $\vec{K}_M$:

$$\vec{A}(\vec{r}) = \frac{\mu_0}{4\pi}\left[\int \frac{\vec{j}_M(\vec{r}')}{|\vec{r} - \vec{r}'|}\, d\vec{r}' + \int \frac{\vec{K}_M \times d\vec{S}'}{|\vec{r} - \vec{r}'|}\right] \tag{9.9}$$

We see that the vector potential produced by a distribution of atomic currents inside matter has the same form as that produced by a distribution of true transport currents.

In principal, if $\vec{M}$ is a known function, $\vec{j}_M$ and $\vec{K}_M$ can be calculated from equation (9.6), and equation (9.9) then gives us the vector potential $\vec{A}$; finally, the field is given by the curl of $\vec{A}$, $\vec{B} = \nabla \times \vec{A}$. In practice, however, it presents some mathematical difficulties.

Example 2: Uniformly Magnetized Rod and Equivalent Air-Filled Solenoid

A uniformly magnetized bar is equivalent to an air-filled solenoid. The realization of this connection is very useful for us. So let us revisit a long bar magnet of uniform circular cross-section and length L. It consists of miniature Amperian current loops (**Figure 9.4**). Each loop is equivalent to a tiny bar magnet, with the moment IA' of each loop equal to the moment $q_m l$ of each magnet. Assuming there are n loops in a single cross-section of the rod, we have

$$nA' = A$$

where A' is the area of Amperian loop and A is the cross-sectional area of the rod. Further, we assume there are N such sets of Amperian loops in the length of the rod. Then,

$$nN = N'$$

where N' is the total number of loops in the rod. It follows that the magnetization M of the rod is given by

$$M = \frac{N'IA'}{LA} = \frac{NI}{L}\frac{nA'}{A} = \frac{NI}{L} = K_M$$

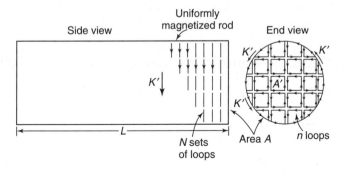

I FIGURE 9.4

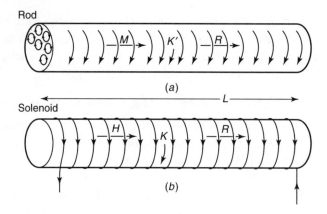

I FIGURE 9.5

where K_M is the equivalent surface current density on the outside surface of the rod, A/m.

Referring to the end view of the rod in Figure 9.4, we note equal and oppositely directed currents wherever loops are adjacent, so the currents cancel out with the exception of currents at the periphery of the rod. As a result, there is the equivalent of a current sheet flowing around the rod, as depicted in Figure 9.4 (end view) and in **Figure 9.5a**. This sheet has a linear current density K_M (A/m). This type of a current sheet is what we also have in the case of a solenoid with many turns of fine wire, as in Figure 9.5b. The actual sheet current density K for the solenoid is

$$K = NI/L \ A/m$$

where N is the number of turns in solenoid, I is the current through each turn, and L is the length of the solenoid.

If the solenoid of Figure 9.5b is the same length and diameter as the magnetized rod and if $K = K_M$, the solenoid is the magnetic equivalent of the rod.

9.3   Ampere's Law in Magnetic Materials and the Auxiliary Field $\vec{H}$

The total current density in magnetic media is now given by

$$\vec{j}_{total} = \vec{j} + \nabla \times \vec{M} \tag{9.10}$$

where $\vec{j}$ is the free (or conduction) current density. It is easy to show that the total current density $\vec{j}_{total}$ satisfies the continuity equation $\partial \rho / \partial t + \nabla \cdot \vec{j}_{total} = 0$, because $\nabla \cdot \nabla \times \vec{M} = 0$.

Ampere's law, the basic equation describing the magnetic effects of current [see equation (3.24)], now takes the form

$$\nabla \times \vec{B} = \mu_0 \vec{j}_{total}$$

$$= \mu_0 (\vec{j} + \nabla \times \vec{M})$$

i.e.,

$$\nabla \times \left(\frac{\vec{B}}{\mu_0} - \vec{M} \right) = \vec{j}$$

In discussing electric polarization, we found it convenient to introduce the displacement vector $\vec{D}$. Similarly here, we introduce a new vector $\vec{H}$ defined by

$$\vec{H} = \vec{B}/\mu_0 - \vec{M} \tag{9.11}$$

Ampere's law now reads

$$\nabla \times \vec{H} = \vec{j} \tag{9.12}$$

Integrating this over a surface S bounded by curve l, and using Stokes' theorem, gives the integral form of Ampere's law

$$\oint \vec{H} \cdot \vec{dl} = \vec{I} \tag{9.13}$$

where $\vec{j}$ is the conduction current density and $\vec{I} = \int \vec{j} \cdot \vec{dS}$. The quantity $\vec{H}$ is measured in units of ampere meter^{-1}, as is $\vec{M}$. It is referred to by different names by different authors, as mentioned in Chapter 4. We call $\vec{B}$ the magnetic field and $\vec{H}$ the auxiliary magnetic vector. We see that the field $\vec{B}$ is the magnetic analog of the electric field $\vec{E}$ and $\vec{H}$ is the magnetic counterpart of the electric displacement vector $\vec{D}$. In free space $\vec{M} = 0$ and equation (9.11) reduces to

$$\vec{H} = \vec{B}/\mu_0 \quad \text{or} \quad \vec{B} = \mu_0 \vec{H}.$$

The divergence equation $\nabla \cdot \vec{B} = 0$ is valid for all magnetic fields that are produced by a current contribution, provided there is no isolated magnetic monopole.

The equation $\nabla \cdot \vec{B} = 0$ and equation (9.12) are the fundamental magnetic field equations in the presence of matter; these two equations, together with appropriate boundary conditions and a relationship between $\vec{B}$ and $\vec{H}$, are sufficient to solve magnetic problems.

9.4 Magnetic Susceptibility and Permeability

The relationship between $\vec{B}$ and $\vec{H}$ or, equivalently, between $\vec{M}$ and one of the magnetic field vectors, depends on the magnetic properties of the material and is usually obtained from experiment. If we except ferromagnetic substances, many materials exhibit an approximately linear relationship between $\vec{M}$ and $\vec{H}$. If the material is isotropic and linear, we may write

$$\vec{M} = \chi_m \vec{H} \tag{9.14}$$

where the dimensionless scalar quantity χ_m is called the magnetic susceptibility of the material. If the material is linear but anisotropic, $\vec{M}$ does not necessarily have the same direction as $\vec{H}$, and equation (9.14) is replaced by a tensor relationship. We limit ourselves to isotropic material.

If $\chi_m > 0$, the magnetic field $\vec{B}$ is strengthened by the presence of the material, so the material is paramagnetic. If $\chi_m < 0$, the field $\vec{B}$ is weakened by the presence of the material, so the material is diamagnetic. Although χ_m is a function of temperature, it is quite small for paramagnetic and diamagnetic materials. Data on χ_m may be obtained from the *Handbook of Chemistry and Physics* (CRC Press, Boca Raton, FL) or other reference books. But χ_m is not listed directly; instead, it is given as the mass susceptibility, $\chi_{m,mass}$, or the molar susceptibility, $\chi_{m,molar}$. These are defined by

$$\chi_{m,mass} = d\chi_m, \qquad \chi_{m,molar} = \chi_m(A/d)$$

where d is the mass density of the material and A is the molecular weight.

From equation (9.11) we have

$$\vec{B} = \mu_0(\vec{H} + \vec{M})$$

Combining this with equation (9.14), we get

$$\vec{B} = \mu_0(1 + \chi_m)\vec{H} = \mu\vec{H} \tag{9.15}$$

where

$$\mu = \mu_0(1 + \chi_m) \tag{9.16}$$

and it is called the magnetic permeability of the material, and the ratio μ/μ_0 is the relative permeability of the material. For paramagnetic materials $\mu > 1$

and for diamagnetic $\mu < 1$. Recall that we have the equivalent results for dielectric materials:

$$\vec{D} = \varepsilon_0(1 + \chi)\,\vec{E} = \varepsilon\vec{E}$$

with the relative permittivity or dielectric constant given by

$$K = \varepsilon/\varepsilon_0 = 1 + \chi$$

In general, the quantities χ_m and μ are not as meaningful as the corresponding quantities χ and ε.

9.5 Boundary Conditions

To solve magnetic problems, we need to know the continuity conditions satisfied by $\vec{B}$ and $\vec{H}$ at the boundary between two media with different permeability. Because the equation $\nabla \cdot \vec{B} = 0$ is unaltered by the presence of magnetic materials, the normal component of $\vec{B}$ should be continuous across the boundary. Let us see why. By divergence theorem we have

$$\int_V \nabla \cdot \vec{B}\,d\tau = \int_S \vec{B} \cdot \hat{n}\,dS$$

But $\nabla \cdot \vec{B} = 0$, and this reduces to

$$\int_S \vec{B} \cdot \hat{n}\,dS = 0$$

We now apply this to a small pillbox-shaped surface S, which intersects the interface as shown in **Figure 9.6**. Because the height of the pillbox is negligibly small in comparison with the diameter of the bases, we find

$$\vec{B_2} \cdot \hat{n}_2\,\Delta S + \vec{B_1} \cdot \hat{n}_1\,\Delta S = 0$$

Now $\hat{n}_2 = -\hat{n}_1$; hence,

$$(\vec{B_2} - \vec{B_1}) \cdot \hat{n}_2 = 0$$

or

$$B_{2n} = B_{1n} \tag{9.17}$$

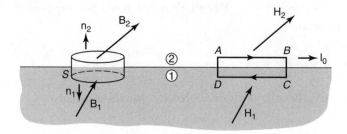

I FIGURE 9.6

The boundary condition on $\vec{H}$ can be found by applying Ampere's law to the rectangular path $ABCD$ in Figure 9.6, where the lengths AB and CD are taken equal to Δl and the sides AD and BC are negligibly small:

$$\oint \vec{H} \cdot d\vec{l} = H_{t1}\Delta l - H_{t2}\Delta l = I$$

If we define a surface current density j by $j\Delta l = I$ this gives

$$H_{t1} - H_{t2} = j \qquad (9.18)$$

i.e., there is a discontinuity in the tangential component of $\vec{H}$-field equal to the surface current density in amperes per meter at the boundary between the two media.

Figure 9.7 show how B and H are refracted at the boundary for the simple case where B and H are collinear in each medium. The diagram is drawn for the case $\mu_2 > \mu_1$. The boundary conditions (9.17) and (9.18) give

$$B_1\cos\theta_1 = B_2\cos\theta_2$$

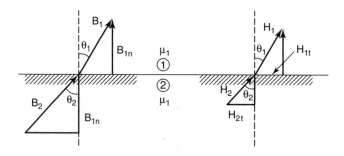

I FIGURE 9.7

and

$$H_1 \sin\theta_1 = H_2 \sin\theta_2$$

From these two relations we obtain the law of refraction, namely

$$\tan\theta_1/\tan\theta_2 = \mu_1/\mu_2.$$

9.6 Boundary-Value Problems

As shown by equations (9.17) and (9.18), $\vec{B}$ and $\vec{H}$ obey boundary conditions similar to those for $\vec{D}$ and $\vec{E}$. The problems of linear media or specified magnetization are similar to the dielectric problems discussed in the preceding chapter. Let us now turn our attention to the calculation of magnetic fields inside magnetic material in which there is no transport current. This is similar to the dielectric with no external charge density.

When $\vec{j} = 0$, the two fundamental magnetic equations, $\nabla \cdot \vec{B} = 0$ and $\nabla \times \vec{H} = \vec{j}$, reduce to

$$\nabla \cdot \vec{B} = 0 \tag{9.19}$$

$$\nabla \times \vec{H} = 0 \tag{9.20}$$

Equation (9.20) implies that the auxiliary magnetic vector $\vec{H}$ can be expressed as the gradient of a scalar field, say V^*:

$$\vec{H} = -\nabla V^* \tag{9.21}$$

where V^* is the magnetic scalar potential due to all sources.

We limit ourselves to magnetic material for which the magnetic field calculation reduces to a simple boundary-value problem. This includes (1) linear or approximately linear magnetic material for which $\vec{B} = \mu\vec{H}$ and (2) a uniformly magnetized piece of material for which $\nabla \cdot \vec{M} = 0$. In both cases, equation (9.19) reduces to

$$\nabla \cdot \vec{H} = 0$$

Combining this with equation (9.21) we get

$$\nabla^2 V^* = 0 \tag{9.22}$$

which is Laplace's equation. We see now that the magnetic problem reduces to finding a solution to Laplace's equation that satisfies the boundary conditions. After finding V_1^*, $\vec{H}$ is may then be obtained from equation (9.21) and $\vec{B}$ from $\vec{B} = \mu\vec{H}$ or $\vec{B} = \mu_0(\vec{H} + \vec{M})$, whichever is appropriate.

We have solved the problem of a polarized sphere in a uniform electric field. We now discuss the corresponding magnetic problem: a sphere of linear magnetic material placed in a uniform magnetic field.

Example 3

Consider a sphere of linear magnetic material of radius a and permeability μ placed in a region of space containing an initially uniform magnetic field $\vec{B_0}$. Find (1) how the magnetic field is modified by the presence of the sphere and (2) the magnetic field in the sphere itself.

Solution

As in the dielectric case, we choose the origin of our coordinate system at the center of the sphere and the direction of $\vec{B_0}$ as the polar direction (z-direction). We assume that the potential outside and inside the sphere to be a sum of zonal harmonics:

$$V_1^*(r, \theta) = A_1 r \cos\theta + C_1 r^{-2} \cos\theta \quad (r > a)$$

$$V_2^*(r, \theta) = A_2 r \cos\theta \qquad (r < a)$$

where θ is the angle that the radius vector makes with the polar direction. The term $r^{-2}\cos\theta$ is not included in V_2^*, for such a term would become infinite at $r = 0$. The constants A_1, C_1, and A_2 must be determined by the boundary conditions.

At distances very far away from the sphere, the magnetic field regains its uniform character, i.e., $\vec{B} = B_0\hat{k}$ and $V_1^* \to -(B_0/\mu_0)r\cos\theta$. Hence, $A_1 = -(B_0/\mu_0)$. We now apply the boundary conditions at the interface $r = a$: $H_{1\theta} = H_{2\theta}$, $B_{1r} = B_{2r}$, or

$$-\left(\frac{B_0}{\mu_0}\right)\sin\theta + \frac{C_1}{a^3}\sin\theta = A_2\sin\theta$$

$$B_0\cos\theta + 2\mu_0\frac{C_1}{a^3}\cos\theta = -\mu A_2\cos\theta$$

Solving these equations yields

$$A_2 = -\frac{3B_0}{\mu + 2\mu_0}, \quad C_1 = [(\mu/\mu_0) - 1]\frac{B_0 a^3}{\mu + 2\mu_0}$$

The magnetic fields inside and outside the sphere are given by, respectively,

$$\vec{B}_2 = \frac{3B_0 \hat{k}}{1 + 2(\mu_0/\mu)}$$

$$\vec{B}_1 = B_0 \hat{k} + \left[\frac{(\mu/\mu_0) - 1}{(\mu/\mu_0) + 2}\right]\left(\frac{a}{r}\right)^3 B_0(2\hat{e}_r \cos\theta + \hat{e}_\theta \sin\theta)$$

Figure 9.8 shows the lines of magnetic field $\vec{B}$ in the vicinity of the sphere.

Example 4

Determine the magnetic field produced by a uniformly magnetized sphere of magnetization $\vec{M}$ and radius a placed in free space where no other magnetic field presented.

Solution

Because $\vec{M}$ is constant, then $\nabla \cdot \vec{M} = 0$, and thus the scalar magnetic potential satisfies Laplace's equation. If we take the magnetization along the z-axis and the origin of our coordinate system at the center of the sphere, then the potential depends on θ as well as on r and we may expand the potential in zonal harmonics:

$$V_1^*(r,\theta) = \sum_{n=0}^{\infty} C_{1,n} r^{-(n+1)} P_n(\cos\theta) \quad (r > a)$$

$$V_2^*(r,\theta) = \sum_{n=0}^{\infty} A_{2,n} r^n P_n(\cos\theta) \quad (r < a)$$

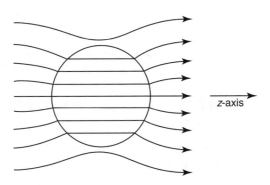

I FIGURE 9.8

Note that we have left out the harmonics with positive power of r in V_1^* because these terms approach infinite as $r \to \infty$. We have also left out the terms of negative power of r in V_2^* because these terms become infinite at the origin. From the boundary conditions at $r = a$:

$$H_{1_\theta} = H_{2_\theta}, \quad B_{1r} = B_{2r}$$

We obtain

$$\sum_{n=0}^{\infty} (C_{1,n} a^{-(n+1)} - A_{2,n} a^n) a^{-1} \frac{d}{d\theta} P_n(\theta) = 0$$

or

$$\sum_{n=0}^{\infty} (C_{1,n} a^{-(n+1)} - A_{2,n} a^n) P_n(\theta) = \text{constant}$$

and

$$\mu_0 C_{1,n} a^{-2} + \mu_0 \sum_{n=0}^{\infty} P_n(\theta) [C_{1,n}(n+1) a^{-(a+2)} + A_{2,n} n a^{n-1}] - \mu_0 M \cos\theta = 0$$

As $P_n(\theta)$ are orthogonal functions, each term of the last two equations vanishes individually. For $n = 0$, we have

$$C_{1,0} a^{-1} - A_{2,0} = \text{constant}, \quad \mu_0 C_{1,0} a^{-2} = 0$$

Therefore, $C_{1,0} = 0$, and $A_{2,0}$ is an arbitrary constant that may be set equal to zero without affecting $\vec{B}$ or $\vec{H}$.

For the $n = 1$ terms, we have

$$C_{1,1} a^{-3} - A_{2,1} = 0, \quad 2C_{1,1} a^{-3} + A_{2,1} - M = 0$$

Solving these two equations yields

$$C_{1,1} = Ma^3/3, \quad A_{2,1} = M/3$$

For all $n \geq 2$, $C_{1,n} = 0$ and $A_{2,n} = 0$. Substituting these back into the equations for V_1^* and V_2^*, we obtain

$$V_1^*(r,\theta) = \frac{1}{3} M(a^3/r^2)\cos\theta, \quad \text{outside the sphere}$$

$$V_2^*(r,\theta) = \frac{1}{3} Mr\cos\theta, \quad \text{inside the sphere}$$

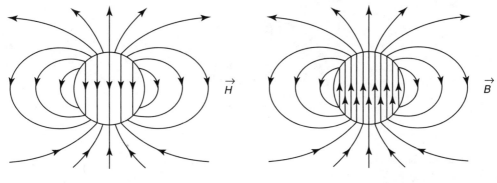

I FIGURE 9.9

From $\vec{H} = -\nabla V^*$ we have

$$\vec{H}_1 = \frac{1}{3} M (a^3/r^3)[a\hat{e}_r \cos\theta + \hat{e}_\theta \sin\theta], \quad \vec{H}_2 = -\frac{1}{3} M \hat{e}_z$$

We see that the external field of the uniformly magnetized sphere is exactly a dipole field, arising from the dipole moment $(4\pi/3)a^3 M$. The field *inside* the sphere is a demagnetizing field, a result that is in accord with the *E*-field inside a uniformly polarized dielectric. We see, therefore, that the magnetized sphere is subjected to its own demagnetizing field. The factor $1/3 = (1/4\pi)(4\pi/3)$ in $\vec{H}_2$ depends explicitly on the spherical geometry. The factor $4\pi/3$ is known as the *demagnetization factor* of a sphere. The demagnetization factors for other geometric shapes have been calculated and tabulated (e.g., see *American Institute of Physics Handbook*, McGraw-Hill, New York). **Figure 9.9** shows the lines of $\vec{H}$ and $\vec{B}$.

9.7 Magnetic Shielding

A sensitive magnetic instrument can be shielded very effectively from outside fields by placing it inside a cylindrical shell made of soft iron of high permeability. We investigate the field in the interior of a cylindrical shell of constant permeability placed in a uniform magnetic field at right angles to the axis of the cylinder, as shown in **Figure 9.10**, where a and b are the inner radius and the outer radius of the infinitely long cylindrical shell, respectively, and the uniform field H_0 is parallel to the x-axis. We need three solutions of Laplace's equation: a solution V_1 to represent the potential outside the shell, a second solution V_2 to represent the potential in the material of the shell, and a third V_3 to represent solution in the cavity. The potential is a function

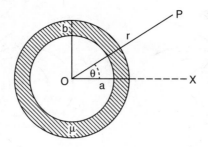

I FIGURE 9.10

of r and θ alone, because the problem has cylindrical symmetry. The boundary conditions that must be satisfied by the solutions are the following:

1. For r very large V_1 must become $-H_0 x = -H_0 r \cos\theta$, provided the potential is taken to be zero on the axis of the cylinder.
2. At each surface of the shell the radial component of the magnetic field $\vec{B}$ and the transverse component of the $\vec{H}$-field must be continuous as we pass from the outside to the inside of the shell.

As the potential at infinity must be $-H_0 r \cos\theta$, it is clear that we are limited to harmonics involving only the first power of $\cos\theta$. The three potentials are of the form

$$\left.\begin{array}{l} V_1 = -H_0 r \cos\theta + B_1 \cos\theta / r \\ V_2 = A_0 r \cos\theta + B_2 \cos\theta / r \\ V_3 = A_2 r \cos\theta \end{array}\right\} \tag{9.23}$$

Note that we have omitted the term in $1/r$ from V_3 because it would become infinite at the origin.

From $V_1 = V_2$ at $r = b$, we have

$$-H_0 b^2 + B_1 = A_2 b^2 + B_2 \tag{9.24}$$

Similarly, the condition $V_2 = V_3$ at $r = a$ gives

$$A_2 a^2 + B_2 = A_3 a^2 \tag{9.25}$$

Continuity of the radial component of the magnetic field $\vec{B}$ at the outer surface gives

$$H_0 b^2 + B_1 = \mu(-A_2 b^2 + B_2) \tag{9.26}$$

Similarly, continuity condition at the inner surface gives

$$\mu(-A_2a^2 + B_2) = -A_3a^2 \tag{9.27}$$

Now we have four equations, from equations (9.24) to (9.27), for four unknowns B_1, B_2, A_2, and A_3. We can eliminate the first three and express A_3 in terms of H_0. Then putting the value of A_3 so found in equation (9.23) for V_3, the result is

$$V_3 = -\frac{4\mu}{(\mu + 1)^2 - a^2(\mu - 1)^2/b^2} H_0 r \cos\theta$$

From this we find

$$H_3 = -\frac{\partial V_3}{\partial x} = \frac{4\mu}{(\mu + 1)^2 - a^2(\mu - 1)^2/b^2} H_0$$

The ratio of H_0 to H_3 in the cavity is called the shielding ratio, and it is given by

$$g = -\frac{H_0}{H_3} = \frac{1}{4\mu}\left\{(\mu + 1)^2 - a^2(\mu - 1)^2/b^2\right\} = 1 + \frac{1}{4}\frac{(\mu - 1)^2}{\mu}\left(1 - \frac{a^2}{b^2}\right) \tag{9.28}$$

If the shell is made of soft iron whose permeability is much larger than unity, the shielding ratio g is then given very closely by

$$g = \frac{1}{4}\mu\left(1 - \frac{a^2}{b^2}\right) = \frac{\mu}{4\pi b^2}\left[\pi(a^2 - b^2)\right] \tag{9.29}$$

where $\pi(b^2 - a^2)$ is the cross-section of the cylindrical shell. Thus, the shielding ratio g for a given amount of soft iron is greater the smaller the outside radius b. The effectiveness of magnetic shielding is somewhat increased by using several coaxial cylindrical shells with air gaps between.

9.8 Cavities

Because neither a compass needle nor a test coil can move freely inside solid bodies, how do we measure the field vectors $\vec{B}$ and $\vec{H}$ in solids? Obviously, the only way to perform a field measurement in a solid is to insert a measuring device into a cavity cut-out in the body. It has been found, however, that such a measurement is affected by the shape and orientation of the cavity. The boundary conditions (9.17) and (9.18) specify the shape and orientation of the cavity for us (**Figure 9.11**). To measure $\vec{H}$ at a point inside a magnetic medium, we cut out a small needle-shaped cavity in this medium at that point such that the axis of the cavity $\vec{H}$ is in the direction of $\vec{H}$, then by equation (9.18)

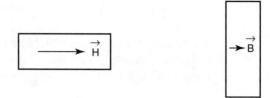

I FIGURE 9.11

$$\vec{H}_{\text{medium}} = \vec{H}_{\text{cavity}}$$

To measure $\vec{B}$ we cut a small disc-shaped cavity oriented so that its axis is in the direction of $\vec{B}$, then by equation (9.17)

$$\vec{B}_{\text{medium}} = \vec{B}_{\text{cavity}}$$

9.9 Origin of Diamagnetism and Paramagnetism

a. Diamagnetism

In the beginning of this chapter, diamagnetic, paramagnetic, and ferromagnetic effects were mentioned briefly. We now revisit these subjects, and in this section we give a semiclassical description of diamagnetic and paramagnetic effects. It begins with a calculation to show how the application of a magnetic field modifies the electronic orbital currents in atoms. A rigorous solution of this problem actually belongs to the domain of quantum mechanics. But a semiclassical treatment gives an explanation of diamagnetic and paramagnetic effects that are correct in general outline though not in details. Most materials present below have been mentioned in previous chapters. For the convenience of the reader, we repeat them again.

In the classical model, atoms in any medium may be considered as consisting of electrons moving in circular obits about fixed nuclei. Now let us consider the motion of an electron (charge $q = -e$, mass m_e) in a circular orbit of radius r about a positive nuclear charge. The equation of motion of the electron is

$$F_e = m_e v^2/r = m_e \omega_0^2 r \qquad (9.30)$$

where F_e is the attractive coulomb force due to the positive nuclear charge.

At any instant, the electron and the positive nuclear charge would appear as an electric dipole, but on the time average the electric dipole moment is zero, producing no steady electric field at a distance. The magnetic field of the system, far away, is not zero on the time average. Instead, it is just the field of a current ring (due to the orbital motion of the electron about the fixed nucleus). Its far field is therefore that of a magnetic dipole $\vec{m}$, and its magnitude is

$$m = IA = \frac{ev}{2\pi r}\,\pi r^2 = \frac{evr}{2}$$

where we assume the electron makes $v/2\pi r$ revolutions per second about the nucleus. We may express the magnetic moment in terms of angular momentum $\vec{L}$ of the electron; its magnitude is $m_e vr$ and pointing opposite to $\vec{m}$

$$\vec{m} = -\frac{e}{2m_e}\vec{L} \qquad (9.31)$$

This relation holds quite generally, not limited to circular orbits.

As $\vec{L}$ is conserved in a central field, and so does $\vec{m}$. The ratio

$$\vec{m}/\vec{L} = -e/2m_e$$

is called the gyromagnetic ratio. The intimate connection between magnetic moment and angular momentum is central to any account of atomic magnetism.

The net magnetic dipole moment of an atom of the medium is simply a sum over all the electrons in the atom. There are two possibilities. First, for an atom with an even number of electrons, close analysis shows an electron with angular momentum $\vec{L}$ for every electron with an angular momentum $-\vec{L}$, that is, the electrons are "paired." Second, for atoms in which the pairing effect is absent, an atom may have a net angular momentum, but it will be randomly oriented and so the net angular momentum of a finite volume of the material, which contains a large number of atoms, vanishes unless some external agency exists to align it.

Now we consider an applied magnetic field $\vec{B}$ as an external agency. By some means we apply a uniform magnetic field $\vec{B}$ over the extent of the atom. We apply the field in such manner that it increases from zero to some finite value in some prescribed manner. The applied magnetic field $\vec{B}$ has two effects: (1) The orbit links with a changing flux and an induced electric field $\vec{E}$ acts tangentially on the electron, and (2) the electron is subject to Lorentz force $-e(\vec{v}\times\vec{B})$, which acts radially.

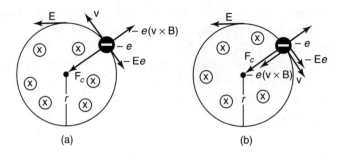

I FIGURE 9.12

Figure 9.12 shows the various quantities of importance to the motion. For the situation shown in Figure 9.12a, where $\vec{B}$ is directed into the paper and is increasing, by Lenz's law $\vec{E}$ is counterclockwise and thus the tangential force on the electron reduces the tangential velocity $\vec{v}$—that is, its effect is to slow the electron down. On the other hand, the magnetic force is directed away from the center and is opposed to the coulomb attraction. We show that under these influences the radius of the orbit remains essentially unchanged but that the electron is either slowed down or speeded up depending on the sense of the vector $(\vec{v} \times \vec{B})$. In Figure 9.12b, the electron velocity has been reversed. Lenz's law still predicts the same sense for the induced field $\vec{E}$, because $\vec{B}$ is still directed into the paper and is increasing. But now the electron is speeded up, and the magnetic force is toward the center.

The rate of magnetic flux through the circular path is

$$\frac{d\Phi}{dt} = \pi r^2 \frac{dB}{dt}$$

and the induced $\vec{E}$ field all around the path is

$$\int \vec{E} \cdot d\vec{l} = \pi r^2 \frac{d\vec{B}}{dt} = 2\pi r E$$

or

$$|E| = \frac{1}{2} r \frac{dB}{dt} \tag{9.32}$$

In deducing equation (9.32), it has been assumed that the electron orbit is unaltered. When the magnetic field is switched on—that is, while the applied field is changing from 0 to its final value B_f—there is a change in the momentum of the electron, which can be described by

$$\frac{d}{dt}(m_e \omega r) = Eq$$

Whence, by the use of equation (9.32),

$$\frac{d\omega}{dt} = \frac{q}{2m_e}\frac{dB}{dt} \tag{9.33}$$

Consequently, the change in the angular velocity from ω_0 to ω_f may be obtained from

$$\Delta\omega = \omega_f - \omega_0 = \int_{\omega_0}^{\omega_f} d\omega = \frac{q}{2m_e}\int_0^{B_f} dB = \frac{q}{2m_e}B_f \tag{9.34}$$

The magnetic force either decreases or increases the force toward the center, as is indicated in Figure 9.12, a and b. The electron can remain in the same orbit if the modification of the central force is just enough to compensate for $\Delta\omega$, the corresponding change in the angular velocity. With the aid of equation (9.31), we can also evaluate $\Delta\omega$ from

$$m_e(\omega_0 + \Delta\omega)^2 r - F_c = \pm\, qvB_f = \pm\, q\omega_f rB_f \tag{9.35}$$

for the situations pictured in Figure 9.12. Even for the highest values of B ($\sim$100 T) obtainable, $\Delta\omega \ll \omega_0$. In solving for $\Delta\omega$ from (9.35), $(\Delta\omega)^2$ may be neglected relative to other terms in the expansion on the left. Also, ω_0 may be considered nearly equal to ω_f. With these approximations, for the numerical value of $\Delta\omega$ we obtain

$$\Delta\omega = \frac{q}{2m_e}B_f \tag{9.36}$$

Consequently, the value of $\Delta\omega$ deduced from the change in the central force does not contradict the result for $\Delta\omega$ as calculated from the change in the orbital momentum on the assumption that the orbital radius remains unaltered. In other words, the magnetic force compensates very nearly for the change in the centripetal force arising from an increase or decrease in the angular velocity. We can conclude that the application of the field either speeds up or slows down the electron in its orbit, depending on the sense of the magnetic force.

The above discussion suggests the following physical picture. To a good approximation (assuming the acceleration due to the induced $\vec{E}$ field is small), the electrons in an atom move in the same orbits for a nonzero $\vec{B}$ field as they do in a zero field; however, for nonzero field, superimposed on the rapid orbital electronic motion is a slow uniform precessional motion of the entire atom about the field direction, the angular frequency of this motion being the Larmor precessional frequency, which is the quantity on the right-hand side of equation (9.36). It is the angular frequency with which the magnetic moment $\vec{m}$ precesses about the field when the field is not normal to the plane of the orbit.

We can calculate the change in the magnetic moment resulting from the change in the angular frequency by the following steps. First recall that

$$m = \left(\frac{qv}{2\pi r}\right)\pi r^2 = qvr/2 = qr^2\omega/2$$

From this we obtain

$$\Delta m = qr^2 \Delta\omega/2 \tag{9.37}$$

Combining this with (36), we get

$$\Delta\vec{m} = -q^2 r^2 \vec{B}/4m_e \tag{9.38}$$

where the subscript of $\vec{B}$ has been dropped and where the minus sign has been included to indicate that $\Delta\vec{m}$ is antiparallel to $\vec{B}$ as we see from the discussion that follows. **Figure 9.13** shows the change in the magnetic moment arising from the application of the field. The figures are drawn for a circulating charge that is negative. In Figure 9.13a the electron is slowed down, the equivalent loop current is reduced, and the initial magnetic moment that is parallel to $\vec{B}$ is reduced. In Figure 9.13b the electron circulation is reversed, the electron is speeded up, and the magnetic moment is increased in magnitude. But in this case the moment is directed opposite to $\vec{B}$. In each

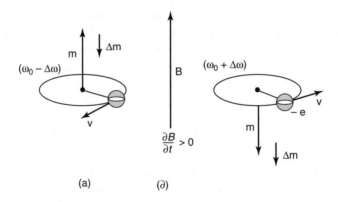

(a)   (∂)

I FIGURE 9.13

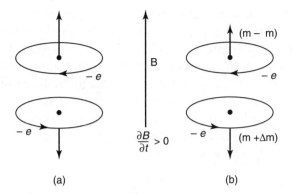

I FIGURE 9.14

case $\Delta \vec{m}$ is such as to give rise to a diminution in the externally applied field. This effect accounts in a general way for the diamagnetic behavior of matter.

We apply the reasoning to an atom of helium, which has two electrons circulating in opposite senses. In the absence of an external field, the net magnetic moment of the atom is zero. If a magnetic field is applied, the individual moments are altered, as shown in **Figure 9.14**. By referring to Figure 9.14b we see that the atom acquires a net magnetic moment of $2\Delta\vec{m}$ in such a direction as to weaken the field of $\vec{B}$. In this simplified treatment we have ignored the spin magnetic moment of the two electrons. In general, orbital and spin moments must be combined to obtain the resultant magnetic moment of the atom.

In metal, due to the spin magnetic moment, the free electrons may constitute a dense collection of permanent magnetic dipoles. According to quantum mechanics, these cannot assume arbitrary orientations but are restricted to one of two possibilities. The component in the direction of any magnetic field can be only ±1 Bohr magneton. These result in a temperature-independent paramagnetism, called Pauli paramagnetism. However, the effect is very small and is often masked by the diamagnetism of the inner electron shells.

Our semiclassical model presented above is only approximately correct, because electronic motions in atoms are not confined to well-defined orbits. But the approach implies correctly that all matter is diamagnetic. This aspect often masks stronger paramagnetic or ferromagnetic behavior that can occur in the material simultaneously. Diamagnetism is particularly prominent in materials that consist entirely of atoms or ions with closed electron shells, because in these cases all paramagnetic contributions cancel out.

This model of a diamagnetic implies that χ_m is independent of temperature, because the core structure of an atom is normally unchanged by thermal vibrations. To calculate χ_m, we have to sum up over all the orbits in an atom and also over all the atoms present per unit volume of the sample. The electron orbits are inclined to the field, the magnetic force that acts at right angles to the angular momentum of the electron gives rise to a precession about the direction of $\vec{B}$ (the Larmor precession). Taking the average of the contributions of all the electrons, it can be shown that the effective induced dipole moment per atom is given by

$$\left\langle \vec{m} \right\rangle = -\frac{q^2}{6m_e} Z r_0^2 \vec{B} \tag{9.39}$$

where Z is the atomic number and r_0^2 is the mean square radius of the electron orbits. The magnetization of the diamagnetic material is then given by

$$\vec{M} = N \left\langle \vec{m} \right\rangle = -\frac{N^2 q^2}{6m_e} Z r_0^2 \vec{B} \tag{9.40}$$

where N is the number of atoms per unit volume. The corresponding susceptibility is

$$\chi_m = \frac{\vec{M}}{\vec{H}} \approx \mu_0 \frac{\vec{M}}{\vec{B}} \approx -\frac{\mu_0 q^2 N}{6m_e} Z r_0^2 \tag{9.41}$$

For a mole of solid diamagnetic occupying 10 cm³, $N = 6 \times 10^{28}$ cm⁻³, and if we put $Z = 29$, $\left\langle r_0^2 \right\rangle = a_0$ the Bohr radius, we find $\chi_m \approx -40 \times 10^{-6}$, which is a similar order of magnitude to χ_m measured for copper (-10×10^{-6}) and glass (-110×10^{-6}).

The field $\vec{B}$ in equations (9.39) and (9.40) is strictly speaking the local magnetic field at the position of the atom, i.e., the total field less the contribution of the atom itself. However, for nonferromagnetic materials, the atoms have only a very small magnetic effect—for most practical purposes the local field can be taken to be the same as the macroscopic field $\vec{B}$ in the absence of the sample.

b. Paramagnetism

According to the discussion above, all materials should exhibit diamagnetic behavior. But there are many materials whose measured susceptibilities are positive. Such materials are called paramagnetic materials. In such materials, each atom has a small magnetic moment composed of the vector sum of the orbital and spin moments of all its electrons; the applied field exerts a torque on the resultant atomic moment. Consequently, there is a tendency to align the moment with the field. This behavior is similar to the rotation experienced by a rigid current loop placed in a field of $\vec{B}$. The torque on the loop is such as to rotate the loop and bring the normal to the loop into alignment with $\vec{B}$. As a result of the orientation effect, the field of the current whirl adds to the external field. Briefly, this is the origin of paramagnetism.

There are other considerations. The circulating electron has angular momentum about the nucleus, and the magnetic torque gives rise to a precessional motion about the field rather than to complete alignment along the field. In addition, according to the laws of atomic physics, the magnetic moment vector, which is proportional to the angular momentum, may assume only a discrete set of orientations relative to the field. Moreover, in an assemblage, the individual atoms possess thermal motions that tend to destroy the alignment effects. However, a magnetic dipole aligned with the field is in stable equilibrium, and the potential energy of the dipole is at a minimum. At room temperature and in a weak field, a relatively small number of atomic moments are aligned. From statistical analysis, it may be shown that the probability of finding any one atom having energy U due to orientation in the field is proportional to $e^{-U/kT}$, where k is Boltzmann's constant and T is the absolute temperature. The torque exerted by the field on an atom of permanent magnetic moment $\vec{m}$ is $\vec{m} \times \vec{B}$ and within an additive constant the potential energy $U = -\vec{m} \cdot \vec{B}$ or $-mB\cos\theta$, where θ is the angle between m and B. Hence, the probability of finding a given moment at an angle θ in a fixed field $\vec{B}$ and at a constant temperature T is given by

$$e^{(mB/kT)\cos\theta} \tag{9.42}$$

From this we can calculate the average value of $(m\cos\theta)$, the component of the moment along $\vec{B}$. If the constant $(mB/kT) \ll 1$, we find that

$$(m\cos\theta)_{ave} = \frac{m^2 B}{3kT} \tag{9.43}$$

The inequality used in obtaining the result in (9.43) is valid for paramagnetic materials in weak fields and at sufficiently high temperatures. The magnetization of a paramagnetic with N atoms per unit volume is therefore

$$\vec{M} = Nm^2 \vec{B}/3kT \tag{9.44}$$

and the corresponding χ_m is

$$\chi_m = \vec{M}/\vec{H} \approx \mu_0 Nm^2/3kT \tag{9.45}$$

This is known as Curie's law and shows that χ_m increases as $1/T$ at low temperature, in contrast to the diamagnetic, where χ_m is independent of temperature.

Again the field $\vec{B}$ should strictly speaking be the local field $\vec{B}_{\text{loc}}$. As for diamagnetic materials, however, the two fields are nearly the same and for most practical purposes they do not need to be distinguished.

The total magnetization of a paramagnetic material must include the induced moments, and adding their contribution to equation (9.43):

$$\vec{M} = N \left\{ \frac{m^2}{3kT} - \frac{q^2}{6m_e} Z r_0^2 \right\} \vec{B} \tag{9.46}$$

9.10 Ferromagnetic Materials

Ferromagnetism is a property of the crystalline state. Isolated atoms of iron, for example, are only paramagnetic. On the other hand, solid formed from irons or from mixtures of aluminum, copper, and manganese exhibit ferromagnetism. We cannot give a semiclassical description of ferromagnetism, because its large magnetization results not from the orbital but from the magnetic moment associated with the unpaired electron spins and so is inherently quantum mechanical.

Ferromagnetic materials not only exhibit abnormally high values of the magnetization, they are also not linear, so $\vec{M} = \chi_m \vec{H}$ and $\vec{B} = \mu \vec{H}$ with constant χ_m and μ do not apply. It is customary to retain the forms of the definitions, however, because χ_m and μ are, in general, functions of $\vec{H}$.

Ferromagnetic materials also display saturation and hysteresis effects. Consider an unmagnetized sample of ferromagnetic material. If the magnetic intensity, initially zero, is increased monotonically, then the $\vec{B} - \vec{H}$ relationship traces out a curve something like that shown in **Figure 9.15**, which is the magnetization curve of the material. It is evident that μ's taken from the magnetization curve, using the expression $\mu = \vec{B}/\vec{H}$, are always of the same sign (positive), but they show a rather large spectrum of values. The maximum permeability occurs at the "knee" of the curve. In some materials, this maximum permeability is as large as $10^5 \mu_0$, but in others it is much lower. The reason for the knee in the curve is that the magnetization $\vec{M}$ approaches a maximum value in the material, and

$$\vec{B} = \mu_0 (\vec{H} + \vec{M})$$

continues to increase at very large $\vec{H}$ only because of the $\mu_0 \vec{H}$ term. The maximum value of $\vec{M}$ is called the saturation magnetization of the material.

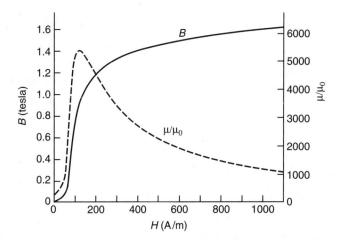

I FIGURE 9.15

In addition to the saturation effect just referred to, ferromagnetic materials display hysteresis effects. If, upon reaching a particular value of B on the B–H curve, H is decreased, B does not follow back down the B–H curve of **Figure 9.16**. Instead, B decreases more slowly and moves along the new curve of Figure 9.16, where it reaches the value B_R when $H = 0$. B_R is called the retentivity. The remarkable feature is that B does not vanish as H vanishes. If H is increased in magnitude in the reverse direction, B is reduced to

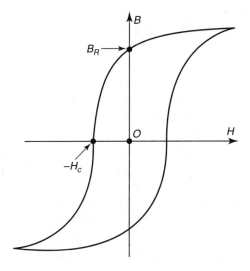

I FIGURE 9.16

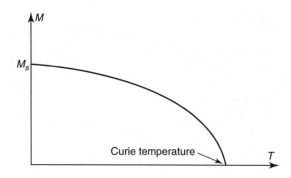

I FIGURE 9.17

zero at $H = -H_c$. H_c is termed the coercive force or coercivity of the material. Further variations in H lead to the values of B shown in the loop of Figure 9.16. This phenomenon is called hysteresis, from a Greek verb meaning "to lag"; the lagging of B behind H for decreasing and increasing values of H. The curve of Figure 9.16 is called the hysteresis loop of the material. The loop is symmetric with respect to the origin O.

The hysteresis loop furnishes information about the properties of ferromagnetic materials. For instance, in making a permanent magnet, one would select a material with a large retentivity and a large coercive force. After the exciting field has been removed, the residual magnetization is high, and large external fields are required to demagnetize the specimen.

Ferromagnetism is a strongly temperature-dependent property. As the temperature is increased, the magnetization decreases gradually, and the material becomes demagnetized at the Curie temperature. Above the Curie temperature, ordinary paramagnetic behavior results. The behavior is shown schematically in **Figure 9.17**.

In electromagnets, transformers, and other applications, ferromagnetic materials are used to enhance the magnetic flux, and the behavior of the material again can be predicted by a study of the loop. In transformer cores, where H varies periodically, heat is generated in the ferromagnetic material as a consequence of the hysteresis effect. The area enclosed by the hysteresis loop is a measure of the heat loss per cycle. To minimize such losses, the material must possess a narrow loop.

A group of substances known as ferrites (mixtures of the oxides of Fe, Co, Ni, Mn, etc.) are of practical importance because they exhibit a large magnetization and at the same time are poor conductors. The ferrites are useful in high-frequency applications where low-resistance magnetic materials cannot be used because of large eddy current losses.

9.11 Maxwell Equations in Matter

After having studied the electric and magnetic properties of matter, we want to know whether the Maxwell equations given in Chapter 6 still apply in matter. They are complete and correct, as long as we take care that the total current density and the total charge density are properly applied. When we are working with materials that are subject to electric and magnetic polarization, there is a more convenient way to write Maxwell equations. To see how to rewrite them, let us first copy down Maxwell equations in vacuum:

$$\nabla \cdot \vec{E} = \rho_t / \varepsilon_0$$

$$\nabla \cdot \vec{B} = 0$$

$$\nabla \times \vec{E} + \frac{\partial \vec{B}}{\partial t} = 0$$

$$\nabla \times \vec{B} - \frac{1}{c^2} \frac{\partial \vec{E}}{\partial t} = \mu_0 \vec{j}_t$$

where, as we have learned in this chapter and in Chapter 6, $\rho_t = \rho_f + \rho_b$ is the total electric charge density in coulomb/meter^3, ρ_f is the free charge density, and $\rho_b = - \nabla \cdot \vec{P}$ is the bound charge density and $\vec{P}$ is the electric polarization in coulomb/m^2. And $\vec{j}_t = \vec{j}_f + \vec{j}_b + \vec{j}_M$ is the total current density in ampere/meter^2; $\vec{j}_f$ is the current density of free charge; $\vec{j}_b = \partial \vec{P} / \partial t$ is the polarization current density; and $\vec{j}_M = \nabla \times \vec{M}$ is the equivalent current density in magnetized matter, $\vec{M}$ is the magnetization in ampere/meter.

The first equation of the Maxwell equations (Gauss' law) can now be written as

$$\nabla \cdot \vec{E} = \frac{1}{\varepsilon_0} (\rho_f - \nabla \cdot \vec{P})$$

or as

$$\nabla \cdot \vec{D} = \rho_f \tag{9.47}$$

where $\vec{D}$, as in the static case, is given by

$$\vec{D} = \varepsilon_0 \vec{E} + \vec{P} \tag{9.48}$$

Meanwhile, the fourth equation (Ampere's law with Maxwell's displacement current) becomes

$$\nabla \times \vec{B} = \mu_0 \left(\vec{j}_f + \nabla \times \vec{M} + \frac{\partial \vec{P}}{\partial t} \right) + \frac{1}{c^2} \frac{\partial \vec{E}}{\partial t}$$

or

$$\nabla \times \vec{H} = \vec{j}_f + \frac{\partial \vec{D}}{\partial t} \tag{9.49}$$

where, as before,

$$\vec{H} = \frac{1}{\mu_0} \vec{B} - \vec{M} \tag{9.50}$$

The second equation (Gauss' law for magnetic field) and the third equation (Faraday's law) remain the same. Thus, in terms of free charges and current, Maxwell equations read

$$
\left.
\begin{aligned}
&(i)\ \nabla \cdot \vec{D} = \rho_f, \qquad (iii)\ \nabla \times \vec{E} = -\frac{\partial \vec{B}}{\partial t}, \\
&(ii)\ \nabla \cdot \vec{B} = 0, \qquad (iv)\ \nabla \times \vec{H} = \vec{j}_f + \frac{\partial \vec{D}}{\partial t}
\end{aligned}
\right\} \tag{9.51}
$$

Note that Maxwell equations now contain both $\vec{E}$ and $\vec{D}$ and both $\vec{B}$ and $\vec{H}$. They must be supplemented by appropriate constitutive relations that give $\vec{D}$ and $\vec{H}$ in terms of $\vec{E}$ and $\vec{B}$. For linear media, we have

$$\vec{P} = \varepsilon_0 \chi \vec{E}, \quad \text{and} \quad \vec{M} = \chi_m \vec{H} \tag{9.52}$$

Hence

$$\vec{D} = \varepsilon \vec{E}, \quad \text{and} \quad \vec{H} = \vec{B}/\mu \tag{9.53}$$

where $\varepsilon = \varepsilon_0(1 + \chi)$ and $\mu = \mu_0(1 + \chi_m)$.

Problems

1. The electron in the lowest energy state in a hydrogen atom moves in such a way that its angular momentum about the nucleus is quantized, i.e., it is equal to $h/2\pi$, where h is Planck's constant. Calculate the magnetic moment.

2. Diamagnetism results from changes in the orbital motions resulting from the force of the local magnetic field on the moving electrons. To get an idea of its magnitude, we consider an electron in an atom

orbiting the nucleus on a circle of constant radius r with angular velocity ω. Show that if a magnetic field $\vec{B}(t)$ is gradually applied, normal to the plane of the orbit, the induced *emf* around the orbit changes ω by an amount $\Delta\omega = eB/2m_e$, assuming $\Delta\omega \ll \omega$. Show that the change in the acceleration of the electron toward the nucleus is given by the Lorentz force $e\vec{v} \times \vec{B}$.

Show that the orbital magnetic moment of the electron changes by

$$\Delta m = (e^2 r^2 / 4m_e)B$$

This result can also be derived from Lenz's law; show it explicitly.

3. Find the magnetization current distribution of a uniformly magnetized sphere, with magnetization M_0.

4. A sphere of radius a is magnetized nonuniformly so that the magnetization is parallel to the z-axis and proportional to the distance from the axis with proportionality constant C. Find the magnetization current distribution.

5. An infinitely long cylinder has uniform magnetization $\vec{M}$ parallel to the axis of the cylinder. Find the $\vec{B}$-field and $\vec{H}$-field outside and inside the cylinder.

6. Given a sphere of radius a that has a constant magnetization $\vec{M}$, find the $\vec{B}$-field and $\vec{H}$-field outside and inside the sphere.

7. A long solenoid is filled with a linear magnetic material. Find $\vec{B}$ and $\vec{H}$ at points on the axis near the middle.

8. In the vicinity of point A (**Figure 9.18**) on a magnetic–vacuum interface, the magnetic field in the vacuum is $\vec{B}_0$, which makes an angle α_0 normal to the interface. The permeability of the magnetic is μ. Find the magnetic field $\vec{B}$ in the magnetic in the vicinity of point A.

9. Equation (9.38) shows the change in angular velocity predicted by Larmor frequency (9.36). To find the magnetization, equation (9.38) must be summed over all electrons in a unit volume. For a

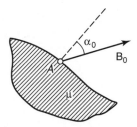

I FIGURE 9.18

substance containing N atoms per unit volume, all the same atomic species, show that

$$\vec{M} = -\frac{Ne^2\mu_0}{4m_e}\vec{B}\sum_i r_i^2$$

10. A cylindrical specimen of iron of length 10 cm and diameter 5 cm is very nearly uniformly magnetized parallel to its axis. It has a dipole moment of 75 $A \cdot m^2$. Estimate the magnetic field $\vec{B}$ inside iron.

11. Estimate the magnetic field $\vec{B}$ at a point P on the axis of a long solenoid of length L, filled with a rod whose relative permeability may be taken to have the constant value μ. The total number of turns on the solenoid is N_t and the current I. The point P is outside the solenoid, and the diameter of the nearest end of the solenoid subtends an angle 2β at P; the diameter of the farthest end subtends an angle 2α at P.

10 Relativity and Electromagnetism

When electric and magnetic fields are static, they can be considered separately. In other cases, electric and magnetic fields must always be considered together as a single total electromagnetic field. Relativity throws much light on the nature of electromagnetic fields. We shall see that magnetic fields result naturally from relativistic Lorentz transformations of electric fields. So let us first review the essentials of special relativity.

A. Elements of Special Theory of Relativity

10.1 Space and Time Before Einstein

Before Einstein, the concept of space and time were those described by Galileo and Newton. In any unaccelerated frame of reference, called the inertial reference frame, Newton's laws of motion are valid, especially the first law, which states that free objects maintain a state of uniform motion. Time was assumed to have an absolute or universal nature, in the sense that any two inertial observers who have synchronized their clocks will always agree on the time of any event. An event is any happening that can be given a space and time coordinate.

In an inertial frame S an event at P has the Cartesian coordinates x, y, z and a time coordinate t, so that (x, y, z, t) locates the event in space and time in S frame. In a different inertial frame S', the same event has space–time coordinates (x', y', z', t'). The physical nature of space and time is expressed by the relation between $((x, y, z, t)$ and (x', y', z', t'). As shown in **Figure 10.1**, S' frame is oriented in the same way as S and is moving relative to S at constant velocity $\vec{v}$ in some arbitrary direction. Without loss of any generality, the origins of the two frames coincide at $t = t' = 0$. An inspection of Figure 10.1 gives

$$\vec{r}' = \vec{r} - \vec{v}t, \quad t' = t \tag{10.1}$$

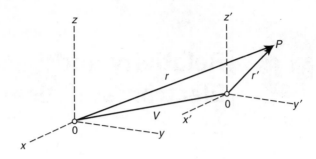

I FIGURE 10.1

This is called the Galilean transformation of coordinates. It is quite symmetric between S and S'. An inversion of the transformational relationship is obtained by simply replacing $\vec{v}$, the velocity of S' relative to S, by $-\vec{v}$, the velocity of S relative to S'.

Differentiating equation (10.1) once, we get

$$\dot{\vec{r}}' = \dot{\vec{r}} - \vec{v} \tag{10.2}$$

Thus, if the velocity $\dot{\vec{r}}'$ of a body at P in S frame is a constant, then the velocity $\dot{\vec{r}}'$ in S' is also a constant. Differentiating the velocity relation once, we get

$$\ddot{\vec{r}}' = \ddot{\vec{r}}$$

which says that the acceleration of P does not change under Galilean transformation, and we say that it is a *Galilean invariant*.

The inertial mass m is also a Galilean invariant. If Newton's laws of motion are valid in all inertial frames of reference, forces must also be Galilean invariants and we then have

$$\vec{F} = m\vec{a} = m\vec{a}' = \vec{F}'$$

The force between two particles is Galilean invariant if it depends only on the relative positions or relative velocities of the particles, because from equations (10.1) and (10.2) $\Delta\vec{r} = \vec{r}_1 - \vec{r}_2$ and $\Delta\dot{\vec{r}} = \dot{\vec{r}}_1 - \dot{\vec{r}}_2$ are both Galilean invariants.

10.2 The Search of Ether

The Galilean transformation asserts that any one inertial frame is as good as another one describing the laws of classical mechanics. However, physicists of the 19th century were not able to grant the same freedom to electromagnetic theory, which did not seem to be Galilean invariant. For example, electromagnetism of Maxwell predicts the velocity of light is a constant, independent of the motion of the source and the observer. Now, a source at rest in an inertial frame S emits a light wave, which travels out as a spherical wave at a constant speed (3×10^5 km/sec). But observed in a frame S' moving uniformly with respect to S, the light wave is no longer spherical and the speed of light is also different. Therefore, for electromagnetic phenomena, inertial frames of reference are not equivalent. Does this suggest the existence of a preferred frame of reference in which Maxwell equations are valid? The idea of a preferred frame of reference is foreign to mechanics. So a number of attempts were proposed to resolve this conflict.

Today we know that Maxwell equations are correct and have the same form in all inertial reference frames. There is some transformation other than the Galilean transformation that makes both electromagnetic and mechanical equations transform in an invariant way. But this proposal was not accepted without resistance. Toward the latter half of the 19th century most physicists were seriously interested in searching for the existence of the ether. Due to the works of Young and Fresnel, light was viewed as a mechanical wave; its propagation therefore required a physical medium that was called the ether. Because light can travel through space, it was assumed that ether must fill all of space and the velocity of light must be measured with respect to the stationary ether.

It is interesting to note that the hypothetical ether has to be assumed to have contradictory mechanical properties: It is the softest and also the hardest substance. It must be assumed softest because all material bodies can pass through it without any resistance from the ether. Otherwise, for example, the earth would have slowed down and fallen into the sun during the billions of years of its traveling around the sun. On the other hand, it must be harder than any material because light (ether vibration) travels with such a high speed that its elastic constant must be highest of all known materials. Such contradictions did not prevent physicists of the 19th century from clinging to their belief in the hypothetical ether.

The existence of such ether would imply a preferred inertial frame of reference, the one at rest relative to the ether. In all other initial frames that move with a constant velocity relative to the ether, measurement and formulation of physical laws would mix both the ether and the effect of motion relative to the ether. Accordingly, the physical laws would appear different in different inertial frames. Only the preferred frame of reference would reveal the true nature of the physical laws.

The existence of ether and the law of velocity addition (according to the Galilean transformation) suggest that it should be possible to detect some variation of the speed of light as emitted by some terrestrial source. As the earth travels through space at 30 km/sec in an almost circular orbit around the sun, it is bound to have some relative velocity with respect to the ether. If this relative velocity is added to that of the light emitted from the source, then light emitted simultaneously in two perpendicular directions should be traveling at different speeds, corresponding to the two relative velocities of the light with respect to the ether.

In one of the most famous and important experiments in physics, in 1887 Michelson set out to detect this variation in the velocity of propagation of light. Michelson's ingenious way of doing this depends on the phenomenon of interference of light to determine whether the time taken for light to pass over two equal paths at right angles was different or not.

10.3 The Michelson-Morley Experiment

The Michelson-Morley experiment was performed with an interferometer, like the one shown schematically in **Figure 10.2**. The interferometer is essentially comprised of a light source S, a half-silvered glass plate A, and two mirrors B and C, all mounted on a rigid base. The two mirrors are placed at equal distances L from the plate A. Light from S enters A and splits into two beams. One goes to mirror B, which reflects it back; the other beam goes to mirror C, also to be reflected back. On arriving back at A, the two reflected beams are recombined as two superposed beams, D and F, as indicated. If the time taken for light from A to B and back equals the time from A to C and back, the two beams D and F will be in phase and will reinforce each other. But if the two times differ slightly, the two beams will be slightly out of phase and interference will result.

We now calculate the two times to see whether they are the same or not. We first calculate the time required for the light to go from A to B and back. If the line AB is parallel to the earth's motion in its orbit and if the earth is moving at a speed u and the speed of light in the ether is c, the time is

$$t_1 = \frac{L_{AB}}{c-u} + \frac{L_{AB}}{c+u} = \frac{2L_{AB}}{c[1-(u/c)^2]} \approx \frac{2L_{AB}}{c}\left(1 + \frac{u^2}{c^2}\right) \tag{10.3}$$

where $(c-u)$ is the upstream speed of light with respect to the apparatus and $(c+u)$ is the downstream speed.

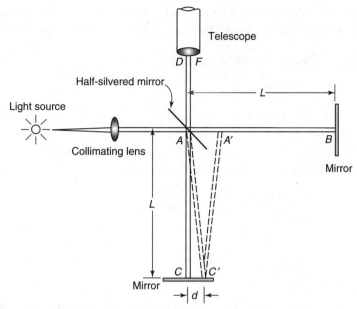

I FIGURE 10.2

Our next calculation is of the time t_2 for the light to go from A to C. We note that while light goes from A to C, the mirror C moves to the right relative to the ether through a distance $d = ut_2$ to the position C'; at the same time the light travels a distance ct_2 along AC'. For this right triangle we have

$$(ct_2)^2 = L_{AC}^2 + (ut_2)^2$$

from which we obtain

$$t_2 = \frac{L_{AC}}{\sqrt{c^2 - u^2}}$$

Similarly, while the light is returning to the half-silvered plate, the plate moves to the right to the position B'. The total path length for the return trip is the same, as can be seen from the symmetry of Figure 10.1. If the return time is therefore also the same, the total time for light to go from A to C and back is then $2t_2$, which we denote by t_3:

$$t_3 = \frac{2L_{AC}}{\sqrt{c^2 - u^2}} = \frac{2L_{AC}}{c\sqrt{1 - (u/c)^2}} \approx \frac{2L_{AC}}{c}\left(1 + \frac{u^2}{2c^2}\right) \tag{10.4}$$

In equations (10.3) and (10.4) the first factors are the same and represent the time that would be taken if the apparatus were at rest relative to the ether.

The second factors represent the modifications in the times caused by the motion of the apparatus. Now the time difference Δt is

$$\Delta t = t_3 - t_1 = \frac{2\,(L_{AC} - L_{AB})}{c} + \frac{L_{AC}}{c}\beta^2 - \frac{2L_{AB}}{c}\beta^2 \tag{10.5}$$

where $\beta = u/c$.

It is most likely that we cannot make $L_{AB} = L_{AC} = L$ exactly. In that case we can rotate the apparatus 90 degrees, so that AC is in the line of motion and AB is perpendicular to the motion. Any small difference in length becomes unimportant. Now we have

$$\Delta t' = t_3{}' - t_1{}' = \frac{2\,(L_{AB} - L_{AC})}{c} + \frac{2L_{AB}}{c}\beta^2 - \frac{2L_{AC}}{c}\beta^2 \tag{10.6}$$

Thus,

$$\Delta t' - \Delta t = \frac{(L_{AB} + L_{AC})}{c}\beta^2 \tag{10.7}$$

This difference yields a shift in the interference pattern across the cross-hairs of the viewing telescope. If the optical path difference between the beams changes by one λ (wavelength), for example, there will be a shift of one fringe. If δ represents the number of fringes moving past the cross-hairs as the pattern shifts, then

$$\delta = \frac{c\,(\Delta t' - \Delta t)}{\lambda} = \frac{L_{AB} + L_{AC}}{\lambda}\beta^2 = \frac{\beta^2}{\lambda/(L_{AB} + L_{AC})} \tag{10.8}$$

In the Michelson-Morley experiment of 1887, the effective length L was 11 m, and sodium light of $\lambda = 5.9 \times 10^{-5}$ cm was used. The orbit speed of the earth is 3×10^4 m/sec, so $\beta = 10^{-4}$. From equation (10.8) the expected shift would be about four-tenths a fringe:

$$\delta = \frac{22m \times (10^{-4})^2}{5.9 \times 10^{-5}} = 0.37 \tag{10.9}$$

A shift of 0.005 fringes can be detected by the Michelson-Morley interferometer. However, no fringe shift in the interference pattern was observed. Thus, no effect at all due to the earth's motion through the ether was found. This null result was very puzzling and most disturbing at the time. How could it be? It was suggested by some, including Michelson, that the ether might be dragged along by the earth, eliminating or reducing the ether wind in the laboratory. This is hard to square with the picture of the ether as an all-pervasive frictionless medium. The ether's status as an absolute reference frame was also gone forever. Many attempts to save the ether failed (see

Basic Concepts in Relativity, by Resnick and Halliday, Wiley, 1985); we just mention one here, namely the contraction hypothesis.

In 1892 George Francis Fitzgerald pointed out that a contraction of bodies along the direction of their motion through the ether by a factor $(1 - u^2/c^2)^{1/2}$ would give the null result. Because equation (10.3) must be multiplied by the contraction factor $(1 - u^2/c^2)^{1/2}$, equation (10.4) reduces to zero. The magnitude of this time difference is completely unaffected by rotation of the apparatus through 90 degrees.

Lorentz obtained a contraction of this sort in his theory of electrons. He found the field equations of electron theory remain unchanged if a contraction by the factor $(1 - u^2/c^2)^{1/2}$ takes places, provided also that a new measure of time is used in a uniformly moving system. The outcome of the Lorentz theory is that an observer will observe the same phenomena, no matter whether he is at rest in the ether or moving with velocity. Thus, different observers are equally unable to tell whether they are at rest or moving in the ether. This means that for optical phenomena, just as for mechanics, ether is unobservable.

Poincare offered another line of approach to the problem. He suggested that the result of the Michelson-Morley experiment was a manifestation of a general principle that absolute motion cannot be detected by laboratory experiments of any kind and the laws of nature must be the same in all inertial frames.

10.4 The Postulates of the Special Theory of Relativity

Einstein realized the full implications of the Michelson-Morley experiment, the Lorentz theory, and Poincare's principle of relativity. Instead of trying to patch up the accumulating difficulties and contradictions connected with the notion of ether, Einstein rejected the ether idea as unnecessary or unsuitable for the description of the physical world and returned to the pre-ether idea of a completely empty space. Along with the exit of ether from the stage of physics, gone also was the notion of absolute motion through space. The Michelson-Morley experiment proves unequivocally that no such special frame of reference exists. All frames of reference in uniform relative motion are equivalent, for mechanical motions and also for electromagnetic phenomena. Einstein extended this as a fundamental postulate, now known as the principle of relativity. Furthermore, he argued that the velocity of light predicted by electromagnetic theory, c, must be a universal constant, the same for all observers. He took an epoch-making step in 1905 and developed the Special Theory of Relativity from these two basic postulates (assumptions), which are rephrased as follows:

1. **The laws of physics are the same in all inertial frames. No preferred inertial frame exists (the principle of relativity).**

    Two people standing in the aisle of an airplane going 700 km/hr can play catch exactly as they would on the ground. If you drop a heavy and light ball together, they fall at the same rate as they would if you drop them on the ground and they hit the cabin floor at the same time. When you are moving uniformly, you do not experience physical sensation of motion. Experiments indicate that the principle of relativity also applies to electromagnetism; it is very general. Radios and tape recorders working the same on an airplane as in the house is a simple example.

2. **The velocity of light in empty space is the same in all inertial frames and is independent of the motion of the emitting body (the principle of the constancy of the velocity of light).**

    According to Einstein, sometime in 1896, after he entered the Zurich Polytechnic Institute to begin his education as a physicist, he asked himself the question of what would happen if he could catch up to a light ray—that is, move at the speed of light. Maxwell's theory says that light is a wave of electric and magnetic fields that move like a water wave through space. But if you could catch up to one of Maxwell's light waves, the way a surfboard rider catches an ocean wave for a ride, then the light wave would not be moving relative to you but instead would be standing still. The light wave would then be a standing wave of electric and magnetic fields, which is not allowed if Maxwell's theory is correct. So, he reasoned, there must be something wrong with the assumption that you can catch a light wave as you can catch a water wave. This idea was a seed from which the fundamental postulate of the constancy of the speed of light and the special theory of relativity grew 9 years later.

All the seemingly very strange results of special relativity came from the special nature of the speed of light. Once we understand this, everything else in relativity makes sense. So we take a brief look at this special nature of the speed of light. The speed of light is very great, 186,000 miles per second. But the bizarre fact of the speed of light is that it is independent of the motion of the observer or the source emitting the light. Michelson hoped to determine the absolute speed of the earth through ether by measuring the difference in times required for light to travel across equal distances that are at right angles to each other. What did he observe? No difference in travel times for the two perpendicular light beams. It was as if the earth were absolutely stationary. The conclusion is that the speed of light does not depend on the motion of the observer.

This bizarre nature is not something we would expect from common sense. The same common sense was once objecting to the idea that the earth is round. Hence, common sense is not always right!

How do we know the speed of light is independent of the motion of the light source? There are many binary star systems in our galaxy, in which two stars revolve about their common center of mass. If the speed of light depended on the motion of the source, then the light emitted by the two stars in a binary system would have different speeds toward the earth, as shown in **Figure 10.3**. The orbit is roughly on the edge of our line of sight, and v is its orbital speed about the center of mass of the system. If the distance to the binary system were right, we could receive light from the star at position A, at the same time as the light sent to us at a slower speed at an earlier time, when the star was at position B. Thus, under some circumstances we could see the same star in a binary system at many different places in its orbit at once, and there would be multiple images or spread out images. But, in fact, we always see binary stars moving in a well-behaved elliptical orbit about each other. We therefore conclude that the motion of its source (the emitter) does not affect the speed of light.

a. Time Is Not Absolute

The constancy of the velocity of light puts an end to the notion of absolute time. We know that Newtonian mechanics abolished the notion of absolute space, but not that of absolute time. Now time is also not absolute anymore, because all inertial observers must agree on how fast light travels but not how far light travels, as space is not absolute. Now time taken is the distance light has traveled divided by the speed of light; thus, they must disagree about the time the journey took. Time had also lost its universal nature. In fact, we see later that moving clocks run slow. This is known as time dilation.

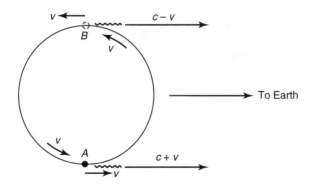

I FIGURE 10.3

10.5 The Lorentz Transformations

Because the Galilean transformation is inconsistent with Einstein's postulate of the constancy of the speed of light, we must modify it in such a way that the new transformation incorporates Einstein's two postulates and makes both mechanical and electromagnetic equations transform in an invariant way.

To this end we consider two inertial frames, S and S'. For simplicity, let the corresponding axes of the two frames be parallel, with frame S' moving at a constant velocity u relative to S along the x_1-axis. The apparatuses used to measure distances and times in the two frames are assumed identical, and both clocks are adjusted to read zero at the moment the two origins coincide. **Figure 10.4** represents the viewpoint of observers in S.

Suppose an event that occurred in frame S at the coordinates (x, y, z, t) is observed at (x', y', z', t') in frame S'. Because of the homogeneity of space and time, we expect the transformation relations between the coordinates (x, y, z, t) and (x', y', z', t') to be linear, for otherwise there would not be a simple one-to-one relation between events in S and S' frames. For instance, a nonlinear transformation would predict acceleration in one system even if the velocity were constant in the other, obviously an unacceptable property for a transformation between inertial systems.

We consider the transverse dimensions first. Because the relative motion of the coordinate systems occurs only along the x-axis, we expect the linear relations are of the forms $y' = k_1 y$, $z' = k_2 z$. The symmetry requires that $y = k_1 y'$ and $z = k_2 z'$. These can both be true only if $k_1 = 1$ and $k_2 = 1$. Therefore, for the transverse direction we have

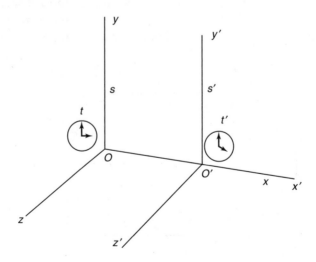

I FIGURE 10.4

$$y' = y, \quad z' = z \tag{10.10}$$

These relations are the same as in Galilean transformations.

Along the longitudinal dimension, the relation between x and x' must depend on the time, so we consider the most general linear relation

$$x' = ax + bt \tag{10.11}$$

Now the origin O', where $x' = 0$, corresponds to $x = ut$. Substituting these into equation (10.11), we have

$$0 = aut + bt$$

from which we obtain

$$b = -au$$

and equation (10.11) simplifies to

$$x' = a(x - ut) \tag{10.12}$$

By symmetry, we also have

$$x = a(x' + ut') \tag{10.13}$$

Now we apply Einstein's second postulate of the constancy of the speed of light. If a pulse of light is sent out from the origin O of frame S at $t = 0$, its position along the x-axis later is given by $x = ct$ and its position along x'-axis is $x' = ct'$. Putting these in equations (10.12) and (10.13), we obtain

$$ct' = a(c - u)t \quad \text{and} \quad ct = a(c + u)t'$$

From these we obtain

$$\frac{t}{t'} = \frac{c}{a(c-u)} \quad \text{and} \quad \frac{t}{t'} = \frac{a(c+u)}{c}$$

Therefore

$$\frac{c}{a(c-u)} = \frac{a(c+u)}{c}$$

Solving for a:

$$a = \frac{1}{\sqrt{1 - (u/c)^2}}$$

then

$$b = -au = -\frac{u}{\sqrt{1 - (u/c)^2}}$$

Substituting these in equations (10.12) and (10.13) gives

$$x' = \frac{x - ut}{\sqrt{1 - \beta^2}} \tag{10.14a}$$

and

$$x = \frac{x' + ut'}{\sqrt{1 - \beta^2}} \tag{10.14b}$$

where $\beta = u/c$. Eliminating either x or x' from equations (10.14a) and (10.14b), we obtain

$$t' = \frac{t - ux/c^2}{\sqrt{1 - \beta^2}} \tag{10.14c}$$

and

$$t = \frac{t' + ux'/c^2}{\sqrt{1 - \beta^2}} \tag{10.14d}$$

Combining all of these results, we obtain the Lorentz transformations:

$$\left.\begin{aligned}
x' &= \gamma(x - ut) & x &= \gamma(x' + ut) \\
y' &= y & y &= y' \\
z' &= z & z &= z' \\
t' &= \gamma(t - ux/c^2) & t &= \gamma(t' + ux/c^2)
\end{aligned}\right\} \tag{10.15}$$

where

$$\gamma (= 1/\sqrt{1 - \beta^2}), \quad \beta = u/c \tag{10.16}$$

is the Lorentz factor.

If $\beta \ll 1$, then $\gamma \cong 1$, and equation (10.15) reduces to the Galilean transformations, which we now see as a first approximation to the Lorentz transformations for $\beta \ll 1$.

When the velocity, $\vec{u}$, of S' relative to S is in some arbitrary direction, equation (10.15) can be given a more general form in terms of the components of $\vec{r}$ and $\vec{r}'$ perpendicular and parallel to $\vec{u}$.

$$\left.\begin{array}{ll} \vec{r}_{\|}{}' = \gamma(\vec{r}_{\|} - \vec{u}t) & \vec{r}_{\|} = \gamma(\vec{r}_{\|}{}' + \vec{u}t) \\[2mm] \vec{r}_{\perp}{}' = \vec{r}_{\perp} & \vec{r}_{\perp}{}' = \vec{r}_{\perp} \\[2mm] t' = \gamma(t - \vec{u} \cdot \vec{r}/c^2) & t = \gamma(t' + \vec{u} \cdot \vec{r}/c^2) \end{array}\right\} \tag{10.17}$$

The Lorentz transformations are valid for all types of physical phenomena at all speeds. As a consequence of this, all physical laws must be invariant under a Lorentz transformation.

The Lorentz transformations that are based on Einstein's postulates contain a new philosophy of space and time measurements. We now examine the various properties of these new transformations. In the following discussion, we still use Figure 10.4.

a. Relativity of Simultaneity: Causality

Two events that happen at the same time but not necessarily at the same place are called simultaneous. Now consider two events in S' that occur at (x_1', t_1') and (x_2', t_2'), and they would appear in frame S at (x_1, t_1) and (x_2, t_2). The Lorentz transformations give

$$t_2 - t_1 = \gamma \left[(t_2' - t_1') + \frac{u(x_2' - x_1')}{c^2} \right] \tag{10.18}$$

It is easy to observe the following:

1. If the two events take place simultaneously in S', then $t_2' - t_1' = 0$. But the events do not occur simultaneously in the S frame, because there is a finite time lapse

$$\Delta t = t_2 - t_1 = \gamma \frac{u(x_2' - x_1')}{c^2}$$

Thus, two spatially separated events that are simultaneous in S' would not be measured to be simultaneous in S. In other words, the simultaneity of spatially separated events is not an absolute property, as it was assumed to be in Newtonian mechanics.

Moreover, depending on the sign of $(x_2' - x_1')$, the time interval Δt can be positive or negative, that is, in the frame S the "first" event in S' can take place earlier or later than the "second" one. The sole exception is the case when two events occur coincidentally in S', then they also occur at the same place and at the same time in frame S.

2. If the order of events in frame S is not reversed in time, then $\Delta t = t_2 - t_1 > 0$, which implies that

$$(t_2' - t_1') + \frac{u(x_2' - x_1')}{c^2} > 0$$

or

$$\frac{x_2' - x_1'}{t_2' - t_1'} < \frac{c^2}{u}$$

which will be true as long as

$$\frac{x_2' - x_1'}{t_2' - t_1'} < c$$

Thus, the order of events remains unchanged if no signal can be transmitted with a speed greater than c, the speed of light.

The preceding discussion also illustrates clearly that the theory of relativity is incompatible with the notion of action at a distance.

b. Time Dilation: Relativity of Colocality

Two events that happen at the same place but not necessarily at the same time are called colocal. Now consider two colocal events in S' taking place at times t_1' and t_2' but at the same place. For simplicity, consider this to be on the x'-axis so that $y' = z' = 0$. These two events would appear in frame S at (x_1, t_1) and (x_2, t_2). According to the Lorentz transformations, we have

$$\Delta x = \frac{u \Delta t'}{\sqrt{1 - \beta^2}} = \gamma u \Delta t', \quad \Delta t = \frac{\Delta t'}{\sqrt{1 - \beta^2}} = \gamma \Delta t' \tag{10.19}$$

where $\beta = u/c$, $\Delta t' = t_2' - t_1'$, and so forth. It is easy to observe the following:

1. The two events that happened at the same place in S' do not occur at the same place in S, and so t_1 and t_2 must be measured by spatially separated synchronized clocks. Einstein's prescription for synchronizing two stationary separated clocks is to send a light signal from clock 1 at a time t_1 (measured by clock 1) and reflected back from clock 2 at a time t_2 (measured by clock 2). If the reflected light returns to clock 1 at a time t_3 (measured by clock 1), then clocks 1 and 2 are synchronous if $t_2 - t_1 = t_3 - t_2$; that is, if the time measured for light to go one way is equal to the time measured for light to go in the opposite direction.

2. The time interval between two events taking place at the same point in an inertial reference is measured by a single clock at that point and it is called the proper time interval between two events. In the second equation of equation (10.19), $\Delta t' = t_2' - t_1'$ is the proper time interval

between the events in S'. Because $\gamma \geq 1$, the time interval $\Delta t = t_2 - t_1$ in S is longer than $\Delta t'$; this is called time dilation, often described by the statement that "moving clocks run slow." This apparent asymmetry between S and S' in time is a result of the asymmetric nature of the time measurement.

Time dilation has been confirmed by experiments on the decay of pions. Pions have a mean lifetime $T_0 = 2.6 \times 10^{-8}$ sec when they are at rest. When they are in fast motion in a synchrotron, their lifetimes become larger according to

$$T = \frac{T_0}{\sqrt{1 - \beta^2}}$$

Time dilation between observers in uniform relative motion is not an artifact of the clock we choose to construct. It is a very real thing. All processes, including atomic and biologic processes, slow down in moving systems.

We often hear the so-called twin paradox. One twin gets on a spaceship and accelerates to 0.866 c, so $\gamma = 2$. If this twin travels for 1 year, as measured by his clock, then heads back at the same speed, the moving twin would report that the trip required 2 years. But his earth-bound twin would report that the time for the journey was 4 years. Can the twin on the spaceship argue that he was at rest and it was the twin on earth who was moving? The answer is no. To make a transition from a resting frame to a moving frame and to turn around and head back home, there must be acceleration. Whichever twin feels acceleration can no longer claim that his frame is the resting frame. Thus, the twin on the spaceship cannot argue that his earth-bound twin was moving, and there is no paradox.

c. Length Contraction

Consider a rod of length L_0 in the S' frame in which the rod lies at rest along the x'-axis: $L_0 = x_2' - x_1'$. L_0 is the proper length of the rod measured in the rod's rest frame S'. Now the rod is moving lengthwise with velocity u relative to the S frame. An observer in the S frame makes a simultaneous measurement of the two ends of the rod. The Lorentz transformations give

$$x_1' = \gamma[x_1 - ut_1], \quad x_2' = \gamma[x_2 - ut_2]$$

from which we get

$$x_2' - x_1' = \gamma[x_2 - x_1] - \gamma u (t_2 - t_1)$$

Let $L(u) = x_2 - x_1$, the length of the rod moving with speed u relative to the observer in S, and remember that $t_1 = t_2$, $\gamma = 1/(1 - \beta^2)^{1/2}$. The result then becomes

$$L(u) = \sqrt{1 - \beta^2}\, L_0 \tag{10.20}$$

Thus, the length of a body moving with velocity u relative to an observer is measured to be shorter by a factor of $(1 - \beta^2)^{1/2}$ in the direction of motion relative to the observer.

Because all inertial frames of reference are equally valid, if $L' = \gamma L$, does the expression $L = \gamma L'$ have to be equally true? The answer is no. The reason is that the measurement was not carried out in the same way in the two frames of reference. The two events of marking the positions of the two ends of the rod were simultaneous in the S frame but not simultaneous in the S' frame. This difference gives the asymmetry of the result. Length L' was equal to length L_0, only because the rod was at rest in the S' frame. As a general expression, $\Delta x' = \gamma \Delta x$ is not true. The full expression relating distances in two frames of reference is $\Delta x' = \gamma (\Delta x - u \Delta t)$. The symmetric inverse relation is $\Delta x = \gamma (\Delta x' + u \Delta t')$. In the case considered earlier, $\Delta t = 0$, so $\Delta x' = \gamma \Delta x$, but $\Delta t' \neq 0$, so $\Delta x \neq \Delta x'$.

A body of proper volume V_0 can be divided into thin rods parallel to u. Each one of these rods is reduced in length by a factor $(1 - \beta^2)^{1/2}$ so that the volume of the moving body measured by an observer in S is $V = (1 - \beta^2)^{1/2} V_0$.

An interesting consequence of the length contraction is the visual apparent shape of a rapidly moving object. This was shown first by James Terrell in 1959 (see *Physics Review*, 1959, 116:1041, and *American Journal of Physics*, 1960, 28:607). The act of seeing involves the simultaneous reception of light from the different parts of the object. For light from different parts of an object to reach the eye or a camera at the same time, light from different parts of the object must be emitted at different times, to compensate for the different distances the light must travel. Thus, taking a picture of a moving object or looking at it does not give a valid impression of its shape. Interestingly, the distortion that makes the Lorentz contraction seem to disappear instead makes an object seem to rotate by an angle $\theta = \sin^{-1}(u/c)$, as long as the angle subtended by the object at the camera is small. If the object moves in another direction or if the angle it subtends at the camera is not small, the apparent distortion becomes quite complex.

Figure 10.5 shows a cube of side l moving with a uniform velocity u with respect to an observer some distance away; the side AB is perpendicular to the line of sight of the observer. For light from corners A and D to reach the

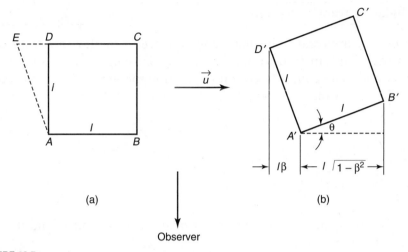

(a)

(b)

Observer

I FIGURE 10.5

observer at the same instant, light from D, which must travel a distance l farther than from A, must have been emitted when D was in position E. The length DE is equal to $(l/c)u = l\beta$. The length of the side AB is foreshortened by Lorentz contraction to $l\sqrt{1-\beta^2}$. The net result corresponds to the view the observer would have if the cube were rotated through an angle $\sin^{-1}\beta$. The cube is not distorted; it undergoes an apparent rotation. Similarly, a moving sphere does not become an ellipsoid; it still appears as a sphere. An interesting discussion of apparent rotations at high velocity is given by V. F. Weisskopf in *Physics Today* (1960, 13:24).

Length contraction opens the possibility of space travel. The nearest star, besides the sun, is Alpha Centauri, which is about 4.3 light years away: light from Alpha Centauri takes 4.3 years to reach us. Even if a spaceship can travel at the speed of light, it would take 4.3 years to reach Alpha Centauri. This is certainly true from the point of view of an observer on earth. But from the point of view of the crew of the spaceship, the distance between the earth and Alpha Centauri is shortened by a factor $\gamma = (1 - \beta^2)^{1/2}$, where $\beta = v/c$ and v is the speed of the spaceship. If v is, say, $0.99c$, then $\gamma = 0.14$ and the distance appears to be only 14% of the value as seen from the earth. The crew therefore deduces that light from Alpha Centauri takes only $0.14 \times 4.3 = 0.6$ year to reach earth. But the crew sees Alpha Centauri coming toward it with a speed of $0.99c$, so they expect to get there in $0.60/0.99 = 0.606$ years, without having to suffer a long tedious journey. In practice, however, the power requirements to launch a spaceship near the speed of light are prohibitive.

10.6 Velocity Transformation

Another very important kinematic consequence of the Lorentz transformation is that Galilean transformation (10.2) for velocity is no longer valid. The new and more complicated transformation for velocities can be deduced easily. By definition, the components of velocity in S and S' are given by, respectively,

$$v_x = \frac{dx}{dt} = \frac{x_2 - x_1}{t_2 - t_1}, \quad v_y = \frac{dy}{dt} = \frac{y_2 - y_1}{t_2 - t_1}$$

$$v_x' = \frac{dx'}{dt} = \frac{x_2' - x_1'}{t_2' - t_1'}, \quad v_y' = \frac{dy'}{dt} = \frac{y_2' - y_1'}{t_2' - t_1'}$$

and so on. Applying the Lorentz transformations to x_1 and x_2 and then taking the difference, we get

$$dx = \frac{dx' + u\,dt'}{\sqrt{1 - \beta^2}}, \quad \beta = \frac{u}{c}$$

Similarly,

$$dx' = \frac{dx - u\,dt}{\sqrt{1 - \beta^2}}$$

Do the same for the time intervals dt and dt'

$$dt = \frac{dt' + u\,dx'/c^2}{\sqrt{1 - \beta^2}}, \quad dt' = \frac{dt - u\,dx/c^2}{\sqrt{1 - \beta^2}}$$

From these we obtain

$$\frac{dx}{dt} = \frac{dx' + u\,dt'}{dt' + u\,dx'/c^2}$$

Dividing both numerator and denominator of the right side by dt' yields the right transformation equation for the x component of the velocity:

$$v_x = \frac{v_x' + u}{1 + uv_x'/c^2} \tag{10.21a}$$

Similarly, we can find the transverse components:

$$v_y = \frac{v_y' \sqrt{1 - \beta^2}}{1 + uv_x'/c^2} = \frac{v_y'}{\gamma(1 + uv_x'/c^2)} \tag{10.21b}$$

$$v_z = \frac{v_z' \sqrt{1 - \beta^2}}{1 + uv_x'/c^2} = \frac{v_z'}{\gamma(1 + uv_x'/c^2)} \tag{10.21c}$$

In these formulas, $\gamma = (1 - \beta^2)^{-1/2}$ as before. We note that the transverse velocity components depend on the x-component. For $v \ll c$, we obtain the Galilean result $v_x = v_x' + u$.

Solving explicitly or merely switching the sign of u would yield (v_x', v_y', v_z') in terms of (v_x, v_y, v_z).

It follows from the velocity transformation formulas that the value of an angle is relative and changes in transition from one reference frame to another. For an object in the S frame moving in the xy-plane with velocity v that makes an angle θ with the x-axis, we have

$$\tan\theta = v_y/v_x, \quad v_x = v\cos\theta, \quad v_y = v\sin\theta$$

In the S' frame, we have

$$\tan\theta' = \frac{v_y'}{v_x'} = \frac{v\sin\theta}{\gamma(v\cos\theta - u)} \tag{10.22}$$

where $\gamma = 1/\sqrt{1 - \beta^2}$, and $\beta = u/c$.

As an application, consider the case of star light, that is, $v = c$; then

$$\tan\theta' = \frac{\sin\theta}{\gamma(\cos\theta - u/c)}$$

Let $\theta = \pi/2$, $\theta' = \pi/2 - \phi$ (**Figure 10.6**), then

$$\tan\phi = \frac{-u/c}{\sqrt{1 - \beta^2}}, \quad \sin\phi = -\frac{u}{c}$$

which is the star aberration formula: To see a star overhead tilt the telescope at angle ϕ.

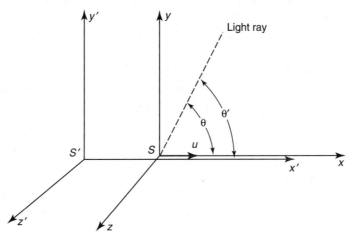

I FIGURE 10.6

10.7 The Doppler Effect

The Doppler effect occurs for light as well as for sound. It is a shift in frequency due to the motion of the source or the observer. Knowledge of the motion of distant receding galaxies comes from studies of the Doppler shift of their spectral lines. Doppler effect is also used for satellite tracking and radar speed traps. We examine the Doppler effect in light only.

Consider a source of light or radio waves moving with respect to an observer or a receiver, at a speed u and at an angle θ with respect to the line between the source and the observer (**Figure 10.7**). The light source flashes with a period τ_0 in its rest frame (the S' frame in which the source is at rest). The corresponding frequency is $\nu_0 = 1/\tau_0$, and the wavelength is $\lambda_0 = c/\nu_0 = c\tau_0$.

While the source is going through one oscillation, the time that elapses in the resting frame of the observer (the S frame) is $\tau = \gamma\tau_0$ because of time dilation, where $\gamma = (1 - \beta^2)^{1/2}$ and $\beta = u/c$. The emitted wave travels at speed c, and therefore its front moves a distance $\gamma\tau_0 c$; the source moves toward the observer with a speed $u\cos\theta$, so a distance $\gamma\tau_0 u\cos\theta$. Then the distance D that separates the fronts of the successive waves (the wavelength) is

$$D = \gamma\tau_0 c - \gamma c_0 u\cos\theta$$

i.e., $\lambda = \gamma\tau_0 c - \gamma\tau_0 u\cos\theta = \gamma\tau_0 c[1 - (u/c)\cos\theta]$, but $c\tau_0 = \lambda_0$, we can rewrite the last expression as

$$\lambda = \lambda_0 \frac{1 - \beta\cos\theta}{\sqrt{1 - \beta^2}} \tag{10.23}$$

In terms of frequency, this Doppler effect formula becomes

$$\nu = \nu_0 \frac{\sqrt{1 - \beta^2}}{1 - \beta\cos\theta} \tag{10.24}$$

Here ν is the frequency at the observer, and θ is the angle measured in the resting frame of the observer. If the source moving directly toward the observer, then $\theta = 0$ and $\cos\theta = 1$. Equation (10.24) reduces to

I FIGURE 10.7

$$v = v_0 \frac{\sqrt{1-\beta^2}}{1-\beta} = v_0 \sqrt{\frac{1+\beta}{1-\beta}} \qquad (10.24a)$$

For a source moving directly away from the observer, $\cos\theta = -1$, equation (10.24) reduces to

$$v = v_0 \frac{\sqrt{1-\beta^2}}{1+\beta} = v_0 \sqrt{\frac{1-\beta}{1+\beta}} \qquad (10.24b)$$

At $\theta = \pi/2$, i.e., the source moving at right angles to the direction toward the observer, equation (10.24) reduces to

$$v = v_0 \sqrt{1-\beta^2} \qquad (10.24c)$$

This transverse Doppler effect is due to time dilation.

10.8 Relativistic Space–Time and Minkowski Space

In our daily experience we think of a world of three dimensions. Objects in space have length, breadth, and height. We tend to think of time as being independent of space. As we have just seen above, however, there is no absolute standard for the measurement of time or of space: The relative motion of observers affects both kinds of measurement. Lorentz transformations treat x^i ($i = 1, 2, 3$) and t as equivalent variables. In 1907 Hermann Minkowski proposed that the three dimensions of space and the dimension of time should be treated together as four dimensions of space–time. Minkowski remarked: "Henceforth space by itself and time by itself, are doomed to fade away into mere shadows, and only a kind of union of the two will preserve an independent reality." He called the four dimensions of space–time "the world space," and the path of an individual particle in space–time a "world line." The four-dimension relativistic space–time is often called the Minkowski space.

It is now a common practice to treat t as a zeroth or a fourth coordinate:

$$x^0 = ct, \quad x^1 = x, \quad x^2 = y, \quad x^3 = z \qquad (10.25a)$$

or

$$x^1 = x, \quad x^2 = y, \quad x^3 = z, \quad x^4 = ix^0 \qquad (10.25b)$$

By analogy with the three-dimensional case, the coordinates of an event (x^0, x^1, x^2, x^3) can be considered as the components of a four-dimensional radius vector, for short, a radius four-vector in a four-dimensional Minkowski space. The square of the length of the radius four-vector is given by

$$(x^1)^2 + (x^2)^2 + (x^3)^2 + (x^4)^2 = -[(x^0)^2 - (x^1)^2 - (x^2)^2 - (x^3)^2]$$

It does not change under Lorentz transformations. The Lorentz transformations now take on the form

$$
\begin{aligned}
x^{1'} &= \gamma(x^1 + i\beta x^4) & x^{0'} &= \gamma(x^0 - \beta x^1) \\
x^{2'} &= x^2 & x^{1'} &= \gamma(-\beta x^0 + x^1) \\
x^{3'} &= x^3 \quad \text{or} & x^{2'} &= x^2 \\
x^{4'} &= \gamma(-i\beta x^1 + x^4) & x^{3'} &= x^3
\end{aligned}
\tag{10.26}
$$

In matrix form, we have

$$
\begin{pmatrix} x^{1'} \\ x^{2'} \\ x^{3'} \\ x^{4'} \end{pmatrix} =
\begin{pmatrix} \gamma & 0 & 0 & i\beta\gamma \\ 0 & 1 & 0 & 0 \\ 0 & 0 & 1 & 0 \\ -i\beta\gamma & 0 & 0 & \gamma \end{pmatrix}
\begin{pmatrix} x^1 \\ x^2 \\ x^3 \\ x^4 \end{pmatrix}
\quad \text{or} \quad
\begin{pmatrix} x^{0'} \\ x^{1'} \\ x^{2'} \\ x^{3'} \end{pmatrix}
\begin{pmatrix} \gamma & -\beta\gamma & 0 & 0 \\ -\beta\gamma & \gamma & 0 & 0 \\ 0 & 0 & 1 & 0 \\ 0 & 0 & 0 & 1 \end{pmatrix}
\tag{10.27}
$$

It is customary to use Greek indices (μ and ν, etc.) to label four-dimensional variables and Latin indices (i and j, etc.) to label three-dimensional variables.

The Lorentz transformations can be distilled into a single equation:

$$x^{\mu'} = \sum_{\nu=1}^{4} L^{\mu}_{\nu} x^{\nu} = L^{\mu}_{\nu} x^{\nu} \mu, \quad \nu = 1, 2, 3, 4 \tag{10.28}$$

where L^{μ}_{ν} is the Lorentz transformation matrix in equation (10.27). The summation sign is eliminated by the Einstein summation convention: The repeated indexes appearing once in the lower and once in the upper position are automatically summed over. The indexes repeated in the lower part or upper part alone, however, are not summed over.

If equation (10.28) reminds you of the orthogonal rotations, it is no accident! The general Lorentz transformations can indeed be interpreted as an orthogonal rotation of axes in Minkowski space. Note that the xt-submatrix of the Lorentz matrix in equation (10.27) is

$$\begin{pmatrix} \gamma & i\beta\gamma \\ -i\beta\gamma & \gamma \end{pmatrix}$$

which is to be compared with the xy-submatrix of the two-dimensional rotation about the z-axis,

$$\begin{pmatrix} \cos\theta & \sin\theta \\ -\sin\theta & \cos\theta \end{pmatrix}$$

Upon identification of matrix elements $\cos\theta = \gamma$, $\sin\theta = i\beta\gamma$, we see that the rotation angle θ (for the rotation in the xt-plane) is purely imaginary.

Some books prefer to use a real angle of rotation ϕ, defining $\phi = -i\theta$. Then note that

$$\cos\theta = \frac{e^{i\theta} + e^{-i\theta}}{2} = \frac{e^{-\phi} + e^{\phi}}{2} = \cosh\phi$$

$$\sin\theta = \frac{e^{i\theta} - e^{-i\theta}}{2i} = \frac{i[e^{\phi} - e^{-\phi}]}{2} = \sinh\phi$$

and the submatrix becomes

$$\begin{pmatrix} \cosh\phi & i\sinh\phi \\ -i\sinh\phi & \cosh\phi \end{pmatrix}$$

We should note that the mathematical form of Minkowski space looks exactly like a Euclidean space; however, it is not physically so because of its complex nature as compared with the real nature of the Euclidean space.

a. Four-Velocity and Four-Acceleration

How do we define four-vectors of velocity and acceleration? It is evident that the set of the four quantities dx^{μ}/dt does not have the properties of a four-vector because dt is not an invariant. But we know that the proper time $d\tau$ is an invariant. Although observers in different frames may disagree about the time interval between events, because each is using ones own time axis, all agree on the value of the time interval that would be observed in the frame moving with the particle. The components of the four-velocity are therefore defined as

$$u^{\mu} = \frac{dx^{\mu}}{d\tau} \tag{10.29}$$

The second equation of equation (10.19) relates the proper time $d\tau$ (was dt' there) to the time dt read by a clock in frame S relative to which the object (S' frame) moves at a constant u:

$$d\tau = dt\sqrt{1 - \beta^2}$$

We can rewrite u^{μ} completely in terms of quantities observed in frame S as

$$u^\mu = \gamma \frac{dx^\mu}{dt} = \frac{1}{\sqrt{1-\beta^2}} \frac{dx^\mu}{dt}$$
(10.30)

where $\gamma = 1/\sqrt{1-\beta^2}$.

In terms of the ordinary velocity components v_1, v_2, v_3, we have

$$u^\mu = (\gamma c, \gamma v_i), \quad i = 1, 2, 3$$
(10.31)

The length of four-velocity must be invariant. We can verify this easily:

$$\sum_{\mu=1}^{4} (u^\mu)^2 = -c^2$$
(10.32)

Similarly, a four-acceleration is defined as

$$w^\mu = \frac{d^2 x^\mu}{d\tau^2} = \frac{du^\mu}{d\tau}$$
(10.33)

Now differentiating equation (10.32) with respect to τ, we obtain

$$w^\mu u^\mu = 0$$
(10.34)

that is, the four-vectors of velocity and acceleration are "mutually perpendicular."

10.9 Equivalence of Mass and Energy

We have learned that Einstein's special relativity drastically revised our concepts about space and time. So, we must follow Einstein and rethink the notions of mass, energy, and other important quantities. The equivalence of mass and energy is the best-known relation Einstein gave in his special relativity in 1905:

$$E = mc^2$$
(10.35)

where E is the energy, m is the mass, and c is the speed of light.

We can get this general idea of the equivalence of mass and energy from the consideration of electromagnetic theory. Electromagnetic field possesses energy E and momentum p, and there is a simple relationship between E and p:

$$p = E/c$$

Thus, if an object emits light in one direction with momentum p, to conserve momentum, the object itself must recoil with a momentum $-p$. If we stick to

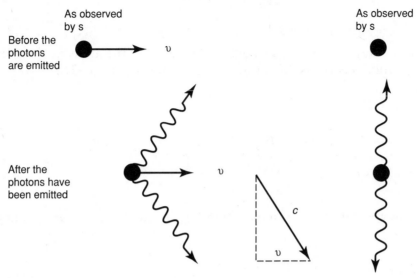

I FIGURE 10.8

the definition of momentum as $p = mv$, we may associate a "mass" with a flash of light:

$$m = \frac{p}{v} = \frac{p}{c} = \frac{E}{c^2}$$

which leads to the famous formula

$$E = mc^2$$

This mass is not merely a mathematical fiction. Let us consider a simple thought experiment. In Figure 10.4, there is an observer S in the S frame and an observer S' and an atom at rest in the S' frame. The atom emits two flashes of light (photons) of equal energy, traveling back-to-back and normal to the direction of the frames' relative motion. **Figure 10.8** depicts what S and S' observe. Before the lights are emitted, the atom is at rest in the S' frame and moves rightward with speed v as observed by observer S in the S system.

We now analyze the emission process, first from the point view of S' and then according to the point view of S. In S' frame, the sum of the momentum of the two lights is zero, and the atom remains at rest after the emission of lights. As observed by observer S, the situation is quite different. The two flashes of light move along diagonal directions, each light's component of velocity parallel to the atom's velocity is v. Thus, the momentum of the two flashes of light parallel to the atom's velocity is

$$2\frac{E}{c}\frac{v}{c}$$

We next study energy and momentum changes for the atom as observed by S. The atom's energy is decreased by an amount equal to $-2E$:

$$\Delta E_{atom} = 2E$$

and its momentum change is

$$\Delta p_{atom} = -2\frac{E}{c}\frac{v}{c} = \Delta E_{atom}\frac{v}{c^2}$$

Now the definition of momentum for the atom ($p_{atom} = m_{atm}v$) implies that

$$\Delta p_{atom} = \Delta m_{atom} \cdot v$$

Comparing the two expressions for the change in the atom's momentum, we obtain

$$\Delta E_{atom}\frac{v}{c^2} = \Delta m_{atom} \cdot v$$

which gives

$$\Delta E_{atom} = \Delta m_{atom} \cdot c^2$$

In short,

$$\Delta E = \Delta m \cdot c^2$$

Einstein also provided a thought experiment some time ago. Consider an emitter and an absorber of light A and B of equal mass M at a distance L, which are rigidly connected together, and the whole system is initially at rest but can move freely (**Figure 10.9**). The center of mass of the system is at a distance $L/2$ from A and B. If the emitter sends out a short light pulse of energy ΔE and momentum $\Delta E/c$ toward the right, the system will recoil toward the left by a small distance d, with a momentum p, where

$$p = 2Mv = -\Delta E/c$$

From which we obtain

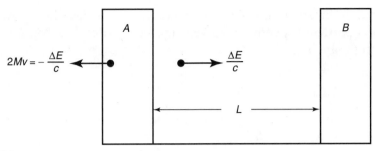

$$v = - \Delta E/2Mc$$

then

$$d = v\Delta t = (- \Delta E/2Mc) \cdot (L/c) = - \Delta EL/2Mc^2$$

The transport of energy ΔE from A to B is accompanied by transport of mass Δm. Everything happened within the system AB and no external forces are involved. The center of mass cannot change. Now the center of mass is $(L/2 - d)$ from B that now has a mass $(M + m)$, and it is $(L/2 + d)$ from A, which has a mass $(M - m)$. Taking moments with respect to the center of mass, we have

$$(M - m)g(L/2 + d) = (M + m)g(L/2 - d)$$

or

$$m = 2Md/L = \Delta E/c^2$$

We should not confuse the notions of equivalence and identity. The energy and mass are different physical characteristics of particles, where here the word "equivalence" established only their proportionality to each other. This is similar to the relation between the gravitational mass and inertial mass of a body: The two masses are indissolubly connected with each other and proportional to each other but at the same time have different characteristics. The equivalence of mass and energy has been beautifully verified by experiments in which matter is annihilated and converted totally into energy. For example, when an electron and a positron, each with a rest mass m_0, come together, they disintegrate and two gamma rays emerge, each with the measured energy of m_0c^2.

Based on Einstein's mass–energy relation $E = mc^2$, we can show that the mass of a particle depends on its velocity. Let a force F act on a particle of momentum mv. Then

$$Fdt = d(mv) \qquad (10.36)$$

If there is no loss of energy by radiation due to acceleration, then the amount of energy transferred in dt is

$$dE = c^2 dm$$

This is equal to the work done by the force F to give

$$Fvdt = c^2 dm$$

Combining this with equation (10.36), we have

$$vd(mv) = c^2 dm$$

Multiply this by m,

$$vmd(mv) = c^2 m dm$$

Integrating

$$(mv)^2 = c^2 m^2 + K$$

where K is the integration constant. Now $m = m_0$ when $v = 0$, we find $K = -c^2 m_0^2$, and

$$m^2 v^2 = c^2(m^2 - m_0^2)$$

where m_0 is known as the rest mass of the particle. Solving for m we obtain

$$m = \frac{m_0}{\sqrt{1 - (v/c)^2}} \qquad (10.37)$$

which is often called the relativistic mass of the particle.

It is now easy to see that a material body cannot have a velocity greater than the velocity of light. If we try to accelerate the body, as its velocity approaches the velocity of light its mass becomes larger and larger, and it becomes increasingly more difficult to accelerate it further. In fact, because the mass m becomes infinite when $v = c$, we can never accelerate the body up to the speed of light.

In the language of relativity theory and high energy physics, however, there is a trend to treat the mass as simply the invariant parameter m_0. We revisit this in the next section.

10.10 Four-Energy and Four-Momentum Vectors

Now it is obvious that Newtonian dynamics cannot hold totally. How do we know what to retain and what to discard? The answer is found in the generalizations that grew from the laws of motion but transcend it in their universality. These are the conservation of momentum and energy.

Thus, we now generalize the definitions of momentum and energy so that in the absence of external forces, the momentum and energy of a system of particles are conserved. In Newtonian mechanics the momentum $\vec{p}$ of a particle is defined as $m\vec{v}$, the product of particle's inertial mass and its velocity. A plausible generalization of this definition is to use the four-velocity u^μ and an invariant scalar m_0 that truly characterize the inertial mass of the particle and define the momentum four-vector (four-momentum, for short) P^μ as

$$P^\mu = m_0 u^\mu \tag{10.38}$$

To ensure that the "mass" of the particle is truly a characteristic of the particle, the "mass" must be that measured in the frame of reference in which the particle is at rest. Thus, the mass of the particle must be its proper mass. We customarily call this mass the rest mass of the particle and denote it by m_0. Using equation (10.31), we write P^μ in component form:

$$P^j = \frac{m_o v_j}{\sqrt{1-\beta^2}} = \gamma m_0 v_j, \quad j = 1, 2, 3$$

$$P^0 = \frac{m_o c}{\sqrt{1-\beta^2}} = \gamma m_0 c \tag{10.39}$$

We see that as $\beta = v/c \to 0$, the first three components of the four-momentum P^μ reduce to $m_o v_j$, the components of the ordinary momentum. This indicates that equation (10.38) appears to be a reasonable generalization.

Let us write the time component of P^0 as

$$P^0 = \frac{m_0 c}{\sqrt{1-\beta^2}} = \frac{E}{c} \tag{10.40}$$

Now what is the meaning of the quantity E? For low velocities, the quantity E reduces to

$$E = \frac{m_0 c^2}{\sqrt{1 - \beta^2}} \cong m_0 c^2 + \frac{1}{2} m_0 v^2$$

The second term on the right is the ordinary kinetic energy of the particle, and the first term can be interpreted as the rest energy of the particle (it is an energy the particle has, even when it is at rest), which must contain all forms of internal energy of the object, including heat energy, internal potential energy of various kinds, or rotational energy, if any. Hence, we can call the quantity E the total energy of the particle (moving at speed v). We now write the four-momentum as

$$P^\mu = \left(\frac{E}{c}, P^j \right) \tag{10.41}$$

The length of the four-momentum must be invariant, just as the length of the velocity four-vector is invariant under Lorentz transformations. We can show this easily:

$$\sum_\mu P^\mu P^\mu = \sum_\mu (m_0 u^\mu)(m_0 u^\mu) = - m_0^2 c^2 \tag{10.42}$$

But equation (10.41) gives

$$\sum_\mu P^\mu P^\mu = P^2 - \frac{E^2}{c^2}$$

Combining this with equation (10.42), we arrive at the relation

$$P^2 - \frac{E^2}{c^2} = - m_0^2 c^2 \quad \text{or} \quad E^2 - P^2 c^2 = m_0^2 c^4 \tag{10.43}$$

The total energy E and the momentum P^μ of a moving body are different when measured with respect to different reference frames. But the combination $P^2 - E^2/c^2$ has the same value for all frames of reference, namely $m_0^2 c^2$. This relation is very useful. Another very useful relation is $\vec{P} = \vec{v}(E/c^2)$. From equation (10.38) we see that $\gamma m_0 = E/c^2$, and combining this with the first equation of equation (10.39) gives the very useful relation $\vec{P} = \vec{v}(E/c^2)$.

The relativistic momentum, however, is not quite the familiar form found in general physics, because its spatial components contain the Lorentz factor γ. We can bring it into the old sense, and the traditional practice was to introduce a "relativistic mass" m as

$$m = m_0 \gamma = \frac{m_0}{\sqrt{1 - \beta^2}}$$

With this introduction of m, P^j takes the old form: $P^j = m v_j$. However, some believe that the introduction of the relativistic mass is not a purely methodologic issue; it often causes misunderstanding and vague interpretations of relativistic mechanics. So some prefer to include the factor γ with v_j forming the proper four-velocity component u_j and treating the mass as simply the invariant parameter m_0. (For details, see the article by Lev B, Okun, The Concept of Mass, *Physics Today*, June 1989.)

10.11 The Conservation Laws of Energy and Momentum

It is now clear that the linear momentum and energy of a particle should not be regarded as different entities but simply as two aspects of the same attributes of the particle, because they appear as separate components of the same four-vector P^μ that transforms according to equation (10.28):

$$P^{\mu'} = L^\mu_\nu P^\nu$$

in matrix form

$$\begin{pmatrix} P^{1'} \\ P^{2'} \\ P^{3'} \\ P^{4'} \end{pmatrix} = \begin{pmatrix} \gamma & 0 & 0 & i\beta\gamma \\ 0 & 1 & 0 & 0 \\ 0 & 0 & 1 & 0 \\ -i\beta\gamma & \gamma & 0 & 0 \end{pmatrix} \begin{pmatrix} P^1 \\ P^2 \\ P^3 \\ P^4 \end{pmatrix} = \begin{pmatrix} \gamma(P^1 + i\beta P^4) \\ P^2 \\ P^3 \\ \gamma(-i\beta P^1 + P^4) \end{pmatrix}$$

Thus,

$$P^{1'} = \gamma(P^1 + i\beta P^4) = \gamma(P^1 - \beta P^0), \quad P^{2'} = P^2, \quad P^{3'} = P^3 \tag{10.44a}$$

$$P^{4'} = \gamma(-i\beta P^1 + P^4), \quad \text{or} \quad P^{0'} = \gamma(P^0 - \beta P^1) \tag{10.44b}$$

which show clearly that what appears as energy in one frame appears as momentum in another frame, and vice versa.

So far we have not discussed explicitly the conservation laws. Because linear momentum and energy are not regarded as different entities but as two aspects of the same attributes of an object, it is no longer adequate to consider linear momentum and energy separately. A natural relativistic generalization of the conservation laws of momentum and energy would be the conservation of the four-momentum. Consequently, the conservation of energy becomes one part of the law of conservation of four-momentum. This is exactly what has been found to be correct experimentally, and, in addition, this generalized conservation law of four-momentum holds for a system of particles, even when the number of particles and their rest energies are different in the initial and final states. It should be emphasized that what we do mean by energy E is the total energy of an object. It consists of rest energy, which contains all forms of internal energy of the body, and kinetic energy. The rest energies and kinetic energies do not need to be individually conserved, but their sum must be. For example, in an inelastic collision, kinetic energy may be converted into some form of internal energy or vice versa, accordingly the rest energy of the object may change.

Energy and momentum conservation go together in special relativity; we cannot have one without the other. This seems a bit puzzling for some readers, because in classical mechanics the conservation laws of energy and momentum are on different footing. That is because energy and momentum are regarded as different entities. Moreover, classical mechanics does not talk about rest energy at all.

10.12 Generalization of Newton's Equation of Motion

A natural relativistic generalization of Newton's equation of motion follows:

$$K^\mu = \frac{dP^\mu}{d\tau} = \frac{d}{d\tau}(m_0 u^\mu) = m_0 \frac{d^2 x^\mu}{d\tau^2} \tag{10.45}$$

where K^μ is the four-force vector, which is also called the Minkowski force.

Using equations (10.19) and (10.39), we obtain the three components for $\mu = 1, 2, 3$:

$$K^j \sqrt{1 - \beta^2} = \frac{d}{dt} \frac{m_0 v_j}{\sqrt{1 - \beta^2}} = F_j \tag{10.46}$$

and $K^4 = iK^0$, and K^0 is given by

$$K^0 = \frac{dP^0}{d\tau} = \frac{\gamma}{c} \frac{dE}{d\tau} = \frac{1}{c\sqrt{1 - \beta^2}} \frac{dE}{dt} \tag{10.47}$$

Thus K^0 is proportional to the time rate of change of the energy.

In practical calculations, we usually do not need to deal with the four-force but prefer to use just the three components given by equation (10.46). In vector form we have

$$\vec{F} = \frac{d}{dt} \frac{m_0 \vec{v}}{\sqrt{1 - \beta^2}} \tag{10.48}$$

This relativistic equation of motion reduces to the Newtonian form $\vec{F} = d\vec{P}/dt$, provided we use the relativistic momentum $\vec{P}$ given by equation (10.38).

We can show that the dot product of the four-force K^μ with the four-velocity u^μ vanishes:

$$\sum_{\mu} K^{\mu} u^{\mu} = \sum_{\mu} \frac{d\,(m_0 u^{\mu})}{d\tau} u^{\mu} = \sum_{\mu} m_0 \frac{d\,(u^{\mu} u^{\mu}/2)}{d\tau}$$

$$= \sum_{\mu} \frac{m_0}{2} \frac{d\,(-c^2)}{d\tau} = 0 \tag{10.49}$$

But the dot product can also be written as

$$\sum_{\mu=1}^{4} K^{\mu} u^{\mu} = \frac{\vec{F} \cdot \vec{v}}{1-\beta^2} + K^4 u^4 = \frac{\vec{F} \cdot \vec{v}}{1-\beta^2} - K^0 u^0 \quad (K^4 = iK^0, u^4 = iu^0)$$

$$= \frac{\vec{F} \cdot \vec{v}}{1-\beta^2} - \frac{cK^0}{\sqrt{1-\beta^2}}$$

Combining this with equation (10.49) we find

$$\frac{\vec{F} \cdot \vec{v}}{1-\beta^2} - \frac{cK^0}{\sqrt{1-\beta^2}} = 0$$

from which we obtain

$$K^0 = \frac{\vec{F} \cdot \vec{v}}{c\sqrt{1-\beta^2}} \tag{10.50}$$

Now

$$K^0 = \frac{dP^0}{d\tau} = \frac{1}{\sqrt{1-\beta^2}} \frac{d}{dt} \frac{m_0 c^2}{\sqrt{1-\beta^2}}$$

Combining this with equation (10.49) gives a very useful relation

$$\frac{d}{dt} \frac{m_0 c^2}{\sqrt{1-\beta^2}} = \vec{F} \cdot \vec{v} \tag{10.51}$$

10.13 The Force Transformation

Consider a particle of rest mass m_0 having velocity $\vec{v}'$ and momentum $\vec{P}'$ relative to frame S' and having velocity $\vec{v}$ and momentum $\vec{P}$ relative to frame S. Then the force acting on the particle measured in S' is, from equation (10.46),

$$F_x' = dP_x'/dt' \tag{10.52}$$

where

$$F_x = F_1', \quad P_x' = P_1' = m_0 v_x / \sqrt{1 - \beta^2} \quad (v_x = v_1)$$

Now, equations (10.44) give, with some modifications in notations,

$$P_x' = \gamma \left(P_x - \frac{u}{c} \frac{E}{c} \right), \quad P_y' = P_y, \quad P_z' = P_z, \quad E' = \gamma (E - u P_x).$$

Hence, equation (10.52) becomes, after making use of the first transformation relation,

$$F_x' = \gamma \frac{d}{dt'} (P_x - uE/c^2) \tag{10.53}$$

But

$$\frac{d}{dt'} = \frac{dt}{dt'} \frac{d}{dt}$$

and

$$\frac{dt'}{dt} = \frac{d}{dt} \gamma \left(t - \frac{ux}{c^2} \right) = \gamma (1 - uv_x/c^2)$$

Substituting these in equation (10.53),

$$F_x' = \frac{\gamma}{\gamma (1 - uv_x/c^2)} \frac{d}{dt} (P_x - uE/c^2)$$

$$= \frac{1}{(1 - uv_x/c^2)} \left(F_x - \frac{u}{c^2} \frac{dE}{dt} \right)$$

Using equations (10.51) and (10.39) we have

$$\frac{dE}{dt} = \vec{F} \cdot \vec{v}$$

Hence

$$F_x' = \frac{1}{(1 - uv_x/c^2)} \left(F_x - \frac{u}{c^2} \vec{F} \cdot \vec{v} \right)$$

Now

$$\vec{F} \cdot \vec{v} = F_x v_x + F_y v_y + F_z v_z$$

Hence

$$F_x' = F_x - \frac{uv_y}{c^2(1 - uv_x/c^2)} F_y - \frac{uv_z}{c^2(1 - uv_x/c^2)} F_z \qquad (10.54)$$

Next, we consider

$$F_y' = \frac{dP_y'}{dt'} = \frac{dP_y}{dt'} = \frac{dP_y}{dt} \frac{dt}{dt'}, \quad (P_y' = P_y)$$

But

$$\frac{dt}{dt'} = 1/\gamma(1 - uv_x/c^2)$$

Hence

$$F_y' = \frac{F_y}{\gamma(1 - uv_x/c^2)} \qquad (10.55)$$

Similarly, because $P_z' = P_z$ we have

$$F_z' = \frac{F_z}{\gamma(1 - uv_x/c^2)} \qquad (10.56)$$

The inverse transformations are

$$F_x = F_x' + \frac{uv_y'}{c^2(1 + uv_x'/c^2)} F_y' + \frac{uv_z'}{c^2(1 + uv_x'/c^2)} F_z' \qquad (10.57)$$

$$F_y = \frac{F_y'}{\gamma(1 + uv_x'/c^2)} \qquad (10.58)$$

$$F_z = \frac{F_z'}{\gamma(1 + uv_x'/c^2)} \qquad (10.59)$$

10.14 Particles of Zero Rest Mass

A surprising consequence of the relativistic energy-momentum generalization is the possibility of "massless" particles, which possess momentum and energy but no rest mass. From the expression for the energy and momentum of a particle,

$$E = \frac{m_0 c^2}{\sqrt{1 - v^2/c^2}}, \quad \vec{P} = \frac{m_0 \vec{v}}{\sqrt{1 - v^2/c^2}} \qquad (10.60)$$

we can define a particle of zero rest mass possessing finite energy and momentum. To this purpose, we allow $v \to c$ in some inertial system S and $m_0 \to 0$ in such a way that

$$\frac{m_0}{\sqrt{1 - v^2/c^2}} = \chi \tag{10.61}$$

remains constant. Then equation (10.60) takes the simple form

$$E = \chi c^2 \quad \vec{P} = \chi c \hat{e}$$

where $\hat{e}$ is a unit vector in the direction of motion of the particle. Eliminating χ from the last two equations, we obtain

$$E = Pc \tag{10.62}$$

which is consistent with equation (10.43): $E^2 - P^2 c^2 = m_0^2 c^4$

Now as $(E/c, \vec{P})$ is a four-vector, $(\chi c, \chi c \hat{e})$ is also a four-vector, the energy and momentum four-vector of a zero rest-mass particle, in frame S and in any other inertial frame such as S'. It can be shown that the transformation of the energy and momentum four-vector $(\chi c, \chi c \hat{e})$ of a zero rest-mass particle is identical with that of a light wave, provided χ is made proportional to the frequency ν. Thus, if we associate a zero rest-mass particle with a light wave in one inertial frame, this association holds in all other inertial frames. The ratio of the energy of the particle to the frequency has the dimensions of action (or angular momentum). This suggests that we can write this association by the following equations

$$E = h\nu \quad \text{and} \quad P = \chi c = h\nu/c$$

where h is Planck's constant. This massless particle of light is called a photon. Einstein introduced it in his pioneering paper on the photoelectric effect published a few months before his work on special relativity, out of concern with the photoelectric effect and consideration of Planck's quantum hypothesis.

B. Relativistic Electrodynamics

It is not necessary to know special relativity in solving problems in electricity and magnetism. But special relativity throws new light on the nature of electromagnetism. We shall see that the division of the electromagnetic field into electric and magnetic fields depends to a very large extent on the reference frame in which the phenomena are considered; in other words, it is of relative nature. As an example, we first demonstrate the relativistic nature of magnetism.

10.15 Relativistic Nature of Magnetism

The relativistic nature of magnetism can be shown clearly by considering the force exerted on a moving charge by another charge moving at the same constant velocity. We again refer to the two reference frames S and S' of Figure 10.4, moving relative to each other with constant velocity u along the x-axis. The two charges q_1 and q_2 are at rest in frame S', q_1 being at the origin O' and q_2 at $(x', y', 0)$. The force on q_2, denoted by $\vec{F}'$, has the components

$$F_{x}' = \frac{q_1 q_2 x'}{4\pi\varepsilon_0 r'^3}, \quad F_{y}' = \frac{q_1 q_2 y'}{4\pi\varepsilon_0 r'^3}, \quad F_{z}' = 0 \tag{10.63}$$

What is the force on q_2, as measured by an observer O in S? For simplicity, we calculate this force at the time $t = 0$. The observer in S sees the two charges moving at the same velocity $u\hat{i}$. Before we can calculate the force, we need to know how the charge transforms. There is exhaustive evidence that electric charge is invariant, and a charged body carries the same charge for all inertial observers. The neutrality of a gas consisting of hydrogen molecules provides one such evidence. Electrons in these molecules move at much higher velocities than protons. Thus, if the charge depended on the velocity, the charges of the electrons and protons would not compensate each other and the gas would be charged. No charge (to 20 decimal places!) has been observed in hydrogen gas. The most direct demonstration of the invariance of charge is the fact that the charge-to-mass e/m for a charged particle moving at a velocity v is found experimentally to agree with the relation

$$e/m = (e/m_0)(1 - v^2/c^2)^{1/2}$$

Now, using equations (10.57) to (10.59) and remembering that $v_x' = v_y' = v_z' = 0$ in this case, we obtain

$$\left.\begin{array}{l} F_x = F_x' = \dfrac{q_1 q_2 x'}{4\pi\varepsilon_0 (x'^2 + y'^2)^{3/2}}, \\[2ex] F_y = \gamma^{-1} F_y' = \dfrac{q_1 q_2 y'}{4\pi\varepsilon_0 \gamma (x'^2 + y'^2)^{3/2}}, \end{array}\right\} \tag{10.64}$$

and $F_z = 0$. Substituting the values of x and y at $t = 0$ from the Lorentz transformation

$$F_x = \frac{\gamma q_1 q_2 x}{4\pi\varepsilon_0 (\gamma^2 x^2 + y^2)^{3/2}},$$

$$F_y = \frac{q_1 q_2 y}{4\pi\varepsilon_0 \gamma (\gamma^2 x^2 + y^2)^{3/2}} = \frac{\gamma q_1 q_2 y}{4\pi\varepsilon_0 (\gamma^2 x^2 + y^2)^{3/2}}(1 - \beta^2) \qquad (10.65)$$

This is the force exerted by q_1 on q_2 at the time $t = 0$ when the charge q_1 is at $x = 0$.

We now can write $\vec{F}$ as follows

$$\vec{F} = \frac{\gamma q_1 q_2 x}{4\pi\varepsilon_0 (\gamma^2 x^2 + y^2)^{3/2}}\hat{i} + \frac{\gamma q_1 q_2 y}{4\pi\varepsilon_0 (\gamma^2 x^2 + y^2)^{3/2}}(1 - \beta^2)\hat{j}$$

At $t = 0$, q_2 is at $\vec{r} = x\hat{i} + y\hat{j}$, so we can put (writing $x\hat{i} = \vec{r} - y\hat{j}$)

$$\vec{F} = \frac{\gamma q_1 q_2 \vec{r}}{4\pi\varepsilon_0 (\gamma^2 x^2 + y^2)^{3/2}} - \frac{\gamma q_1 q_2 u^2 y}{4\pi\varepsilon_0 c^2 (\gamma^2 x^2 + y^2)^{3/2}}\hat{j}$$

$$= \frac{\gamma q_1 q_2}{4\pi\varepsilon_0 (\gamma^2 x^2 + y^2)^{3/2}}\left\{\vec{r} - \frac{u^2}{c^2}y\hat{j}\right\} \qquad (10.65a)$$

$$= q_2\left\{\frac{\gamma q_1 \vec{r}}{4\pi\varepsilon_0 (\gamma^2 x^2 + y^2)^{3/2}} + \left[\vec{u} \times \frac{\gamma q_1 uy\hat{k}}{4\pi\varepsilon_0 c^2 (\gamma^2 x^2 + y^2)^{3/2}}\right]\right\}$$

If frame S is fixed with respect to the laboratory, then the above force and the above coordinates are those measured by an observer in the laboratory. The first term between the braces in equation (10.65a) is the electric field $\vec{E_1}$ of charge q_1 at the point x, y, z and at the time $t = 0$, when the origins of the two systems coincide:

$$\vec{E_1} = \frac{\gamma q_1 \vec{r}}{4\pi\varepsilon_0 (\gamma^2 x^2 + y^2)^{3/2}} \qquad (10.66)$$

What is the limiting case where u is small and $\gamma \cong 1$? This should throw some light on the subject for us. To see this, let us first rewrite equation (10.65a) in the form of the Lorentz force law:

$$\vec{F} = q_2(\vec{E_1} + \vec{u} \times \vec{B_1}) \qquad (10.67)$$

where $\vec{E_1}$ is given by equation (10.66), and $\vec{B_1}$ is given by

$$\vec{B}_1 = \frac{\gamma q_1 u y \hat{k}}{4\pi\varepsilon_0 c^2 (\gamma^2 x^2 + y^2)^{3/2}} = \frac{\mu_0 \gamma q_1 u y \hat{k}}{4\pi (\gamma^2 x^2 + y^2)^{3/2}} \tag{10.68}$$

where $\mu_0 = 1/\varepsilon_0 c^2$, the permeability of free space. $\vec{B}_1$ is the magnetic field due to q_1 at the position of q_2. When u is small and $\gamma \cong 1$, equations (10.66) and (68) are seen to have the familiar form

$$\vec{E} = \frac{q}{4\pi\varepsilon_0 r^2}\hat{r} \quad \text{and} \quad \vec{B} = \frac{\mu_0}{4\pi}\frac{\vec{j}\times\hat{r}}{r^2}, \quad \vec{j} = q\vec{u} \tag{10.69}$$

The vectors $\vec{E}_1$ and $\vec{B}_1$, and the corresponding forces are shown in **Figure 10.10**.

We see that for an observer in S, the charge q_1 exerts on q_2 both an electric force $q_2\vec{E}_1$ and a magnetic force $q_2\vec{u}\times\vec{B}_1$. For an observer in S' of q_1, however, the charge q_2 is stationary and the force exerted by q_1 on q_2 is purely electric. Thus, the appearance of the magnetic field is a relativistic effect. Moreover, it follows from the existence in nature of the limiting velocity c equal to the velocity of light in vacuum. If this velocity were infinite, no magnetism would exist at all.

The relativistic nature of magnetism is a universal physical fact, and its origin is associated with the absence of magnetic charges.

Unlike most relativistic effects, magnetism can be easily observed in many cases, for example, the magnetic field of a current-carrying conductor. The reason behind such favorable circumstances is that the magnetic field may be created by a very large number of moving electric charges under the conditions of the almost complete vanishing of the electric field due to practically ideal balance of the numbers of electrons and protons in conductors. In this case, the magnetic interaction is predominant.

The almost complete compensation of electric charges made it possible for physicists to study relativistic effects (i.e., magnetism) and discover

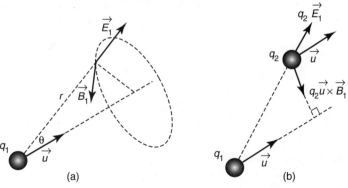

(a)   (b)

I FIGURE 10.10

correct laws. For this reason, the laws of electromagnetism, unlike Newton's laws, have not been refined after the introduction of the theory of relativity.

Because the relations between electric and magnetic fields vary on a transition to another reference system, the fields $\vec{E}$ and $\vec{B}$ must be treated with care. While going from one reference system to another, they are transformed in a certain way. The laws of this transformation can be established in various ways in the context of special relativity. We must first cast the electromagnetic variables into four-vector forms.

10.16 The Four-Current Density J^μ

The current density $\vec{j}$ and the charge density ρ form the components of the four-current density vector $\vec{j}^\mu$: $\vec{j}^\mu = (c\rho, \vec{j})$. It is easy to show this: If charge is invariant, then charge density cannot be, because volume is subject to length-contraction factor γ^{-1}. We refer again to the two frames S and S' of Figure 10.4, moving relative to each other with a constant velocity u along the x-axis. A charge Q of volume V_0 is located at the origin O' of the frame S', and the charge density is $\rho_0 = Q/V_0$. In frame S, Q is moving with a velocity u, and its volume is now given by $V = V_0(1 - u^2/c^2)^{1/2}$. So the charge density ρ is given by

$$\rho = Q/V = \rho_0(1 - u^2/c^2)^{1/2} = \gamma\rho_0.$$

We now rewrite this expression in a form more suitable for our further discussion. For this purpose, let us consider two nearby points A and B on the world line of charge Q in frame S, separated by the four-vector

$$dx^\mu = (d\vec{r}, c\,dt)$$

with $d\vec{r} = \vec{u}\,dt$. The proper time interval $d\tau$ between A and B is, according to equation (10.19),

$$d\tau = dt\,\sqrt{1 - u^2/c^2} = dt/\gamma$$

or $\gamma = dt/d\tau$. Then ρ becomes

$$\rho = \gamma\rho_0 = \frac{dt}{d\tau}\,\rho_0 \tag{10.70}$$

The current density in S is

$$\vec{j} = \rho\vec{u} = \rho\,\frac{d\vec{r}}{dt} = \gamma\rho_0\,\frac{d\vec{r}}{dt} = \rho_0\,\frac{dt}{d\tau}\frac{d\vec{r}}{dt} = \rho_0\,\frac{d\vec{r}}{d\tau} \tag{10.71}$$

From equations (10.70) and (10.71) we have

$$J^\mu = (\vec{j}, c\rho) = \rho_0 \frac{d}{d\tau}(\vec{r}, ct) = \rho_0 \frac{dx^\mu}{d\tau} \tag{10.72}$$

which shows that $\vec{j}$ and $c\rho$ form the components of the four-vector, the four-current density vector. Then they should transform according to equation (10.28), i.e.,

$$J'^\mu = L^\mu_\nu J^\nu$$

Or, written in full

$$c\rho' = \gamma(c\rho - \beta j^1), \quad j'^1 = \gamma(j^1 - u c\rho), \quad j'^2 = j^2, \quad j'^3 = j^3 \tag{10.73}$$

Alternatively, we can obtain the above transformation properties of the four-current density J^μ through the continuity equation

$$\nabla \cdot \vec{j} + \frac{\partial \rho}{\partial t} = \frac{\partial j_1}{\partial x} + \frac{\partial j_2}{\partial y} + \frac{\partial j_3}{\partial z} + \frac{\partial \rho}{\partial t} = 0 \tag{10.74}$$

Let us look into the transformation of this equation. For this purpose, we first derive the transformation equation for the partial derivative operators. From the chain rule of differential calculus, we have

$$\frac{\partial}{\partial t} = \frac{\partial t'}{\partial t}\frac{\partial}{\partial t'} + \frac{\partial x'}{\partial t}\frac{\partial}{\partial x'}$$

The coefficients can be read off from equations (10.15):

$$\frac{\partial}{\partial t} = \gamma \frac{\partial}{\partial t'} + (-\beta\gamma c)\frac{\partial}{\partial x'} = \gamma\left(\frac{\partial}{\partial t'} - \beta c \frac{\partial}{\partial x'}\right) \tag{10.75a}$$

Similarly,

$$\frac{\partial}{\partial x} = \frac{\partial t'}{\partial x}\frac{\partial}{\partial t'} + \frac{\partial x'}{\partial x}\frac{\partial}{\partial x'} = \gamma\left(\frac{\partial}{\partial x'} - \frac{\beta}{c}\frac{\partial}{\partial t'}\right) \tag{10.75b}$$

Now we substitute for the differential operators in equation (10.74) from equations (10.75), obtaining

$$\gamma\left(\frac{\partial}{\partial x'} - \frac{\beta}{c}\frac{\partial}{\partial t'}\right)j_1 + \frac{\partial j_2}{\partial y} + \frac{\partial j_3}{\partial z} + \gamma\left(\frac{\partial}{\partial t'} - u\frac{\partial}{\partial x'}\right)\rho = 0$$

On rearranging terms, we have

$$\frac{\partial}{\partial x'}(\gamma j_1 - \gamma u \rho) + \frac{\partial j_2}{\partial y} + \frac{\partial j_3}{\partial z} + \frac{\partial}{\partial t'}\left(\gamma\rho - \frac{\gamma\beta}{c}j_1\right) = 0$$

This can be written in the same form as equation (10.74), namely

$$\frac{\partial j_1'}{\partial x'} + \frac{\partial j_2'}{\partial y'} + \frac{\partial j_3'}{\partial z'} + \frac{\partial \rho'}{\partial t'} = 0 \tag{10.76}$$

provided $c\rho'$, j_1', j_2' and j_3' are given by equation (10.73).

We can now write the continuity equation (10.74) in terms of the four-current density J^μ

$$\partial_\mu J^\mu = 0 \tag{10.77}$$

where $\partial_\mu = \partial/\partial x^\mu$ is the four-dimensional operator

$$\partial_\mu = \left(\nabla, -\frac{\partial}{c\partial t} \right) \tag{10.78}$$

Because the current continuity equation takes the same form in all inertial frames, so it is Lorentz covariant. Covariance is a requirement of any valid physical laws in view of the principle of relativity.

The d'Alembertian operator $\square$ is the scalar product of the four-dimensional operator ∂_μ with itself:

$$\square = \partial_\mu \partial^\mu = \frac{\partial^2}{\partial x^2} + \frac{\partial^2}{\partial y^2} + \frac{\partial^2}{\partial z^2} - \frac{1}{c^2}\frac{\partial^2}{\partial t^2} = \nabla^2 - \frac{1}{c^2}\frac{\partial^2}{\partial t^2} \tag{10.79}$$

The d'Alembertian is invariant, like the square of the norm of a four-vector.

10.17 The Contravariant Four-Vector $A^\mu = (\phi/c, \vec{A})$

With the Lorentz gauge condition, we obtained, in Chapter 6, the following two inhomogeneous equations:

$$\nabla^2 \vec{A} - \frac{1}{c^2}\frac{\partial^2 \vec{A}}{\partial t^2} = -\mu_0 \vec{j} \quad \text{or} \quad -\mu_0^{-1} \square \vec{A} = \vec{j}$$

$$\nabla^2 \phi - \frac{1}{c^2}\frac{\partial^2 \phi}{\partial t^2} = -\rho/\varepsilon_0 \quad \text{or} \quad -\varepsilon_0 c^2 \square (\phi/c) = c\rho$$

Because the d'Alembertian operator, $\square$, is an invariant and because the coefficients on the left are equal, it is clear that $(\phi/c, \vec{A})$ transform like $(c\rho, \vec{j})$; in other words, it is another four-vector. Explicitly, the transformation equations

$$\phi'/c = \gamma\left(\phi/c - \beta A^1\right), \quad A'^1 = \gamma\left(-\beta\phi/c + A^1\right), \quad A'^2 A^2, \quad A'^3 = A^3 \tag{10.80}$$

The Lorentz gauge condition is now simply

$$\partial_\mu A^\mu = 0 \tag{10.81}$$

where ∂_μ is the four-dimensional operator. Similarly, we can express the two inhomogeneous equations in terms of four-vectors J^μ and A^μ:

$$\partial_\nu \partial^\nu A^\mu = -\mu_0 J^\mu \tag{10.82}$$

So far we have dealt only with four-vectors, and a four-vector has four components. But electromagnetic field is a new quantity that has 16 components, and it is called a tensor of rank two. We can consider a four-vector is a tensor of rank one. We now digress for a moment to introduce the ideas of tensor and covariance.

10.18 Covariance and Tensors

Covariance refers to the form invariance of the laws of physics (or any mathematical equation) under a general coordinate transformation. If some physical law in S frame is expressed by the statement $C_l(q) = 0$, where q is a generalized coordinate (e.g., r, θ, or charge ρ), viewed from S' frame in which the corresponding coordinate is q', the same law is expressed by $C'_m(q') = 0$, and C and C' are related by the transformation

$$C'_m = Q^l_m C_l, \quad |Q^l_m| \neq 0$$

where Einstein summation convention is implied, then the relation $C_l(q) = 0$ is said to be invariant under the transformation. If the Q^l_m is a Lorentz transformation, the relation $C_l(q) = 0$ is said to be a Lorentz covariant.

Before we introduce the idea of tensor of rank two, we review the transformation properties of four-vector first; in particular, the radius four-vector x^μ, equation (10.82), describes its law of transformation:

$$x^{\mu'} = L^\mu_\nu x^{\nu'} \tag{10.82}$$

where the matrix L^μ_ν is given by equation (10.27).

We define any set of four quantities A^μ ($\mu = 0, 1, 2, 3,$ or $= 1, 2, 3, 4$) that transform under the Lorentz transformations like the components of the radius four-vector x^μ

$$A^{\mu'} = L^\mu_\nu A^\nu \tag{10.83}$$

to be a (contravariant) four-vector, where Einstein's summation convention is implied. If we set $\mu = 0, 1, 2, 3$, then we must use the second L_ν^μ in equation (10.27) and equation (10.83) gives

$$A'^0 = \gamma(A^0 - \beta A^1), \quad A'^1 = \gamma(-\beta A^0 + A^1), \quad A'^2 = A^2, \quad A'^3 = A^3 \tag{10.83a}$$

We can construct tensor of rank two from four-vectors (tensors of rank one). The set of 16 quantities $A^\mu B^\nu$ that transform according to

$$\left(A^\mu B^\nu\right)' = L_\sigma^\mu L_\rho^\nu A^\sigma B^\rho \tag{10.84}$$

form a (contravariant) tensor of rank two. The extension to higher ranks is obvious.

Any tensor relation, if valid in frame, holds in any other frame.

10.19 Electromagnetic Field Tensor $F^{\mu\nu}$

We may now proceed to the transformation of the field themselves because

$$\vec{E} = -\nabla\phi - \frac{\partial \vec{A}}{\partial t}, \quad \vec{B} = \nabla \times \vec{A}$$

It follows from the four-vector nature of A^μ that the field components form a tensor of rank two, $F^{\mu\nu}$, which must be antisymmetric because they are in all just six field quantities: the three components of $\vec{E}$ and the three components of $\vec{B}$. Thus

$$F^{\mu\nu} = \partial^\mu A^\nu - \partial^\nu A^\mu \tag{10.85}$$

and $F^{\mu\nu} = -F^{\nu\mu}$; $F^{\mu\mu} = 0$. Then, in terms of the field components

$$F^{12} = -F^{21} = \partial^1 A^2 - \partial^2 A^1 = \frac{\partial A_y}{\partial x^1} - \frac{\partial A_x}{\partial x^2} = B_z \tag{10.86a}$$

$$F^{13} = -F^{31} = \partial^1 A^3 - \partial^3 A^1 = \frac{\partial A_z}{\partial x^1} - \frac{\partial A_x}{\partial x^3} = -B_y \tag{10.86b}$$

$$F^{14} = -F^{41} = \partial^1 A^4 - \partial^4 A^1 = \frac{\partial}{\partial x^1}\left(\frac{\phi}{c}\right) - \frac{\partial A_x}{\partial (ct)} = -E_x/c \tag{10.86c}$$

Similarly, we find

$$F^{23} = -F^{32} = B_x, \quad F^{24} = -F^{42} = -E_y/c, \quad F^{34} = -F^{43} = -E_z/c \tag{10.86d}$$

The electromagnetic field tensor $F^{\mu\nu}$ is therefore

$$F^{\mu\nu} = \begin{bmatrix} 0 & B_z & -B_y & -E_x/c \\ -B_z & 0 & B_x & -E_y/c \\ B_y & -B_x & 0 & -E_z/c \\ E_x/c & E_y/c & E_z/c & 0 \end{bmatrix} \tag{10.87}$$

where "μ" is the row number and "ν" the column number.

10.20 Covariant Form of Maxwell Equations

Maxwell equations may now be expressed in terms of the field tensor. Consider first the two inhomogeneous equations

$$\nabla \cdot \vec{E} = \rho/\varepsilon_0, \quad \nabla \times \vec{B} = \mu_0 \vec{j} + \frac{1}{c^2} \frac{\partial \vec{E}}{\partial t} = 0; \quad c = 1/\sqrt{\mu_0 \varepsilon_0}$$

From equation (10.85), forming the (covariant) derivative of the field tensor, we get

$$\partial_\nu F^{\mu\nu} = \partial_\nu \partial^\mu A^\nu - \partial_\nu \partial^\nu A^\mu$$

which becomes, after applying the Lorentz gauge (10.81),

$$\partial_\nu F^{\mu\nu} = -\partial_\nu \partial^\nu A^\mu$$

from which we obtain, using equation (10.82)

$$\partial_\nu F^{\mu\nu} = \mu_0 J^\mu \tag{10.88}$$

It is easy to show that both of the inhomogeneous Maxwell equations are contained in equation (10.86). For example, with $\mu = 1$ we have

$$\partial_2 F^{12} + \partial_3 F^{13} + \partial_4 F^{14} = \mu_0 J^1$$

i.e.,

$$\frac{\partial B_z}{\partial y} - \frac{\partial B_y}{\partial z} - \frac{1}{c} \frac{\partial E_x}{\partial (ct)} = \mu_0 j_x$$

which is the x component of

$$\nabla \times \vec{B} = \mu_0 \vec{j} + \frac{1}{c^2} \frac{\partial \vec{E}}{\partial t} = 0$$

Similarly, with $\mu = 4$ we find

$$\partial_1 F^{41} + \partial_2 F^{42} + \partial_3 F^{43} = \mu_0 J^4$$

i.e.,

$$\frac{1}{c}\left(\frac{\partial E_x}{\partial x} + \frac{\partial E_y}{\partial y} + \frac{\partial E_z}{\partial z}\right) = \mu_0 c\rho$$

or $\nabla \cdot \vec{E} = \rho/\varepsilon_0$.

The two homogeneous Maxwell equations

$$\nabla \cdot \vec{B} = 0, \quad \nabla \times \vec{E} + \frac{\partial \vec{B}}{\partial t} = 0$$

may be shown to correspond to

$$\partial_\lambda F^{\mu\nu} + \partial_\mu F^{\nu\lambda} + \partial_\nu F^{\lambda\mu} = 0 \tag{10.89}$$

with (λ, μ, ν) to be $(1, 2, 3)$, and equation (10.89) gives

$$\nabla \cdot \vec{B} = 0$$

and $(4, 2, 3)$ gives

$$c^{-1}\left(\partial B_x/\partial t + \partial E_z/\partial y - \partial E_y/\partial z\right) = 0$$

which is the x component of

$$\nabla \times \vec{E} + \frac{\partial \vec{B}}{\partial t} = 0$$

Thus, in covariant form Maxwell equations become

$$\partial_\nu F^{\mu\nu} = \mu_0 J^\mu \tag{10.88}$$

$$\partial_\lambda F^{\mu\nu} + \partial_\mu F^{\nu\lambda} + \partial_\nu F^{\lambda\mu} = 0 \tag{10.89}$$

10.21 Transformation of $\vec{E}$ and $\vec{B}$

Any four-vector transforms in the same way. Each index of a tensor transforms like a four-vector. Hence, the Lorentz transformation of $F^{\mu\nu}$ is, in accordance with equation (10.84),

$$F^{\mu\nu\,\prime} = L^\mu_\sigma L^\nu_\rho F^{\sigma\rho} \tag{10.90}$$

From this we find that

$$\left.\begin{array}{ll} \vec{E}_{\parallel}{}' = \vec{E}_{\parallel}, & \vec{B}_{\parallel}{}' = \vec{B}_{\parallel} \\[6pt] \vec{E}_{\perp}{}' = \gamma(\vec{E} + \vec{u} \times \vec{B})_{\parallel}, & \vec{B}_{\perp}{}' = \gamma(\vec{B} - c^{-2}\vec{u} \times \vec{E})_{\perp} \end{array}\right\}$$ (10.91)

We leave the detail for a homework. The inverse transformations are

$$\left.\begin{array}{ll} \vec{E}_{\parallel} = \vec{E}_{\parallel}{}', & \vec{B}_{\parallel} = \vec{B}_{\parallel}{}' \\[6pt] \vec{E}_{\perp} = \gamma(\vec{E} - \vec{u} \times \vec{B}')_{\parallel}, & \vec{B}_{\perp} = \gamma(B' + c^{-2}\vec{u} \times \vec{E}')_{\perp} \end{array}\right\}$$ (10.92)

From these transformation formulas we see again that electric and magnetic fields are relative. It is possible for either $\vec{E}$ or $\vec{B}$ to vanish in the frame S and to be present in another frame S'.

10.22 Field of a Uniformly Moving Point Charge

To demonstrate the utility of the transformation of $\vec{E}$ and $\vec{B}$, let us calculate the fields of a point charge in uniform motion. The charge q moves with uniform velocity $\vec{u}$ in the positive x-axis in the frame S. Then in the frame S', which moves with the charge q, the charge is at rest and so

$$\vec{E}' = \frac{q\vec{r}'}{4\pi\varepsilon_0 r'^3}, \quad \vec{B}' = 0$$ (10.93)

For convenience we have made the origin of the frame S' coincide with the position of the charge. The fields in the laboratory frame S may be obtained only by using equation (10.92), i.e.,

$$E_x = E_x{}' = \frac{qx'}{4\pi\varepsilon_0 r'^3}, \quad B_x = B_x{}' = 0$$ (10.94a)

$$\vec{E}_{\perp} = \gamma \vec{E}_{\perp}{}' = \frac{q\gamma \vec{r}_{\perp}'}{4\pi\varepsilon_0 r'^3}, \quad \vec{B}_{\perp} = \gamma c^{-2}(\vec{u} \times \vec{E}') = \frac{\gamma q\,(\vec{u} \times \vec{E}')}{4\pi\varepsilon_0 r'^3 c^2}$$ (10.94b)

where $\gamma = 1/\sqrt{1 - \beta^2}$, $\beta = u/c$; $x' = \gamma(x - ut)$, $y' = y$, $z' = z$ and t is the time elapsed in the frame S (the laboratory frame) from the moment at which the two origins coincide. Thus the vector $\vec{r}'$ is given by

$$\vec{r}' = \{\gamma(x - ut), y, z\}$$

Defining vectors

$$\gamma \vec{R}^{*} = \vec{r}\,' = \{\gamma(x - ut), y, z\} \tag{10.95}$$

and $\vec{R}$, the position vector of the charge relative to the point of observation at time t by

$$\vec{R} = \{(x - ut), y, z\} \tag{10.96}$$

Then, equations (10.94a) and (10.94b) become

$$E_x = \frac{q}{4\pi\varepsilon_0}\frac{\gamma(x - ut)}{\gamma^3(\vec{R}^{*})^3}, \quad E_y = \frac{q}{4\pi\varepsilon_0}\frac{\gamma y}{\gamma^3(\vec{R}^{*})^3}, \quad E_z = \frac{q}{4\pi\varepsilon_0}\frac{\gamma z}{\gamma^3(\vec{R}^{*})^3} \tag{10.97}$$

or

$$\vec{E} = \frac{q}{4\pi\varepsilon_0}\frac{\vec{R}}{\gamma^2(\vec{R}^{*})^3} \tag{10.98}$$

and

$$\vec{B} = \frac{q\mu_0}{4\pi}\frac{\vec{u} \times \vec{R}}{\gamma^2(\vec{R}^{*})^3} \tag{10.99}$$

The electric field is no longer spherically symmetric. To see this more clearly, let us rewrite it in terms of $\theta = \cos^{-1}(\vec{u} \cdot \vec{R}/uR)$:

$$\vec{E} = \frac{q\vec{R}(1 - \beta^2)}{4\pi\varepsilon_0 R^3(1 - \beta^2\sin^2\theta)^{3/2}} \tag{10.100}$$

In this form it is immediately clear that the electric field is no longer spherically symmetric. For $\theta = 0$, i.e., along the line of motion the field is reduced from the Coulomb value by a factor $(1 - \beta^2)$,

$$\vec{E} = \frac{q}{4\pi\varepsilon_0}\frac{\vec{R}}{R^3}(1 - \beta^2) \tag{10.101}$$

Perpendicular to the line of motion ($\theta = \pi/2$), the field is enhanced over the Coulomb value by a factor $(1 - \beta^2)^{-1/2}$:

$$\vec{E} = \frac{q}{4\pi\varepsilon_0}\frac{\vec{R}}{R^3}(1 - \beta^2)^{-1/2} \tag{10.102}$$

From equations (10.101) and (10.102) we see that for a fast-moving charge, the field is strongly concentrated in the plane at right angles to its motion (**Figure 10.11b**), no longer spherically symmetric as for a stationary charge (Figure 10.11a).

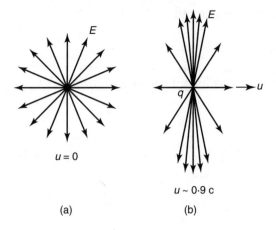

$u = 0$

$u \sim 0{\cdot}9\,c$

(a)   (b)

I FIGURE 10.11

If the charge is moving with a small uniform velocity, then $u \ll c$, $\gamma \to 1$ and the field given by equations (10.98) and (10.99) reduce to

$$\vec{E} = \frac{q}{4\pi\varepsilon_0}\frac{\vec{r}}{r^3}, \quad \vec{B} = \frac{q\mu_0}{4\pi}\frac{\vec{u} \times \vec{r}}{r^3} \tag{10.103}$$

We can easily derive these results directly. Because the charge is moving with a constant velocity u, the charge density must stay constant. That is, if the charge density at point P is ρ, then at a later time Δt the charge density at point Q that is $\vec{u}\Delta t$ from P, is also ρ. In other words, the charge density satisfies the relation

$$f(x + u_x\Delta t,\, y + u_y\Delta t,\, z + u_z\Delta t,\, t + \Delta t) = f(x, y, z, t)$$

Likewise, $\vec{E}$ and $\vec{B}$ also satisfy the same relation. From this relation we have

$$(\vec{u} \cdot \nabla) f + \partial f/\partial t = 0$$

which reduces to $u(\partial f/\partial x) + \partial f/\partial t = 0$, provided we let $\vec{u}$ parallel to x-axis. Applying this to ρ, $\vec{E}$, and $\vec{B}$ we obtain

$$\frac{\partial \rho}{\partial t} = -u\frac{\partial \rho}{\partial x}, \quad \frac{\partial \vec{E}}{\partial t} = -u\frac{\partial \vec{E}}{\partial x}, \quad \frac{\partial \vec{B}}{\partial t} = -u\frac{\partial \vec{B}}{\partial x} \tag{10.104}$$

Because when $u = 0$, we have

$$\vec{E} = -\nabla\phi, \quad \vec{B} = 0, \quad \phi(x, y, z) = \int \frac{\rho(\xi, \eta, \zeta)}{r}\, d\xi\, d\eta\, d\zeta$$

we expect that, when u is small,

$$\vec{E} = -\nabla\phi + 0\,(\vec{u}/c) \tag{10.105}$$

As for the $\vec{B}$ field, we go back to Ampere's law,

$$\nabla \times \vec{B} = \mu_0\rho\vec{u} + \mu_0\varepsilon_0\frac{\partial\vec{E}}{\partial t}$$

which can be rewritten, with the help of equation (10.104) and $\nabla \cdot \vec{E} = \rho/\varepsilon_0$, as

$$\nabla \times \vec{B} = \mu_0\varepsilon_0\vec{u}\nabla\cdot\vec{E} - \mu_0\varepsilon_0(\vec{u}\cdot\nabla)\vec{E} = \mu_0\varepsilon_0\nabla\times(\vec{u}\times\vec{E})$$

Thus

$$\vec{B} = \mu_0\varepsilon_0\vec{u}\times\vec{E} - \nabla\phi' \tag{10.106}$$

where ϕ' is to be determined. Because $\vec{B} = 0$ when $\vec{u} = 0$, ϕ' is of the order u/c. Now substituting equation (10.106) into Gauss' law for magnetic field, we get

$$\nabla \cdot \vec{B} = \nabla \cdot (\mu_0\varepsilon_0\vec{u}\times\vec{E} - \nabla\phi') = 0$$

from which it follows that

$$-\mu_0\varepsilon_0\nabla\cdot\vec{E} - \nabla^2\phi' = 0$$

Combining this with equation (10.105) we see that

$$\nabla\phi' = 0(u^2/c^2)$$

and then equation (10.106) becomes

$$\vec{B} = \mu_0\varepsilon_0\vec{u}\times\vec{E} + 0\,(u^2/c^2) \tag{10.107}$$

Now $\phi = q/4\pi\,\varepsilon_0 r$
Substituting this into equations (10.105) and (10.107) we get the results as given by equation (10.103).

10.23 Electromagnetic Field Invariants

Quantities that remain invariant when transformations are made from one inertial frame to another are particularly important. Because field vectors $\vec{E}$

and $\vec{B}$ characterizing electromagnetic field depend on the frame of reference, a natural question arises concerning the electromagnetic field invariants, and they are independent of the reference frame. It is fairly simple to show that there exist two such invariants:

$$B^2 - E^2/c^2 = B'^2 - E'^2/c^2 \quad \text{and} \quad \vec{E} \cdot \vec{B} = \vec{E'} \cdot \vec{B'} \tag{10.108}$$

We leave the proof as homework.

The invariance of these two quantities allows us to establish the following conclusions or predictions:

1. The invariance of $\vec{E} \cdot \vec{B}$ implies that if $\vec{E} \perp \vec{B}$ in some reference frame, in all other inertial frames $\vec{E'} \cdot \vec{B'}$ as well.
2. The invariance of $B^2 - E^2/c^2$ that if $E = cB$ in some reference frame, in all other inertial frames $\vec{E'} = c\vec{B'}$ as well.
3. If $\theta = \cos^{-1}(\vec{E} \cdot \vec{B}/EB)$ is acute (obtuse) in some frame, then θ will be acute (obtuse) in all other inertial frames.
4. If $E > cB$ (or $E < cB$) in some frame, then $E > cB$ (or $E < cB$) in all other inertial frames.
5. If both invariants are equal to zero, then $\vec{E} \perp \vec{B}$ and $E = cB$ in all inertial frames. It will be shown later that this is precisely observed in electromagnetic waves.
6. If the invariant $\vec{E} \perp \vec{B}$ alone is equal to zero, we can find a reference frame in which either $\vec{E'} = 0$ or $\vec{B'} = 0$. The sign of the other invariant determines which of these vectors is equal to zero.

It should be borne in mind that generally $\vec{E}$ and $\vec{B}$ depend both on coordinates and on time. Consequently, each invariant of (10.108) refers to the same point in space and time of the field, the coordinates and time of the point in different systems of frame being connected through the Lorentz transformations.

Problems

1. Observer O notes that two events are separated in space and time by 600 m and 8×10^{-7} sec. How fast must an observer to O' be moving relative to O in order that the events be simultaneous to O'?
2. A meter stick makes an angle of 30 degrees with respect to x'-axis of O'. What must be the value of v if the meter stick makes an angle of 45 degrees with respect to the x-axis of O?

3. Find the speed of a particle that has a kinetic energy equal to exactly twice its rest mass energy.

4. Find the law of transformation of the components of a symmetric four-tensor $T^{\mu\nu}$ under Lorentz transformations.

5. A man on a station platform sees two trains approaching each other at the rate $7/5\ c$, but the observer on one of the trains sees the other train approaching him with a velocity $35/37\ c$. What are the velocities of the trains with respect to the station?

6. The equation for a spherical pulse of light starting from the origin at $t = t' = 0$ is

$$c^2 dt^2 - x^2 - y^2 - z^2 = 0$$

Show from the Lorentz transformations that O' will observe this same pulse as spherical, in accordance with Einstein's postulate stating that the velocity of light is the same for all observers.

7. Referring to Figure 10.4, frame S' moves with a velocity u relative to frame S along the x axis. A pair of oppositely charged plates is at rest in S' frame in a direction parallel to x'-axis, the field E between the plates is perpendicular to the plates (i.e., $\perp$ to the x' axis) and has a value that depends on the charge density σ on the plates: $E = \sigma/\varepsilon_0$. Show that view from frame S in which the plates are now moving in the x-direction with a velocity u, the field E' is given by

$$E' = \frac{1}{\sqrt{1 - u^2/c^2}} E$$

8. Referring to previous problem, now the plates are at rest in frame S' along the y'-axis. Find the electric field in both frames.

9. A large metallic plate moves at a constant velocity $\vec{v}$ perpendicular to a uniform magnetic field $\vec{B}$. Find the surface charge density induced on the surface of the plate.

10. A point charge q moves at a constant velocity $\vec{v}$. Using the transformation formulas, find the magnetic field of this charge at a point whose radius vector is $\vec{r}$.

11. In a region where there exists only a uniform magnetic field $\vec{B}$, a charged particle is placed, at rest, in the origin of an inertial frame Σ. The charge remains at rest because there is no force acting on it. Now there exists a second inertial frame Σ' that moves relative to Σ with a constant velocity $\vec{v}$. Describe the motion of the charge in Σ'.

11

Electromagnetic Waves in Matter

11.1 The Wave Equation

We have seen that Maxwell equations admit electromagnetic waves. The propagation of electromagnetic waves in free space was discussed in Chapter 6. In this chapter, we discuss the propagation of electromagnetic waves in homogeneous, isotropic linear, and stationary media. We start again from Maxwell equations. Inside matter they become

$$
\left.
\begin{array}{ll}
(i)\ \ \nabla \cdot \vec{D} = \rho_f & (iii)\ \ \nabla \times \vec{E} = -\dfrac{\partial \vec{B}}{\partial t} \\[2mm]
(ii)\ \ \nabla \cdot \vec{B} = 0 & (iv)\ \ \nabla \times \vec{H} = \dfrac{\partial \vec{D}}{\partial t} + \vec{j}_f
\end{array}
\right\}
\tag{11.1}
$$

For isotropic and linear media

$$
\vec{D} = \varepsilon \vec{E}, \quad \vec{H} = \vec{B}/\mu, \quad \text{and} \quad \vec{j}_f = \sigma \vec{E}
\tag{11.2}
$$

If the medium is homogeneous, ε and μ do not vary from point to point, and equation (11.1) reduces to

$$
\left.
\begin{array}{ll}
(a)\ \ \nabla \cdot \vec{E} = \rho_f/\varepsilon & (c)\ \ \nabla \times \vec{E} = -\dfrac{\partial \vec{B}}{\partial t} \\[2mm]
(b)\ \ \nabla \cdot \vec{B} = 0 & (d)\ \ \nabla \times \vec{B} = \mu\varepsilon \dfrac{\partial \vec{E}}{\partial t} + \mu \vec{j}_f
\end{array}
\right\}
\tag{11.3}
$$

We call a medium stationary if it is at rest with respect to the coordinate system used. We further assume that the medium is infinite in extent. This avoids the problems of reflection and refraction, dealt with in the next chapter.

Taking the curl of equation (11.3c)

$$
\nabla \times (\nabla \times \vec{E}) + \frac{\partial}{\partial t}(\nabla \times \vec{B}) = 0
$$

Using the vector identity $\nabla \times (\nabla \times \vec{E}) = \nabla (\nabla \cdot \vec{E}) - \nabla^2 \vec{E}$ and equations (11.3d), (11.3a), and (11.2), the above equation becomes

$$\nabla^2 \vec{E} - \mu\varepsilon \frac{\partial^2 \vec{E}}{\partial t^2} - \sigma\mu \frac{\partial \vec{E}}{\partial t} = \nabla (\rho_f / \varepsilon) \qquad (11.4a)$$

Similarly, we have

$$\nabla^2 \vec{H} - \mu\varepsilon \frac{\partial^2 \vec{H}}{\partial t^2} - \sigma\mu \frac{\partial \vec{H}}{\partial t} = 0 \qquad (11.4b)$$

The right side of equation (11.4b) is zero because of the absence of the magnetic monopole.

Equations (11.4a) and (11.4b) describe an attenuated electromagnetic wave and $\vec{E}$ and $\vec{H}$ vectors are transverse. To see this, we consider again a plane electromagnetic wave propagating along z-direction. $\vec{E}$ and its components are not functions of x and y. Thus

$$\nabla \cdot \vec{E} = \frac{\partial E_x}{\partial x} + \frac{\partial E_y}{\partial y} + \frac{\partial E_z}{\partial z} = \rho_f / \varepsilon$$

reduces to

$$\frac{\partial E_z}{\partial z} = \rho_f / \varepsilon$$

From this we obtain

$$\frac{\partial^2 E_z}{\partial z^2} \hat{k} = \frac{\partial}{\partial z} (\rho_f / \varepsilon) \hat{k} \qquad (11.5)$$

We now write the wave equation (11.4a) in rectangular coordinate

$$\frac{\partial^2}{\partial z^2} (E_x \hat{i} + E_y \hat{j} + E_z \hat{k}) = \mu \left(\varepsilon \frac{\partial^2}{\partial t^2} + \sigma \frac{\partial}{\partial t} \right)(E_x \hat{i} + E_y \hat{j} + E_z \hat{k}) + \left(\frac{\partial}{\partial x} \hat{i} + \frac{\partial}{\partial y} \hat{j} + \frac{\partial}{\partial z} \hat{k} \right)\left(\frac{\rho_f}{\varepsilon} \right)$$

which reduces to, with the help of equation (11.5),

$$\frac{\partial^2}{\partial z^2} (E_x \hat{i} + E_y \hat{j}) = \mu \left(\varepsilon \frac{\partial^2}{\partial t^2} + \sigma \frac{\partial}{\partial t} \right)(E_x \hat{i} + E_y \hat{j} + E_z \hat{k}) + \left(\frac{\partial}{\partial x} \hat{i} + \frac{\partial}{\partial y} \hat{j} \right)\left(\frac{\rho_f}{\varepsilon} \right)$$

From this equation we obtain

$$\left(\varepsilon \frac{\partial^2 E_z}{\partial t^2} + \sigma \frac{\partial E_z}{\partial t} \right) = 0 \qquad (11.6)$$

It has solution of the form

$$E_z = a + be^{-\sigma t/\varepsilon}$$

where a and b are integration constants, independent of t. We see that E_z decreases exponentially with time and so there is no E_z wave. We may set $E_z = 0$ or a constant. Then from

$$\nabla \cdot \vec{E} = \frac{\partial E_z}{\partial z} = \frac{\rho_f}{\varepsilon}$$

we find that $\rho_f = 0$ for a plane wave in a conducting medium. Now in a non-conducting medium, $\sigma = 0$; then equation (11.6) reduces to

$$\varepsilon \frac{\partial^2 E_z}{\partial t^2} = 0$$

Integrating twice, we find

$$E_z = a + bt$$

which is not a function of z. And there is no wave. In conclusion, we have found for plane electromagnetic waves in matter

1. The $\vec{E}$ vector is transverse (normal to the direction of propagation),
2. The wave equation in matter takes the general form

$$\nabla^2 \vec{E} - \mu\varepsilon \frac{\partial^2 \vec{E}}{\partial t^2} - \sigma\mu \frac{\partial \vec{E}}{\partial t} = 0 \tag{11.7}$$

Similarly, we can show that the vector $\vec{H}$ is also transverse ($\nabla \cdot \vec{H} = 0 \rightarrow \partial H_z/\partial z = 0 \rightarrow H_z = 0$).

As in free space, we can show that the $\vec{E}$ and $\vec{H}$ vectors are mutually perpendicular, and $\vec{E} \times \vec{H}$ points in the direction of propagation.

11.2 Propagation of Plane Electromagnetic Waves in Nonconducting Media

In a uniform nonconducting medium, the term involving σ can be neglected and we have

$$\nabla^2 \vec{E} - \mu\varepsilon \frac{\partial^2 \vec{E}}{\partial t^2} = 0 \tag{11.8}$$

which differs from the vacuum analog only in the replacement of $\mu_0\varepsilon_0$ by $\mu\varepsilon$. Evidently, electromagnetic waves propagate through such a medium at a speed

$$v = \frac{1}{\sqrt{\varepsilon\mu}} = \frac{c}{n} \tag{11.9}$$

where

$$n = \sqrt{\varepsilon\mu/\varepsilon_0\mu_0} \tag{11.10}$$

is the index of refraction of the medium. For magnetic medium, μ is very close to μ_0, and so

$$n \approx \sqrt{\varepsilon/\varepsilon_0} = \sqrt{K} \tag{11.11}$$

Note that both n and K are functions of the frequency.

Because the dielectric constant K is almost always greater than 1, light travels more slowly through matter. This observation is trivial mathematically, but the physical implication is not trivial at all, as pointed out by D. J. Griffiths (see *Introduction to Electrodynamics*, third edition, 1999, p. 383, Prentice Hall).

As the wave passes through, the fields busily polarize and magnetize all the molecules, and the resulting (oscillating) dipoles create their own electric and magnetic fields. These combine with the original fields in such a way as to create a *single* wave with the same frequency but a different speed. This extraordinary conspiracy is responsible for the phenomenon of transparency. It is a distinctly nontrivial consequence of the *linearity* of the medium. More detailed discussion may be found in an article by M. B. James and D. J. Griffiths (see *American Journal of Physics*, 1992, 60:309).

As in free space, $\vec{E}$ and $\vec{H}$ are in phase. To see this, let us assume that there is a phase difference ϕ between $\vec{E}$ and $\vec{H}$, i.e.,

$$\vec{E} = \vec{E_0}e^{i(kz-\omega t)}, \quad \vec{H} = \vec{H_0}e^{i(kz-\omega t + \phi)}$$

The fields $\vec{E}$ and $\vec{H}$ have to satisfy the Maxwell equation

$$\nabla \times \vec{E} = -\frac{\partial\vec{B}}{\partial t}$$

Now because $\vec{E}$ travels in z-direction, $\frac{\partial}{\partial x}$ and $\frac{\partial}{\partial y}$ of $\vec{E}$ is zero, and the above equation reduces to

$$\frac{\partial E_y}{\partial z} = \frac{\partial B_x}{\partial t}, \quad \frac{\partial E_x}{\partial z} = -\frac{\partial B_y}{\partial t}$$

From the second equation,

$$ikE_{0x}e^{i(kz-\omega t)} = -i\omega\mu H_{0y}e^{i(kz-\omega t+\phi)}$$

Taking the real parts

$$kE_{0x}\cos(kz-\omega t) = \omega\mu H_{0y}\cos(kz-\omega t+\phi)$$

This is true for all z and t. The only value of ϕ that satisfies this condition is zero. Therefore, $\vec{E}$ and $\vec{H}$ are in phase. For the propagation of the wave in any arbitrary direction, we can write

$$\vec{E} = \vec{E}_0 e^{i(\vec{k}\cdot\vec{r}-\omega t)}, \quad \vec{H} = \vec{H}_0 e^{i(\vec{k}\cdot\vec{r}-\omega t)}$$

Substituting these into Maxwell equations

$$\nabla\cdot\vec{E} = 0 \,(\rho_f = 0 \text{ in non-conductor}), \quad \nabla\cdot\vec{B} = 0$$

we obtain

$$\vec{k}\cdot\vec{E} = 0 \quad \text{and} \quad \vec{k}\cdot\vec{H} = 0$$

These show that both $\vec{E}$ and $\vec{H}$ are perpendicular to the propagation vector $\vec{k}$. Furthermore, we can show that $\vec{H}$ is perpendicular both to $\vec{k}$ and $\vec{E}$. (You can see this by substituting $\vec{E} = \vec{E}_0 e^{i(\vec{k}\cdot\vec{r}-\omega t)}$ into Maxwell equation $\nabla\times\vec{E} = -\partial\vec{B}/\partial t$.)

All the previous results for free space may carry over with the simple replacements: $\varepsilon_0 \to \varepsilon$, $\mu_0 \to \mu$, and $c \to v$. For examples, the energy density is

$$u = \frac{1}{2}\left(\varepsilon E^2 + \frac{1}{\mu}B^2\right) \tag{11.12}$$

and the electric and magnetic energy densities are equal,

$$\frac{\varepsilon E^2/2}{\mu H^2/2} = \frac{\varepsilon}{\mu}\left(\frac{\mu}{\varepsilon}\right) = 1$$

the Poynting vector is

$$\vec{S} = \frac{1}{\mu}\left(\vec{E}\times\vec{B}\right) \tag{11.13}$$

and the intensity I, which is the average power per unit area transported by the electromagnetic wave, is now given by

$$I = \frac{1}{2} v \varepsilon E_0^2 \qquad (11.14)$$

It should be noted that in dielectrics there can be losses, associated with the damping of the electronic oscillators in the simple atomic model for dielectrics. In this simple atomic model, electrons are bounded to their nuclei. Any displacement by an applied electric field is balanced by a restoring force proportional to the displacement. For this model, application of a linearly polarized electromagnetic wave with electric vector of magnitude

$$E_x = E_0 e^{i(kz - \omega t)}$$

would induce a damped, simple, harmonic motion in the bound electrons given by

$$-eE_x = m(\ddot{x} + b\dot{x} + \omega_0^2 x)$$

where x is the displacement parallel to E_x and b is the damping constant of the electronic oscillators of natural frequency ω_0. The effect of the magnetic component of the electromagnetic wave is assumed to be negligible. The steady-state displacement is the dynamic response to the incident wave, after initial transients:

$$x = x_0 e^{i\omega t}$$

and so

$$\dot{x} = i\omega x, \quad \ddot{x} = -\omega^2 x$$

Hence

$$\vec{x} = \frac{-e}{m(\omega_0^2 - \omega^2 + ib\omega)} \vec{E}$$

Now the instantaneous dipole moment due to the displacement of the electron is

$$\vec{P} = -e\vec{x} = \alpha \vec{E}$$

From the last two equations, we have the polarizability

$$\alpha(\omega) = \frac{e^2}{m(\omega_0^2 - \omega^2 + ib\omega)}$$

Clearly, the polarizability $\alpha(\omega)$ is complex. For electromagnetic waves this is seen as an attenuation of the wave as it penetrates a dielectric.

11.3 Propagation of Plane Electromagnetic Waves in Conducting Media

Now the wave equations take the general form

$$\nabla^2 \vec{E} - \mu\varepsilon \frac{\partial^2 \vec{E}}{\partial t^2} - \sigma\mu \frac{\partial \vec{E}}{\partial t} = 0 \tag{11.15a}$$

$$\nabla^2 \vec{H} - \mu\varepsilon \frac{\partial^2 \vec{H}}{\partial t^2} - \sigma\mu \frac{\partial \vec{H}}{\partial t} = 0 \tag{11.15b}$$

Again, we consider a plane-polarized wave $\vec{E}$ parallel to the x-axis

$$\vec{E} = E_0 e^{i(kz - \omega t)} \hat{\imath} \tag{11.16}$$

It is easy to show that the wave number (or propagation constant) k is complex. Substituting equation (11.16) into equation (11.15a) we obtain

$$\left(k^2 - \varepsilon\mu\omega^2 - i\sigma\mu\omega\right)\vec{E} = 0$$

which requires

$$k^2 = \varepsilon\mu\omega^2\left(1 + i\frac{\sigma}{\varepsilon\omega}\right) \tag{11.17}$$

where the first term corresponds to the displacement current and the second to the conduction current. Thus, in a conducting medium, the wave number k is complex. We may express it as

$$k = \alpha + i\beta \tag{11.18}$$

Squaring thus and comparing with equation (11.17), we find

$$\alpha^2 - \beta^2 = \varepsilon\mu\omega^2 \quad \text{and} \quad 2\alpha\beta = \sigma\mu\omega$$

Solving these equations simultaneously, we find

$$\alpha = \omega \sqrt{\varepsilon\mu/2} \left\{ 1 + \left[1 + (\sigma/\varepsilon\omega)^2 \right]^{1/2} \right\}^{1/2} \tag{11.19}$$

where the positive root has been chosen to make the solutions yield the proper form for k in free space, and

$$\beta = \omega \sqrt{\varepsilon\mu/2} \left\{ -1 + \left[1 + (\sigma/\varepsilon\omega)^2 \right]^{1/2} \right\}^{1/2} \tag{11.20}$$

Now equation (11.16) becomes, after k is replaced by $\alpha + i\beta$,

$$\vec{E} = E_0 e^{-\beta z} e^{i(\alpha z - \omega t)} \hat{\imath} \tag{11.21}$$

where α and β are given by equations (11.19) and (11.20). Because $\beta > 1$, equation (11.21) shows that a plane wave cannot propagate in a conducting medium without attenuation. As a plane wave is propagating in a conducting medium, the oscillating electric field in the wave sets up currents. Work must be done to drive the currents, and some of the energy is dissipated as heat in the medium. This results in the attenuation of the wave. Thus, the quantity β is called the absorption coefficient, and it is a measure of the attenuation.

a. The Skin Depth

For a good conductor, $\sigma/\varepsilon\omega \gg 1$ for frequencies up to at least the microwave range, equations (11.19) and (11.20) reduce to

$$\alpha = \beta = \sqrt{\omega\sigma\mu/2} = 1/\delta \tag{11.22}$$

where

$$\delta = \sqrt{2/(\sigma\omega\mu)} \tag{11.23}$$

and equation (11.21) becomes

$$\vec{E} = \hat{\imath} E_0 e^{-z/\delta} e^{i(z/\delta - \omega t)} \tag{11.24}$$

which indicates that when $\delta = z$, the amplitude decreases in magnitude to $1/e$ times its value at the surface. The quantity δ therefore is a measure of the distance of penetration of an electromagnetic wave into a good conductor before its magnitude drops to $1/e$ times its value at the surface. The distance δ is called the skin depth, and it is much less than the wavelength $2\pi k$ of the electromagnetic wave in the conductor.

Equation (11.23) shows that $\delta \to 0$ as $\sigma \to \infty$, and δ is very small for good conductors at high frequency currents. For example, for copper $\sigma = 58 \times 10^6$ mhos per meter (one mho is reciprocal of ohm), the skin depth at various frequencies is

Frequency	δ
60 Hz	8.5×10^{-3} m
1 MHz	6.6×10^{-5} m
30 MHz	3.8×10^{-7} m

Radio communication on submarines becomes difficult at depths of several meters because of the relatively higher skin depth of seawater ($\delta = 10^{-1}$ m at 30 kHz).

Because in high frequency circuits current flows only on the surface of the conductor, a poor conductor can be made into a good conductor with a thin coating of copper or silver.

b. The Poynting Vector

The attenuated plane waves, equation (11.21), satisfy the modified wave equation (11.15a). Maxwell equations (11.3) impose further constraints; as before, equations (11.3a) and (11.3b) rule out any z-component, so the fields are transverse. But we will see that $\vec{E}$ and $\vec{H}$ are not in phase. So we have to handle the Poynting vector with care. To see these, let us go back to equation (11.16) and the associated $\vec{H}$ waves:

$$\vec{E} = E_0 e^{i(kz - \omega t)}\hat{\imath} = E_0 e^{i[(\alpha + i\beta)z - \omega t]}\hat{\imath}$$
$$= E_0\left(e^{i\alpha z} e^{-\beta z}\right) e^{-i\omega t}\hat{\imath} = (\vec{E}_{0r} + i\vec{E}_{0i}) e^{-i\omega t} \tag{11.25}$$

where

$$\vec{E}_{0r} = E_{0r}\hat{\imath}, \qquad \vec{E}_{0i} = E_{0i}\hat{\imath}$$
$$E_{0r} = \cos(\alpha z)\, e^{-\beta z}, \quad E_{0i} = \sin(\alpha z)\, e^{-\beta z} \tag{11.26}$$

Similarly,

$$\vec{H} = (\vec{H}_{0r} + i\vec{H}_{0i})\, e^{-i\omega t}\hat{\jmath} \tag{11.27}$$

We first calculate the Poynting vector, averaged over time:

$$\vec{S}_{av} = \left(\text{Re}\,\vec{E} \times \text{Re}\,\vec{H} \right)_{av}$$

$$= \left[\left(\vec{E}_{0r}\cos\omega t - \vec{E}_{0i}\sin\omega t \right) \times \left(\vec{H}_{0r}\cos\omega t - \vec{E}_{0i}\sin\omega \right) \right]_{av}$$

$$\vec{S}_{av} = \left[\left(\vec{E}_{0r} \times \vec{H}_{0r}\cos^2\omega t + \vec{E}_{0i} \times \vec{H}_{0i}\sin^2\omega t \right) - \vec{E}_{0r} \times \vec{H}_{0i}\cos\omega t \sin\omega t - \vec{E}_{0i} \times \vec{H}_{0r}\sin\omega t \cos\omega t \right]_{av}$$

But

$$[\cos^2\omega t]_{av} = [\sin^2\omega t]_{av} = \frac{1}{2} \quad \text{and} \quad [\sin\omega t\,\cos\omega t]_{av} = 0$$

and the expression for $\vec{S}_{av}$ now reduces to

$$\vec{S}_{av} = \frac{1}{2}\left(\vec{E}_{0r} \times \vec{H}_{0r} + \vec{E}_{0i} \times \vec{H}_{0i} \right) \tag{11.28}$$

This can be rewritten in a more convenient form. To this end, we observe that

$$\frac{1}{2}\,\text{Re}\left(\vec{E} \times \vec{H}^* \right) = \frac{1}{2}\,\text{Re}\left[\left(\vec{E}_{0r} + i\vec{E}_{0i} \right) e^{i\omega t} \times \left(\vec{H}_{0r} - i\vec{H}_{0i} \right) e^{-i\omega t} \right]$$

$$= \frac{1}{2}\,\text{Re}\left[\left(\vec{E}_{0r} + i\vec{E}_{0i} \right) \times \left(\vec{H}_{0r} - i\vec{H}_{0i} \right) \right]$$

$$= \frac{1}{2}\left(\vec{E}_{0r} \times \vec{H}_{0r} + \vec{E}_{0i} \times \vec{H}_{0i} \right).$$

Thus, we write $\vec{S}_{av}$ as

$$\vec{S}_{av} = \frac{1}{2}\,\text{Re}\left(\vec{E} \times \vec{H}^* \right) \tag{11.29}$$

This is the most convenient expression for calculating $\vec{S}_{av}$. Because both $\vec{E}$ and $\vec{H}$ are proportional to $e^{-z/\delta}$, $\vec{S}_{av}$ is proportional to $e^{-2z/\delta}$, and the power density decreases very rapidly with distance:

$$\vec{S}_{av} = \frac{1}{2}\,\text{Re}\left(\vec{E} \times \vec{H}^* \right) = \frac{1}{2}\left(\frac{\sigma}{2\omega\mu} \right)^{1/2} e^{-2z/\delta} E_0^2\,\hat{k}$$

c. $\vec{E}$ and $\vec{H}$ Are Not in Phase

Unlike the nonconductor case, $\vec{E}$ and $\vec{H}$ are no longer in phase. To see this, let us go back to Maxwell equations that are now written as

$$\nabla \cdot \vec{E} = 0 \quad \nabla \times \vec{E} + \mu \frac{\partial \vec{H}}{\partial t} = 0$$

$$\nabla \cdot \vec{H} = 0 \quad \nabla \times \vec{H} - \varepsilon \frac{\partial \vec{E}}{\partial t} = \sigma \vec{E}.$$

Substituting the plane-polarized waves

$$\vec{E} = E_0 e^{i(kz - \omega t)}\hat{i}, \quad \vec{H} = H_0 e^{i(kz - \omega t)}\hat{j}$$

into Maxwell equations, we obtain

$$ik\vec{E} \cdot \hat{k} = 0 \quad -ik\vec{E} \times \hat{k} = i\omega\mu\vec{H}$$

$$ik\vec{H} \cdot \hat{k} = 0 \quad -ik\vec{H} \times \hat{k} = \sigma\vec{E} - i\omega\varepsilon\vec{E}$$

or

$$\hat{k} \cdot \vec{E} = 0, \quad \vec{H} = -\frac{k}{\omega\mu} \hat{k} \times \vec{E}$$

$$\hat{k} \cdot \vec{H} = 0, \quad \vec{E} = -\frac{k}{\omega\varepsilon - i\sigma} \hat{k} \times \vec{H}$$

Note that the quantity k is the wave number, whereas $\hat{k}$ is a unit coordinate vector along the z-axis.

We see that, as in nonconductors, $\vec{E}$ and $\vec{H}$ are transverse and orthogonal, and $\vec{E} \times \vec{H}$ points in the direction of propagation. The ratio of E/H is a complex quantity:

$$\frac{E}{H} = \frac{k}{\omega\varepsilon - i\sigma} = \frac{\omega\mu}{k} \tag{11.30}$$

This shows that $\vec{E}$ and $\vec{H}$ are not in phase. We need the value of k to determine the phase relation between $\vec{E}$ and $\vec{H}$. Now we go back to equation (11.17), and we see that the imaginary part of k^2 is much larger than unity, and so we write

$$k^2 \cong \omega^2\varepsilon\mu \left(-i \frac{\sigma}{\omega\varepsilon}\right) = -i\omega\sigma\varepsilon$$

from which we find k

$$k = \sqrt{-i}\sqrt{\omega\sigma\mu}$$

What is $\sqrt{-i}$? Using polar form we have

$$-i = e^{-i\pi/2} = \cos(\pi/2) - i\sin(\pi/2)$$

hence

$$\sqrt{-i} = \left(e^{-i\pi/2}\right)^{1/2} = e^{-i\pi/4}$$

and accordingly

$$k = \sqrt{\omega\varepsilon\sigma}\, e^{-i\pi/4}$$

Substituting this into equation (11.30) we find

$$\frac{E}{H} = \frac{\omega\mu}{k} = \sqrt{\frac{\omega\mu}{\sigma}}\, e^{-i\pi/4} \tag{11.31}$$

Equation (11.31) indicates clearly that $\vec{H}$ lags $\vec{E}$ by 45 degrees (or $\pi/4$ radians) in good conductors. Now we can write, taking into account the phase difference between $\vec{E}$ and $\vec{H}$:

$$\vec{E} = E_0 e^{-z/\delta} e^{i(z/\delta - \omega t)}\hat{i}, \quad \vec{H} = \sqrt{\sigma/\omega\mu}\, e^{i(z/\delta - \omega t - \pi/4)}\hat{j} \tag{11.32}$$

What is the speed of propagation v? The value of α (the real part of k) supplies the answer:

$$v = \frac{\omega}{\mathrm{Re}(k)} = \frac{\omega}{\alpha} = \sqrt{\frac{2\omega}{\sigma\mu}} \tag{11.33}$$

It is easy to show that the energy density is almost in the magnetic form:

$$\frac{\varepsilon E_0^2/2}{\mu H_0^2/2} = \frac{\varepsilon}{\mu}\frac{\omega\mu}{\sigma} = \frac{\omega\varepsilon}{\sigma} \tag{11.34}$$

where we have used the relation $H_0 = \sqrt{\sigma/(\omega\varepsilon)}E_0$. For good conductors $\sigma/\omega\varepsilon \gg 1$ (or $\omega\varepsilon/\sigma \ll 1$); thus the above equation indicates that the energy density is essentially all in the magnetic form.

11.4 Propagation of Plane Electromagnetic Waves in a Uniform Plasma

Plasma is an ionized gas that contains free electrons and massive positive ions. We have studied plane electromagnetic waves in linear dielectrics, in which the current is entirely of the displacement type, and in good conducts, in which current of the conduction type is predominant; and we have left the permittivity ε and the electrical conductivity σ unspecified in these earlier discussions (to determine them takes us beyond the limit of classical electromagnetic theory). We now study the propagation of plane electromagnetic waves in a uniform plasma, in which the current arises from the presence of free electrons and ions in the electric and magnetic fields of the incident electromagnetic waves, and see how it is possible to obtain an expression for ε and for σ. For simplicity we consider a cold plasma in which the thermal motion of electrons and ions may be neglected.

Consider a plane electromagnetic wave traveling in $+z$ direction with $\vec{E}$ and $\vec{H}$ vectors, respectively, parallel to the x- and y-axes. An electron of charge q, mass m, and velocity $\vec{v}$ is subjected to a Lorentz force

$$\vec{F} = q\,(\vec{E} + \vec{v} \times \vec{B})$$

We are not concerned with the motion of the positive ions because these particles, being massive in comparison with the electrons, are correspondingly sluggish in their response to an electric field. When the period of the field alternations becomes comparable with or shorter than the mean free time τ, collisions must be taken into account and the full equation of motion has the form

$$m\frac{d\vec{v}}{dt} + \frac{m}{\tau}\vec{v} = q\,(\vec{E} + \vec{v} \times \vec{B}) \tag{11.35}$$

where the term $m\vec{v}/\tau$ represents the momentum loss due to collisions. We may neglect the magnetic force in the Lorentz force for non-relativistic particle motion in the field of a plane wave, because

$$\frac{|\vec{v} \times \vec{B}|}{|\vec{E}|} \approx \frac{|\vec{v} \times (\vec{k} \times \vec{E})|}{\omega E} \approx \frac{v}{c}$$

Then the equation of motion becomes

$$m\frac{d\vec{v}}{dt} + \frac{m}{\tau}\vec{v} = q\vec{E} \tag{11.36}$$

Assuming a time dependence of the electric field

$$\vec{E} = \vec{E}_0 e^{i\omega t}$$

the velocity $\vec{v}(t)$ then also has the same time frequency: $\vec{v} = \vec{v}_0 e^{i\omega t}$. Equation (11.36) then gives

$$-im\omega\vec{v}_0 + m\vec{v}_0/\tau = q\vec{E}_0$$

or

$$\vec{v_0} = \frac{q\tau/m}{1 - i\omega\tau} \vec{E_0} \qquad (11.37)$$

From this we may write for the corresponding current density $\vec{j}$

$$\vec{j} = Nq\vec{v} = \frac{Nq^2\tau}{m} \frac{1}{1 - i\omega\tau} \vec{E} \qquad (11.38)$$

Now plasma in many instances may, to a very good approximation, be assumed to be collisionless. Setting $\tau \to \infty$ in equation (11.38) we find

$$\vec{j} = i \frac{Nq^2}{m\omega} \vec{E} \qquad (11.39)$$

and we see that there is a phase difference between $\vec{j}$ and $\vec{E}$.

The electric conductivity σ is

$$\begin{aligned}
\sigma = \frac{j}{E} &= \frac{Nq^2\tau}{m} \frac{1}{1 - i\omega\tau} \\
&= \frac{Nq^2\tau}{m(1 + \omega^2\tau^2)} + i \frac{Nq^2\tau^2\omega}{m(1 + \omega^2\tau^2)}
\end{aligned} \qquad (11.40)$$

The conductivity σ is complex and approaches a very small purely imaginary value as $\omega\tau \gg 1$. For collisionless plasma, $\tau \to \infty$, and equation (11.40) reduces to

$$\sigma = i \frac{Nq^2}{m\omega} \qquad (11.40a)$$

The meaning of this imaginary conductivity is made evident by consideration of both the conduction and displacement currents. We do so a little later.

Now the electron current lags the electric field intensity by 90 degrees. The electron current can therefore be said to be inductive. Because $\vec{j}$ and $\vec{E}$ are 90 degrees out of phase, the scalar product $\vec{j} \cdot \vec{E}$ is also purely imaginary, and there is no energy loss in the medium. This means that, on the average, the oscillating electrons in the current do not gain energy from the field once it is established.

Because the displacement current $\partial \vec{D}/\partial t = i\varepsilon_0 \omega \vec{E}$ leads the electric field $\vec{E}$ by 90 degrees, whereas the electric current lags by the same angle, so the displacement and electric currents are 180 degrees out of phase, and the total current is less than if there were no electrons present. As an example, consider a plasma-filled parallel-plate capacitor, where the total current density is

$$\vec{j}_{total} = \frac{\partial \vec{D}}{\partial t} + \vec{j} = i\omega\varepsilon_0 \vec{E} - i\frac{Nq^2}{m\omega}\vec{E}$$

$$= i\varepsilon_0\omega\left(1 - \frac{\omega_p^2}{\omega^2}\right)$$

(11.41)

where the second term in the parentheses is the ratio of the convection to the displacement current density and where

$$\omega_p = \left(\frac{Nq^2}{m\varepsilon_0}\right)^{1/2}$$

(11.42)

is the plasma frequency.

$\vec{j}_{total} = 0$ when $\omega = \omega_p$. We have here a resonance phenomenon that makes the total current zero at a certain critical value of ω. This is mathematically similar to that in which an alternating voltage V is applied to a capacitor and an inductor connected in parallel, as shown in **Figure 11.1**. The total current is given by

$$I = Vi\omega C + \frac{V}{i\omega L} = i\omega C\left(1 - \frac{1}{\omega^2 LC}\right)V$$

We see that $I = 0$ at $\omega = 1/(LC)^{1/2}$. At resonance, both currents are finite and equal in magnitude but opposite in sign so that the total current is zero.

The plasma frequency depends solely on the properties of the ionized gas considered. It corresponds to a frequency

$$f_p = 8.94N^{1/2} \text{ hertz}$$

(11.43)

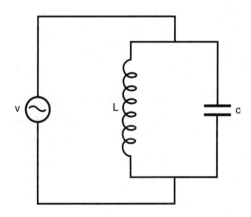

I FIGURE 11-1

In the ionosphere N is of the order of 10^{11} electrons/meter^3, f_p is about 3 megahertz, whereas in plasma where N is of the order of 10^{18} electrons/meter^3, f_p is about 10^4 megahertz.

a. Effective Dielectric Constant

Let us now try to understand the meaning of the complex conductivity. To this end, we rewrite the total current as

$$\vec{j}_{total} = \vec{j} + \partial\vec{D}/\partial t = \sigma\vec{E} + \varepsilon\partial\vec{E}/\partial t = (\sigma - i\omega\varepsilon)\,\vec{E}$$

$$= \left[\sigma_{re} + i\sigma_{im} - i\omega\left(\varepsilon_{re} + i\varepsilon_{im}\right)\right]\vec{E}$$

or

$$\vec{j}_{total} = i\omega\left[\left(\varepsilon_{re} - \frac{\sigma_{im}}{\omega}\right) + i\left(\varepsilon_{im} + \frac{\sigma_{re}}{\omega}\right)\right]\vec{E} \tag{11.44}$$

We see that σ_{im}, the imaginary part of the conductivity, makes a negative contribution to ε and to the dielectric constant K ($= \varepsilon/\varepsilon_0$) of the material. Equation (11.44) suggests that we can define an effective dielectric constant K_{eff}

$$K_{eff} = K\,(1 - \sigma_{im}/\varepsilon_0\,\omega) \tag{11.45}$$

where σ_{im} is the imaginary part of the right side of equation (11.40). Thus, for $\omega\tau \gg 1$,

$$K_{eff} = K\left[1 - \frac{Nq^2}{m\omega}\frac{1}{\varepsilon_0\,\omega}\right] = K\,[1 - (\omega_p/\omega)^2] \tag{11.46}$$

It is seen that the effective dielectric constant is negative for frequencies below the plasma frequency (as long as the assumption $\omega\tau \gg 1$ remains valid). This has profound effects on the propagation of electromagnetic waves through material.

b. Wave Propagation at Frequencies Higher Than ω_p

We consider a collisionless plasma, and then the conductivity σ is given by equation (11.40a). Substituting this into equation (11.17) for the wave number k, we obtain

$$k^2 = \frac{\omega^2}{c^2}\left[1 - \frac{\omega_p^2}{\omega^2}\right] \tag{11.47}$$

or

$$\omega^2 = \omega_p^2 + k^2 c^2 \tag{11.48}$$

Thus, for frequencies above the plasma frequency, k is real and waves propagate in the plasma without attenuation. The phase velocity

$$v_{ph} = \frac{\omega}{k} = \frac{c}{\left[1 - (\omega_p^2/\omega^2)\right]^{1/2}} \tag{11.49}$$

is always greater than the velocity of light in vacuum. Because the phase velocity v_{ph} depends on frequency, plasma is disperse; equation (11.48) is the corresponding dispersion relation. The velocity of propagation of a wave packet, which disperses as it propagates, is given by the group velocity v_g,

$$v_g = \frac{d\omega}{dk} = \frac{kc^2}{\omega} = \frac{c^2}{v_{ph}} c \tag{11.50}$$

For very high frequencies where $\omega \gg \omega_p$, the wave is unaware of the presence of the plasma and behaves as in free space, i.e., $\omega^2 \cong k^2 c^2$.

To obtain some inside picture, let us consider a wave propagating in $+z$-direction and with its $\vec{E}$ vector parallel to the x-axis,

$$\vec{E} = E_0 e^{i(kz - \omega t)} \hat{i}$$

From the general relation for E/H, equation (11.30), we find

$$H = \frac{k}{\omega\mu} E = \left(\frac{\varepsilon_0}{\mu_0}\right)^{1/2} \left[1 - \left(\frac{\omega_p}{\omega}\right)^2\right]^{1/2} E_0 e^{i(kz - \omega t)}$$

When $\omega > \omega_p$, the wave number k is real and the $\vec{E}$ and $\vec{H}$ vectors in phase, the ratio E/H being larger than in free space by a factor of $1/[1 - (\omega_p/\omega)^2]^{1/2}$. As ω approaches ω_p, H tends to zero. This is expected, because the total current density $\vec{j} + (\partial\vec{D}/\partial t)$ tends to zero. The Poynting vector $\vec{S}$ also tends to zero.

Problems

1. Two waves, having the same frequency, amplitude, and direction of polarization, travel in opposite directions in a medium. Find the resultant field (it is a standing wave).

2. Two waves travel in the same direction, having the same frequency and in phase with each other but having different amplitudes and their planes of polarization being perpendicular to each other. What is the resultant field?

3. If the two waves of the previous problem have equal amplitude but are 90 degrees out of phase with each other, find the resultant field.

4. Consider a plane wave (time-dependence omitted)

$$\vec{H} = H_0 e^{\pm kz} \hat{e}_H$$

where $\hat{e}_H$ is a fixed unit vector. Show that $\hat{k} \cdot \hat{e}_H = 0$, where $\hat{k}$ is a unit in +z-direction.

5. Examine the field

$$\vec{E}(z,t) = 10 \sin(kz + \omega t)\,\hat{i} + 10 \cos(kz + \omega t)\,\hat{j}$$

in the $z = 0$ plane, for $\omega t = 0$, $\pi/4$, $\pi/2$, $3\pi/4$, and π. Is $\vec{E}(z,t)$ circularly polarized? What is the direction of propagation of the wave?

6. Consider the plane wave

$$\vec{E} = \hat{i} E_{0x} \cos[\omega t - kz + \alpha] + \hat{j} E_{0y} \cos[\omega t - kz + \beta]$$

Show that it is elliptically polarized when $\beta = \alpha \pm \pi/2$.

7. Estimate the skin depth and wave velocity in copper ($\sigma = 6 \times 10^7$ mho/m) at a frequency of 1 gigahertz (0^9 Hz).

8. Show that the skin depth in a poor conductor ($\sigma \ll \omega\varepsilon$) is $(2/\sigma)(\varepsilon/\mu)^{1/2}$. And it is $\lambda/2\pi$ for a good conductor ($\sigma \gg \omega\varepsilon$), where λ is the wavelength in the conductor.

9. Given $f = f_0 e^{-i\omega t}$ and $g = g_0 e^{-i\omega t}$, show that

$$\langle \mathrm{Re}(f)\,\mathrm{Re}(g) \rangle = \frac{1}{2}\,\mathrm{Re}(fg^*)$$

where < > denotes time average over any integral number of cycles and the asterisk denotes complex conjugate.

10. Calculate the time-average energy density and energy flux in a plane-polarized plane wave in a dielectric medium, assuming a real frequency-independent dielectric response function. What is the velocity of energy propagation?

11. A circularly polarized wave results from the superposition of two waves that are (a) of the same frequency and amplitude, (b) plane-polarized in perpendicular directions, and (c) 90 degrees out of phase. Show that the average value of the Poynting vector for such a wave is equal to the sum of the average values of the Poynting vectors of the two waves.

12 Electromagnetic Waves in Bounded Media

12.1 Introduction

In Chapter 11 we studied the propagation of electromagnetic waves in linear, homogeneous, isotropic, and stationary media. The medium was silently assumed to be infinite in extent, so we avoided the problems of reflection and refraction. In this chapter we extend our discussion in bounded media, and so we can no longer avoid the problems of reflection and refraction. But we limit our discussion at plane boundaries between two different media that are the simplest to treat. Curved boundaries tend to scatter incident plane waves into different directions simultaneously. Such problems are, in principle, solvable by superposing an infinite series of uniform plane waves; obviously they are difficult to handle.

Thus, we assume an ideally thin, infinite, plane interface between two linear, homogeneous, isotropic media. An incident wave along $\hat{k}_i$ give rises to a reflected wave along $\hat{k}_r$ and a refracted (or transmitted) wave along $\hat{k}_t$. The three waves satisfy the appropriate conditions of continuity for the tangential components of $\vec{E}$ and $\vec{H}$ at the interface. For future convenience, these conditions are summarized in **Table 12.1**. For the time-dependent case the boundary conditions on the normal components of $\vec{D}$ and $\vec{B}$ are not

Table 12-1

Dielectric	Conductors
$E_{t1} = E_{t2}$	$E_{t1} = E_{t2}$ (=0 for perfect conductors)
$D_{n1} = D_{n2}$	$D_{n1} = D_{n2} = \rho$
$H_{t1} = H_{t2}$	$H_{t1} - H_{t2} = 0$ (or $j_{s\perp}$ for perfect conductors)
$B_{n1} = B_{n2}$	$B_{n1} = B_{n2}$

independent but are contained already in Maxwell field equations. The boundary conditions on D_n and B_n are automatically satisfied provided the conditions on E_t and H_t are met.

For perfect conductors, the skin depth becomes vanishingly small. Thus, electric fields are excluded from inside a perfect conductor and the tangential component of the magnetic field also vanishes within the conductor. It is not zero outside, so that $H_{t1} - H_{t2} = j_{s\perp}$, where $j_{s\perp}$ is the current per unit width, assumed to flow as a vanishingly thin current sheet. In general, the concept of perfect conductivity is not very useful at optical frequencies because for most metals it is a poor approximation.

12.2 Reflection and Refraction of Plane Waves at a Dielectric Boundary

a. Law of Reflection and Snell's Law (Law of Refraction)

Both media are assumed to be dielectrics (so $\sigma = 0$) and no energy losses occur. The incident and reflected waves in medium 1 and the refracted wave in medium 2 are shown in **Figure 12.1** ($\vec{E}$ is normal to the incident plane). To avoid the problems of multiple reflections, we assume the media extend to infinity on either side of the interface. The origin is at some convenient point on the interface (the plane $x = 0$).

The incident wave is of the form

$$\vec{E}_i = \vec{E}_{0i} e^{i(k_i \cdot \vec{r} - \omega_i t)}, \quad \vec{H}_i = \frac{\vec{k}_i \times \vec{E}_i}{\omega_i \mu_1} \tag{12.1}$$

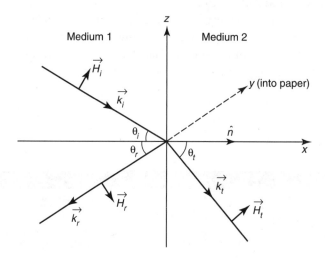

I FIGURE 12.1

The vector $\vec{k}_i$ is normal to the wave fronts and points in the direction of propagation. We expect the reflected and refracted waves to be of the form:

$$\vec{E}_r = \vec{E}_{0r} e^{i(\vec{k}_r \cdot \vec{r} - \omega_r t)}, \quad \vec{H}_r = \frac{\vec{k}_r \times \vec{E}_r}{\omega_r \mu_1} \qquad (12.2)$$

and

$$\vec{E}_t = \vec{E}_{0t} e^{i(\vec{k}_t \cdot \vec{r} - \omega_t t)}, \quad \vec{H}_t = \frac{\vec{k}_t \times \vec{E}_t}{\omega_t \mu_2} \qquad (12.3)$$

We have made no assumptions regarding the amplitude, phases, frequencies of the waves, or the directions of the reflected and refracted waves.

We now apply boundary conditions to these waves:

1. The tangential components of both $\vec{E}$ and $\vec{H}$ must be continuous across the interface, i.e.,

$$(\vec{E}_i + \vec{E}_r)_{\text{tang}} = (\vec{E}_t)_{\text{tang}} \qquad (12.4a)$$

$$(\vec{H}_i + \vec{H}_r)_{\text{tang}} = (\vec{H}_t)_{\text{tang}} \qquad (12.4b)$$

2. The normal components of $\vec{D}$ and $\vec{B}$ must be continuous, i.e.,

$$(\vec{D}_i + \vec{D}_r)_{\text{nor}} = (\vec{D}_t)_{\text{nor}} \qquad (12.5a)$$

$$(\vec{B}_i + \vec{B}_r)_{\text{nor}} = (\vec{B}_t)_{\text{nor}} \qquad (12.5b)$$

To obtain continuity of the tangential components of $\vec{E}$ and $\vec{H}$ at the interface, we have

(a) All three vectors $\vec{E}_i, \vec{E}_r,$ and $\vec{E}_t$ are identical functions of the time t;
(b) All three vectors $\vec{E}_i, \vec{E}_r,$ and $\vec{E}_t$ are identical functions of position on the interface;
(c) There exist certain relations between $\vec{E}_{0i}, \vec{E}_{0r},$ and $\vec{E}_{0t}$.

From (a) we have

$$\omega_i = \omega_r = \omega_t \equiv \omega \qquad (12.6)$$

i.e., all three waves have the same frequency. This is quite obvious, because they are superposition of the wave emitted by the source and of those waves emitted by the electrons executing forced vibrations in media 1 and 2. The forced vibrations have the same frequency as the applied one.

From (b) we have at all points on the interface (the plane $x = 0$) at all time

$$\vec{k}_i \cdot \vec{r} = \vec{k}_r \cdot \vec{r} = \vec{k}_t \cdot \vec{r} \tag{12.7}$$

where $\vec{r}$ is on the interface. This shows that the projections of the three $\vec{k}$'s on the interface must be equal. Because the y-component of $\vec{k}_i$ is zero, both $\vec{k}_r$ and $\vec{k}_t$ have zero y-components, and the three propagation vectors are coplanar. So if we choose $\vec{k}_i$ in the Oxz plane, then $\vec{k}_i \cdot \vec{r} = \vec{k}_r \cdot \vec{r}$ gives

$$k_i \sin \theta_i = k_r \sin \theta_r, \quad \text{or} \quad \theta_i = \theta_r \tag{12.8}$$

where we have made use of the fact that $k_i = k_r = \omega(\varepsilon_1 \mu)^{1/2}$. Equation (12.8) says that the angle of reflection is equal to the angle of incidence.

Similarly, the relation $\vec{k}_i \cdot \vec{r} = \vec{k}_t \cdot \vec{r}$ gives

$$k_i \sin \theta_i = k_t \sin \theta_t$$

or

$$\frac{\sin \theta_i}{\sin \theta_t} = \frac{k_t}{k_i}$$

Now

$$k_i = \frac{2\pi}{\lambda} = \frac{2\pi f}{\lambda f} = \frac{\omega}{v_1} = \frac{\omega}{cv_1/c} = \frac{\omega}{cn_1}$$

And

$$k_t = \frac{\omega}{cn_2}$$

Substituting these into the last expression, we obtain

$$\frac{\sin \theta_i}{\sin \theta_t} = \frac{k_t}{k_i} = \frac{n_2}{n_1} \tag{12.9}$$

which is Snell's law, where n_1 and n_2 are the indices of refraction of the media 1 and 2, respectively. Equations (12.8) and (12.9) are the simple laws of geometric optics that we are familiar with. They are general and apply to any two linear, homogeneous, isotropic, and stationary media, whether conducting or not.

When $(n_1/n_2)\sin\theta_i > 1$, then $\sin\theta_t > 1$. This is not absurd at all, and we revisit this in the section on total reflection.

b. The Fresnel Equations

If we want to obtain information about the reflected and refracted waves, we must apply the boundary conditions at $x = 0$. Omitting the factor $e^{-i\omega t}$, the tangential components of $\vec{E}$ and $\vec{H}$ give, respectively,

$$(\vec{E}_i + \vec{E}_r) \times \hat{n} = \vec{E}_t \times \hat{n} \tag{12.10}$$

and

$$(\vec{H}_i + \vec{H}_r) \times \hat{n} = \vec{H}_t \times \hat{n} \tag{12.11}$$

where $\hat{n}$ is the unit normal vector to the interface directed along Ox. For nonmagnetic materials, we may take $\mu_1 = \mu_2$, and then equation (12.11) can be rewritten as

$$(\vec{k}_i \times \vec{E}_i + \vec{k}_r \times \vec{E}_r) \times \hat{n} = (\vec{k}_t \times \vec{E}_t) \times \hat{n} \tag{12.12}$$

We now must consider two cases separately.

1. TE polarization. The $\vec{E}$ vector is normal to the incident plane, the plane defined by the propagation vector $\vec{k}_i$ and the unit normal vector $\hat{n}$, and the $\vec{H}$ vector lies in the incident plane (Figure 12.1).
2. TM polarization. The $\vec{H}$ vector is normal to the incident plane, and the $\vec{E}$ vector lies in the incident plane.

The TE and TM polarizations are also known as *s*- and *p*-polarization, respectively. The general case of arbitrary polarization is an appropriate linear combination of these two cases.

Case 1. TE polarization (s-polarization)

We find, from equations (12.10) and (12.12)

$$E_{0i} + E_{0r} = E_{0t} \tag{12.13}$$

and

$$k_i E_{0i} \cos\theta_i - k_r E_{0r} \cos\theta_i = k_t E_{0t} \cos\theta_t \tag{12.14}$$

In applying equation (12.12) we have made use of the vector identity

$$(\vec{A} \times \vec{B}) \times \vec{C} = \vec{B}(\vec{A} \cdot \vec{C}) - \vec{A}(\vec{B} \cdot \vec{C})$$

Solving for E_{0r}/E_{0i}

$$\frac{E_{0r}}{E_{0i}} = \frac{k_i \cos \theta_i - k_t \cos \theta_t}{k_r \cos \theta_i + k_t \cos \theta_t}.$$

Now

$$k_i = \omega/v_1 = \omega \sqrt{\varepsilon_1 \mu_1} = k_r, \quad k_t = \omega/v_2 = \omega \sqrt{\varepsilon_2 \mu_2}, \quad \mu_1 = \mu_2 \tag{12.15}$$

Accordingly, we can rewrite E_{0r}/E_{0i} as

$$\frac{E_{0r}}{E_{0i}} = \frac{\sqrt{\varepsilon_1} \cos \theta_i - \sqrt{\varepsilon_2} \cos \theta_t}{\sqrt{\varepsilon_1} \cos \theta_i + \sqrt{\varepsilon_2} \cos \theta_t} \tag{12.16}$$

Similarly, solving equations (12.13) and (12.14) for E_{0t}/E_{0i}

$$\frac{E_{0t}}{E_{0i}} = \frac{2\sqrt{\varepsilon_1} \cos \theta_i}{\sqrt{\varepsilon_1} \cos \theta_i + \sqrt{\varepsilon_2} \cos \theta_t} \tag{12.17}$$

These two equations are known as the Fresnel relations, which are usually given in a more familiar form that we show a little later.

It is obvious that E_{0t}/E_{0i} is always positive; the sign of E_{0r}/E_{0i}, on the other hand, depends on the nature of the dielectrics. If the dielectric 2 has a larger refractive index ($= c/v = c\sqrt{\varepsilon\mu}$) than that in medium 1, then, according to equation (12.16), $E_{0r}/E_{0i} < 0$. Reflection of an electromagnetic wave from such an interface results in a phase change of π in the electric field. In $n_2 < n_1$, there is no change in sign in $\vec{E}$ on reflection.

Using Snell's law equation (12.9), we may rewrite the Fresnel relations in a familiar form given in most textbooks:

$$\frac{E_{0r}}{E_{0i}} = \frac{\sin(\theta_t - \theta_i)}{\sin(\theta_t + \theta_i)} \tag{12.18}$$

$$\frac{E_{0t}}{E_{0i}} = \frac{2\sin \theta_t \cos \theta_i}{\sin(\theta_t + \theta_i)} \tag{12.19}$$

Reflection and Transmission Coefficients

To obtain information about the power transmitted and reflected at the interface, we can use the Poynting vector

$$\left\langle \vec{S} \right\rangle = \frac{1}{2} \operatorname{Re} \left(\vec{E} \times \vec{H}^* \right)$$

because the reflection coefficient is defined as the rate of the reflected energy flux from the interface to that incident on the interface, i.e.,

$$R_\perp = \frac{\hat{n} \cdot \left\langle \vec{S_r} \right\rangle}{\hat{n} \cdot \left\langle \vec{S_i} \right\rangle} = \frac{|\vec{E_r} \times \vec{H_r}^*|}{|\vec{E_i} \times \vec{H_r}^*|} = \frac{|E_{0r}|^2}{|E_{0i}|^2} \tag{12.20}$$

where the subscript $\perp$ refers to the TE polarization and the transmission coefficient $T_\perp$ is defined as

$$T_\perp = \frac{\hat{n} \cdot \left\langle \vec{S_t} \right\rangle}{\hat{n} \cdot \left\langle \vec{S_i} \right\rangle} = \frac{n_2 \, \cos\theta_t |\, E_{0t}|^2}{n_1 \, \cos\theta_i |\, E_{0i}|^2} \tag{12.21}$$

Using equations (12.18) and (12.19) and Snell's law, we can rewrite these two expressions as

$$R_\perp = \frac{\sin^2(\theta_t - \theta_i)}{\sin^2(\theta_t + \theta_i)} \tag{12.22}$$

$$T_\perp = \frac{\sin 2\theta_i \sin 2\theta_t}{\sin^2(\theta_t + \theta_i)} \tag{12.23}$$

It is easy to show that

$$R_\perp + T_\perp = 1 \tag{12.24}$$

As we expect, at the outset there is no energy losses.

For normal incidence, $\theta_I = \theta_r = 0$ and $\theta_t = 0$ (from Snell's law), $R_\perp$ and $T_\perp$ take on simple form

$$R_\perp = \left(\frac{n_2 - n_1}{n_2 + n_1} \right)^2, \quad T_\perp = \frac{n_2}{n_1} \left(\frac{2n_1}{n_1 + n_2} \right)^2 \tag{12.25}$$

Note that it is a straightforward matter to obtain equations (12.25) from equations (12.16) and (12.20) and from equations (12.21) and (12.17), respectively.

Case 2. TM polarization (p-polarization)

In this case, the $\vec{E}$ vector lies in the plane of incidence, as shown in **Figure 12.2**. Equations (12.10) and (12.12) now become

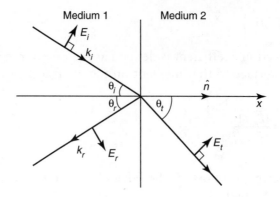

I FIGURE 12.2

$$(E_{0i} - E_{0r})\cos\theta_i = E_{0t}\cos\theta_t \tag{12.26}$$

$$k_i E_{0i} + k_r E_{0r} = k_t E_{0t} \tag{12.27}$$

From these two equations, we obtain, after some manipulations,

$$\frac{E_{0r}}{E_{0i}} = \frac{\tan(\theta_i - \theta_t)}{\tan(\theta_i + \theta_t)} \tag{12.28}$$

and

$$\frac{E_{0t}}{E_{0i}} = \frac{2\sin\theta_t\cos\theta_i}{\sin(\theta_i + \theta_t)\cos(\theta_i - \theta_t)} \tag{12.29}$$

Equations (12.28) and (12.29) are the Fresnel relations for TM polarization. From equations (12.18) and (12.28) we can see easily that there is an important distinction between the TE and TM polarizations. From equation (12.18) we see that $(E_0 r/E_{0i})_\perp \neq 0$ for any angle of incidence θ_i lying between 0 and $\pi/2$ other than in the trivial case when $n_2 = n_1$. But equation (12.28) shows that $(E_0 r/E_{0i})_\parallel$, and hence $R_\parallel$, vanishes when $\theta_i + \theta_t = \pi/2$.

This means, if the wave is incident at an angle $\theta_i = \pi/2 - \theta_t$, it crosses the interface without reflection. This angle of incidence is known as Brewster's angle and is usually denoted by the symbol θ_B. We can express it in terms of indices of refraction n_1 and n_2 by applying Snell's law:

$$\frac{\sin\theta_B}{\sin(\pi/2 - \theta_B)} = \frac{\sin\theta_B}{\cos\theta_B} = \tan\theta_B = \frac{n_2}{n_1}$$

or

$$\theta_B = \tan^{-1}\left(\frac{n_2}{n_1}\right) \tag{12.30}$$

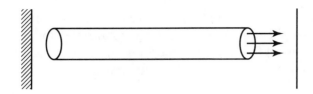

I FIGURE 12.3

It is clear that if the incident wave is unpolarized, the only component of $\vec{E}$ normal to the plane of incidence is reflected and, hence, the reflected wave is plane polarized perpendicular to the plane of incidence. Thus, θ_B is also known as the polarizing angle.

Brewster's angle finds its important application in the construction of gas lasers. In a gas laser we usually have mirrors outside the glass windows, as shown in **Figure 12.3**. Normally, about 92% of the incident intensity is transmitted through a glass window, i.e., about 8% is lost in each traverse. In a laser there are a large number of traverses, and hence very little light is left after a few traverses. To overcome this difficulty, the windows are arranged at Brewster's angle. The electric field component polarized parallel to the plane of incidence is transmitted perfectly and suffers negligible loss even after many traverses. The component polarized perpendicular to the plane of incidence is partly reflected and partly transmitted each time it strikes the surface; after a number of traverses it is completely eliminated and, hence, the emerging beam is 100% linearly polarized.

c. Total (Internal) Reflection

So far we considered the incident from medium 1 on medium 2 that generally has a large index of refraction, i.e., $n_2 > n_1$. We now show the opposite case in which a wave is incident from a higher refractive index on a medium of lower refractive index, i.e., $n_1 > n_2$, and consider what takes place when θ_i is gradually increased from zero. First let us rewrite Snell's law as

$$\sin\theta_t = \frac{n_1}{n_2}\sin\theta_i$$

Obviously, as θ_i increases so does θ_t until the point is reached at which $\theta_t = \pi/2$, and

$$\sin\theta_i = \frac{n_2}{n_1}\sin(\pi/2) = \frac{n_2}{n_1}$$

We call this angle of incidence the critical angle and denote it by θ_C:

$$\sin\theta_C = \frac{n_2}{n_1} \tag{12.31}$$

At this point there is only reflected wave and no transmitted wave in medium 2. This phenomenon is known as total reflection.

What happens if θ_i is increased beyond the critical angle θ_C? We see that θ_t becomes imaginary. To this end, let us first express θ_t in terms of θ_i and θ_C:

$$\sin\theta_t = \frac{n_1}{n_2}\sin\theta_i = \frac{\sin\theta_i}{\sin\theta_C}$$

Now

$$\cos\theta_t = \sqrt{1 - \sin^2\theta_t} = \sqrt{1 - \sin^2\theta_i/\sin^2\theta_C}$$

and we see that the value of $\cos\theta_t$ decreases as θ_i is increased; it vanishes at $\theta_i = \theta_C$ and for $\theta_i > \theta_C$, $\cos\theta_t$ becomes an imaginary number. This is not inconsistent with the analysis of the previous section. To see this, let us calculate the amplitude of the reflected electric vector when $\theta_i > \theta_C$. We first write

$$\cos\theta_t = \sqrt{1 - \sin^2\theta_i/\sin^2\theta_C} = iQ \tag{12.32}$$

or

$$Q = \sqrt{\sin^2\theta_i/\sin^2\theta_C - 1}$$

For TE polarization, we have, from equation (12.16),

$$\frac{E_{0r}}{E_{0i}} = \frac{n_1\cos\theta_i - n_2\cos\theta_t}{n_1\cos\theta_i + n_2\cos\theta_t} = \frac{n_1\cos\theta_i - n_2 iQ}{n_1\cos\theta_i + n_2 iQ}$$

and so

$$\left|\frac{E_{0r}}{E_{0i}}\right|^2 = 1, \quad \text{or} \quad |E_{0r}| = |E_{0i}|$$

Similarly for TM polarization, we have, from equation (12.28),

$$\left|\frac{E_{0r}}{E_{0i}}\right|^2 = 1, \quad \text{or} \quad |E_{0r}| = |E_{0i}|$$

Thus, the wave is totally reflected. However, there is a change of phase on reflection. If the incident wave is polarized in plane intermediate between the plane of incidence and the plane normal to it, the two components are not in phase after reflection and the wave is elliptically polarized.

When the angle of incidence θ_i is greater than the critical angle θ_C, there is no refracted wave and we can show that there is no energy flow across the interface, but the field does penetrate to the other side. Let us first calculate the rate of energy flow, which is

$$\left(\vec{S}\right) \cdot \hat{n} = \frac{1}{2}\operatorname{Re}\left(\vec{E}_t \times \vec{H}_t^*\right) \cdot \hat{n} = \frac{1}{2}\operatorname{Re}\left(\vec{E}_t \times \frac{\vec{k}_t \times \vec{E}_t^*}{\omega\mu_2}\right) \cdot \hat{n}$$

$$= \frac{1}{2\omega\mu_2}\operatorname{Re}\left[(\vec{E}_t \cdot \vec{E}_t^*)\vec{k}_t - (\vec{E}_t \cdot \vec{k}_t)\vec{E}_t^*\right] \cdot \hat{n}$$

$$= \frac{1}{2\omega\mu_2}\operatorname{Re}(\vec{E}_t \cdot \vec{E}_t^*)\vec{k}_t \cdot \hat{n} \quad (\because \vec{E}_t \perp \vec{k}_t)$$

$$= \frac{1}{2\omega\mu_2}\operatorname{Re}|E_{0t}|^2 k_t \cos\theta_t$$

$$= \frac{1}{2\omega\mu_2}iQ|E_{0t}|^2 k_t$$

Because this expression is purely imaginary, there is no energy flow across the interface. However, the field does penetrate to the other side to certain distance. To this end, let consider, for example, the TE polarization case. Then the incident and reflected wave amplitudes are described by equation (12.16). The electric field of the refracted wave is

$$\vec{E}_t = \hat{j}E_{0t}e^{i(k_t x \cos\theta_t - k_t x \sin\theta_t - \omega t)}$$

Now $k_t\sin\theta_t$ [$= k_i\sin\theta_I$, by equation (12.9)] is real. But $k_t\cos\theta_t$ is purely imaginary, and we can see this easily with the help of equation (12.32):

$$k_t\cos\theta_t = ik_t Q$$

The refracted wave may now be written

$$\vec{E}_t = \hat{j}E_{0t}e^{-qx}e^{-i(k_i x \sin\theta_i + \omega t)} \qquad (12.33)$$

where $q = k_t Q$. Equation (12.33) shows that the transmitted wave travels only a small distance into the second medium, decaying in amplitude by $1/e$ within $\lambda/2\pi$. The penetration distance d_0 is approximately given by

$$d_0 \sim q^{-1} = \frac{1}{k_t}\left(\frac{\sin^2\theta_i}{\sin^2\theta_C} - 1\right)^{-1/2} \qquad (12.34)$$

Consider for example light going from glass to air. The refractive index of glass is 1.5, and then the critical angle is

$$\sin\theta_C = n_2/n_1 \quad \text{or} \quad \theta_C = \sin^{-1}(2/3) = 42°$$

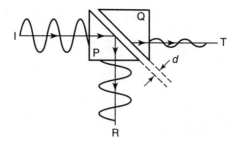

I FIGURE 12.4

If $\theta_i = 45$ degrees ($>\theta_C$) there will be total reflection, and

$$d_0 = \frac{1}{k_t}\left[\frac{(1/\sqrt{2})^2}{(2/3)^2} - 1\right]^{-1/2} = \frac{\lambda}{2\pi}\left(\frac{1}{2} \times \frac{9}{4} - 1\right)^{-1/2} = 0.45\lambda$$

Thus, the field becomes negligible beyond distances of the order of a few wavelengths. The existence of this evanescent wave can be demonstrated most conveniently with microwaves (say, $\lambda = 3$ cm), as illustrated in **Figure 12.4,** where an incident microwave beam I is internally reflected by the prism p to the receiver R, but a small transmitted signal can be observed at T when a second prism Q is at a distance $d < A$, the microwave wavelength. This demonstrates that an evanescent wave accompanies total internal reflection.

What would be the microscopic explanation of this attenuation? The molecular charges in the medium oscillate due to interaction with the "incident" wave and give rise to a radiation field (this is shown later in Chapter 13). The forward wave in this field interferes destructively with the original wave and gives rise to a small transmission. Then why is there no transport of energy across the boundary? Energy does flow in the second medium because the component of the field in the medium is finite, but during the later part of the cycle the flow is in the opposite direction and the energy is returned to the first medium.

12.3 Reflection from the Surface of a Conductor: Normal Incidence

We now consider reflection from the surface of a uniform conductor. This can be done easily by extending the arguments of the preceding section. The case of oblique incident is more involved, so we limit it to the simple case of normal incidence. As in the dielectric case, for the incident, reflected, and refracted waves we have

$$\vec{E}_I = \vec{E}_{0I}\,e^{i(\vec{k}_I \cdot \vec{r} - \omega t)}; \quad \vec{H}_I = \frac{\vec{k}_I \times \vec{E}_I}{\omega\mu_1} \tag{12.35a}$$

$$\vec{E}_R = \vec{E}_{0R}\, e^{i\,(\vec{k}_I \cdot \vec{r} - \omega t)}; \quad \vec{H}_R = \frac{\vec{k}_I \times \vec{E}_R}{\omega \mu_1} \tag{12.35b}$$

$$\vec{E}_T = \vec{E}_{0T}\, e^{i\,(\vec{k}_T \cdot \vec{r} - \omega t)}; \quad \vec{H}_T = \frac{\vec{k}_T \times \vec{E}_T}{\omega \mu_2} \tag{12.35c}$$

As medium 2 is a conductor, the propagation vector is given by equations (12.10) to (12.17):

$$k_T^2 = \varepsilon_2 \mu_2 \omega^2 \left(1 + i\,\frac{\sigma}{\varepsilon_2 \omega}\right) \tag{12.36}$$

Applying the boundary conditions as before gives

$$E_{0I} - E_{0R} = E_{0T} \tag{12.37}$$

$$k_I(E_{0I} - E_{0R}) = k_T E_{0T} \tag{12.38}$$

Because k_T is complex, E_{0R} and E_{0T} cannot both be real. Thus, phase shifts other than 0 and π are to be expected from the reflected and refracted waves. Solving equations (12.37) and (12.38) we have

$$E_{0R} = \frac{k_T - k_I}{k_T + k_I}\, E_{0I} \tag{12.39}$$

$$E_{0T} = \frac{2k_I}{k_T + k_I}\, E_{0I} \tag{12.40}$$

Now $k_I = \omega(\varepsilon_1 \mu_1)^{1/2}$ and k_T is given by equation (12.36). Equations (12.39) and (12.40) now become

$$E_{0R} = \frac{\sqrt{\varepsilon_2 \mu_2 \omega^2}\left(1 + \dfrac{i\sigma}{\varepsilon_2 \omega}\right)^{1/2} - \omega\,(\varepsilon_1 \mu_1)^{1/2}}{\sqrt{\varepsilon_2 \mu_2 \omega^2}\left(1 + \dfrac{i\sigma}{\varepsilon_2 \omega}\right)^{1/2} + \omega\,(\varepsilon_1 \mu_1)^{1/2}}\, E_{0I} \tag{12.41}$$

and

$$E_{0T} = \frac{2\omega\,(\varepsilon_1 \mu_1)^{1/2}}{\sqrt{\varepsilon_2 \mu_2 \omega^2}\left(1 + \dfrac{i\sigma}{\varepsilon_2 \omega}\right)^{1/2} + \omega\,(\varepsilon_1 \mu_1)^{1/2}}\, E_{0I} \tag{12.42}$$

For a perfect conductor, $\sigma = \infty$, we have $E_{0R} = E_{0I}$ and $E_{0T} = 0$; the reflection is complete. For a very good conductor, we may neglect displacement current compared with conduction current, i.e., $\sigma/\varepsilon_2 \omega \gg 1$. The approximation adopted in the previous chapter gives

$$k_T = \alpha + i\beta = (1 + i)\sqrt{\omega\sigma\mu_2/2} = (1 + i)/\delta$$

where δ is the skin depth. Equation (12.39) can now be rewritten as

$$E_{0R} = \frac{(1 + i)/\delta - \omega(\varepsilon_1\mu_1)^{1/2}}{(1 + i)/\delta + \omega(\varepsilon_1\mu_1)^{1/2}} E_{0I} = \frac{[1/\delta - \omega(\varepsilon_1\mu_1)^{1/2}] + i/\delta}{[1/\delta + \omega(\varepsilon_1\mu_1)^{1/2}] + i/\delta} E_{0I}$$

from which we find the reflection coefficient R:

$$R = \frac{|E_{0R}|^2}{|E_{0I}|^2} = \frac{[1 - \omega(\varepsilon_1\mu_1)^{1/2}\delta] + 1}{[1 - (\varepsilon_1\mu_1)^{1/2}\delta] + 1}$$

Now

$$\sigma/\varepsilon_2\omega \gg 1, \quad \omega(\varepsilon_1\mu_1)^{1/2}\delta \ll 1$$

Accordingly, R reduces to

$$R \cong \frac{2 - 2\omega(\varepsilon_1\mu_1)^{1/2}\delta}{2 + 2\omega(\varepsilon_1\mu_1)^{1/2}\delta} \cong 1 - 2\omega(\varepsilon_1\mu_1)^{1/2}\delta$$

$$= 1 - 2\sqrt{2\omega\varepsilon_1\mu_1/\sigma\mu_2}$$

which reduces to, when $\mu_1 = \mu_2$,

$$R \cong 1 - 2\sqrt{2\omega\varepsilon_1/\sigma}$$

The transmission coefficient T is given by

$$T = 1 - R = 2\sqrt{2\omega\varepsilon_1/\sigma}$$

which is very small for a good conductor. Thus, the energy transmitted to good conductors is very small. For example, for copper we have $\sigma = 6 \times 10^7 (\text{ohm·m})^{-1}$, $\nu = 10^{10} \text{ sec}^{-1}$, then

$$T = 2\sqrt{(2 \times 2\pi \times 10^{10} \times 8.85 \times 10^{-12})/6 \times 10^7} \approx 3 \times 10^{-4}$$

which is very small for measurement to be made.

As the reflection coefficient is almost unity for good conductors, the field in medium 1 is represented by a standing wave, as shown in **Figure 12.5**.

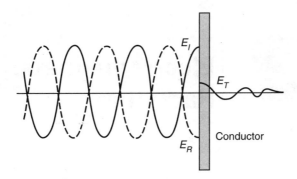

I FIGURE 12.5

Problems

1. When both media are nonmagnetic, find the conditions under which there is no reflected wave.
2. Show that conservation of energy holds in reflection and refraction.
3. Calculate the electric field amplitude in medium 2 for the case $n_2 < n_1$. Take $\mu_1 = \mu_2$ for simplicity.
4. Calculate the time-average Poynting vector for the evanescent wave in medium 2.
5. Calculate the phase shift of the reflected wave under total reflection conditions.
6. Calculate the Brewster angle for each of the following cases:
   a. Light incident on a glass whose index of refraction is 1.6
   b. Light emerging from the same type of glass
   c. A radiofrequency wave incident on water ($n = 9$ at radiofrequency). This concerns communication with submarines.
7. Show that the reflection coefficient from a dielectric to a vacuum is the same as the reflection coefficient from a vacuum to the dielectric (take $\mu_r = 1$ in the dielectric).
8. Light falls at normal incidence in a dielectric from a vacuum. Show that the refracted wave is retarded in phase by an angle $\tan^{-1}[2\kappa/(n^2 + \kappa^2 - 1)]$, where $\sqrt{\varepsilon_r} = kc/\omega = n + i\kappa$ (for vacuum, $\varepsilon_r = 1$, then we have the vacuum relation $\omega = kc$).

Electromagnetic Radiation

We have studied the propagation of electromagnetic waves in free space, in dielectrics, and in conductors. In this chapter we study the generation of electromagnetic waves. We will see that any distribution of changing charge and current acts as a source of electromagnetic radiation.

13.1 Retarded Potentials

It is convenient to start the study of electromagnetic radiation from the vector and scalar potentials satisfying the inhomogeneous wave equation with source. These two inhomogeneous equations with sources can be obtained from Maxwell equations with the help of the Lorentz condition (see Chapter 6):

$$\nabla^2 \vec{A} - \frac{1}{c^2}\frac{\partial^2 \vec{A}}{\partial t^2} = -\mu_0 \vec{j} \tag{13.1}$$

$$\nabla^2 \phi - \frac{1}{c^2}\frac{\partial^2 \phi}{\partial t^2} = -\rho/\varepsilon_0 \tag{13.2}$$

Equations (13.1) and (13.2) are formally simple generalizations of the corresponding time-independent equations $\nabla^2 \phi = -\rho/\varepsilon_0$ and $\nabla^2\vec{A} = -\mu_0\vec{j}$, which have solutions

$$\phi = \frac{1}{4\pi\varepsilon_0}\int_V \frac{\rho}{R}\, d\tau \quad \text{and} \quad \vec{A} = \frac{\mu_0}{4\pi}\int_V \frac{\vec{j}}{R}\, d\tau$$

We expect the physical solutions of equations (13.1) and (13.2) are the generalizations of the above two integrals. For simplicity, let us consider equation (13.2) for the time harmonic case $e^{i\omega t}$. Then equation (13.2) becomes

$$\nabla^2 \phi + k_0^2 \phi = -\rho/\varepsilon_0 \tag{13.3}$$

where $k_0 = \omega/c$.

Now ϕ has spherical symmetry and so depends on r only, and we note

$$\frac{\partial \phi}{\partial x} = \frac{d\phi}{dr}\frac{\partial r}{\partial x} = \frac{x}{r}\frac{d\phi}{dr}, \quad \frac{\partial^2 \phi}{\partial x^2} = \frac{1}{r}\frac{d\phi}{dr} + \frac{x^2}{r}\frac{1}{r}\frac{d}{dr}\left(\frac{i}{r}\frac{d\phi}{dr}\right)$$

Similarly, expressions for the y- and z-components. Thus,

$$\nabla^2 \phi = \frac{3}{r}\frac{d\phi}{dr} + r\frac{d}{dr}\left(\frac{1}{r}\frac{d\phi}{dr}\right) = \frac{1}{r}\frac{d^2}{dr^2}(r\phi)$$

We consider the homogeneous equation first [i.e., set the right side of equation (13.3) to zero], which is now of the form

$$\frac{1}{r}\frac{d^2}{dr^2}(r\phi) + k_0^2 \phi = 0$$

or

$$\frac{d^2}{dr^2}(r\phi) + k_0^2 r\phi = 0 \tag{13.4}$$

This is a simple harmonic equation for $r\phi$ and has solutions $\exp(\pm ik_0 r)$. Hence, the solution sought for ϕ is some linear combination of

$$e^{-ik_0 r}/r \quad \text{and} \quad e^{ik_0 r}/r$$

Now $e^{-ik_0 r}/r$ represents an outgoing spherical wave, and we are concerned with the generation of radiation rather than absorption. Hence, the solution of equation (13.2) is therefore proportional to $e^{-ik_0 r}/r$. Moreover, when $r \to 0$ the solution must agree with that for the case $k_0 = 0$; otherwise, the differential operator $\nabla^2 + k_0^2$ acting on it will not match the source singularity. Hence, the constant of proportionality is $1/(4\pi\varepsilon_0)$. Thus, the solution for a unit point source at the origin is

$$\frac{1}{4\pi\varepsilon_0}\frac{e^{-ik_0 r}}{r}$$

For any arbitrary charge density ρ we have

$$\phi(\vec{r}) = \frac{1}{4\pi\varepsilon_0}\int \rho(\vec{r}')\frac{e^{-ik_0 R}}{R}\, d\tau \tag{13.5}$$

where R is the distance of the field point at $\vec{r}$ from the volume element $d\tau$ at $\vec{r}'$, as shown in **Figure 13.1**.

The next task is to extend this result for monochromatic fields to fields with arbitrary time variation. To this end, we note that with the time factor $e^{i\omega t}$ made explicit the integrand of equation (13.5) is the product of $1/R$ with

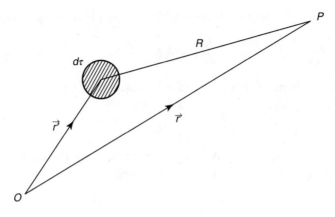

I FIGURE 13.1

$$\rho e^{i\omega(t - R/c)}$$

Because $\rho e^{i\omega t}$ is the charge density evaluated at time t, $\rho e^{i\omega(t - R/c)}$ is the charge density evaluated at time $t - R/c$. Thus, equation (13.5) can be written as

$$\phi(\vec{r}, t) = \frac{1}{4\pi\varepsilon_0} \int \frac{\rho(\vec{r}', t - R/c)}{R} d\tau \tag{13.6}$$

where ρ and ϕ stand for the actual quantities. Similarly, the corresponding solution for $\vec{A}$ is

$$\vec{A}(\vec{r}, t) = \frac{\mu_0}{4\pi} \int \frac{\vec{j}(\vec{r}', t - R/c)}{R} d\tau \tag{13.7}$$

Equations (13.6) and (13.7) reflect the fact that electromagnetic disturbances travel in vacuum with speed c, the effect of charge and current variations at each volume element taking time R/c to travel to a field point at distance R from $d\tau$. The time interval $t - R/c$ is commonly called the retarded time and the expressions for ϕ and $\vec{A}$ the retarded potentials.

13.2 Radiation from an Oscillating Electric Dipole

In general, the calculation of $\vec{A}$ and ϕ is complicated for finite sources, and so we first solve a simple case, an oscillating electric dipole, sometimes called a Hertzian oscillator. It is just an electric charge oscillating sinusoidally with time. We show that such a system radiates electromagnetic waves.

Many practical radiating systems may be considered to be made up by putting together a large number of such dipoles. Furthermore, the radiation field of an oscillating electric dipole contains many features useful in quantum theory of emission of radiation by atoms, molecules, and nuclei.

Consider two small spheres at the ends of a short wire of length l, and the charge is transferred harmonically from one sphere to the other:

$$q = q_0 e^{i\omega t} \tag{13.8}$$

where q_0 is the amplitude of the oscillating charge and ω the angular frequency of oscillation. The charge oscillation between the two spheres is equivalent to an oscillating dipole moment $\vec{p}$ (**Figure 13.2**):

$$\vec{p} = q\vec{l} = \vec{p}_0 e^{i\omega t} \tag{13.9}$$

where $\vec{p}_0 = q_0\vec{l}$. Let us consider the situation where the wavelength λ of the radiation is large compared with the length of the wire l, the scale length of the system, i.e.,

$$\lambda = \frac{2\pi c}{\omega} \gg l \quad \text{or} \quad \frac{2\pi}{\omega} = T \gg \frac{l}{c}$$

This means that the time l/c taken for a signal to propagate along the wire from one end to the other is much less than that over which the source

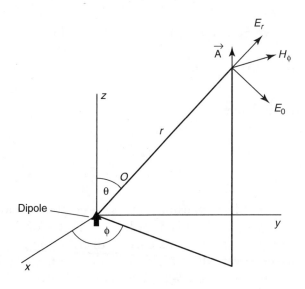

I FIGURE 13.2

current changes appreciably. So we may take the current I to be the same at all points along the length, where

$$I = dq/dt = i\omega q_0 e^{i\omega t} \tag{13.10}$$

We are interested to know how the radiation field of the dipole is distributed in space and the total power radiated. We first calculate the field value at a point P at position $\vec{r}$. In the near zone, defined by $r \ll \lambda$, the electric field is exactly that calculated in Chapter 2 for the electrostatic dipole. In the far-field or radiation zone, characterized by

$$r \gg \lambda \gg l,$$

we find fields that vary as $1/r$. To determine the fields, we have to find the retarded potentials. The vector potential $\vec{A}$ is given by equation (13.11):

$$\vec{A}(\vec{r},t) = \frac{\mu_0}{4\pi} \int \frac{\vec{j}(\vec{r}',t - |\vec{r} - \vec{r}'|c|)}{|\vec{r} - \vec{r}'|} \, d\tau' \tag{13.11}$$

The integration is over the volume occupied by the current, which is along the z-axis.

At large distance r, $r' (= |\vec{r}'|$, which may be replaced by z) is negligible compared with r and

$$\frac{1}{|\vec{r} - \vec{r}'|} = \frac{1}{r}\left(1 + \frac{r'^2}{r^2} - \right)^{-1/2} \cong \frac{1}{r}\left(1 + \frac{\vec{r} \cdot \vec{r}'}{r^2}\right) \cong \frac{1}{r}$$

Hence, at large r, we may replace equation (13.11) by

$$\vec{A}(\vec{r},t) = \frac{\mu_0}{4\pi r} \int \vec{j}(\vec{r}',t - |\vec{r} - \vec{r}'|c|) \, d\tau' \tag{13.12}$$

In the integrand, we have

$$t - \frac{|\vec{r} - \vec{r}'|}{c} = t - \frac{1}{c}r\left(1 + \frac{r'^2}{r^2} - \frac{2\vec{r} \cdot \vec{r}'}{r^2}\right) \cong t - \frac{r}{c} + \frac{\vec{r} \cdot \vec{r}'}{cr} + \dots.$$

Higher order terms in this expansion $\to 0$ as $r \to \infty$, but it may be necessary to retain the term $\vec{r} \cdot \vec{r}'/cr$, which is of order r'/c. This is of the magnitude of the time taken for light to cross the current distribution and may be comparable with the time over which the current distribution changes appreciably. However, for our case, $r'/c = l/c$, which is much less than that over which the source current changes appreciably, and so we may ignore the term $\vec{r} \cdot \vec{r}'/cr$. Then equation (13.12) becomes

$$\vec{A}(\vec{r},t) = \frac{\mu_0}{4\pi r} \int \vec{j}(\vec{r}',t-r/c)\, d\tau'$$
$$= \frac{\mu_0}{4\pi r} \int_{-l/2}^{l/2} I(z,t-r/c)\hat{z}\, dz = \frac{\mu_0 I}{4\pi r} I(t-r/c)\hat{z} \tag{13.13}$$

which shows that the vector potential $\vec{A}$ is parallel to l. Thus, in rectangular coordinates the components of $\vec{A}$ are

$$\vec{A}_x(\vec{r},t) = 0, \quad \vec{A}_y(\vec{r},t) = 0$$

and

$$\vec{A}_z(\vec{r},t) = \frac{\mu_0 I}{4\pi r} I(t-r/c)\hat{z}$$
$$= \frac{\mu_0 I}{4\pi r} i\omega q_0 e^{i\omega(t-r/c)}\hat{z} = \frac{\mu_0}{4\pi r} i\omega p_0 e^{i\omega t} e^{-i\omega r/c}\hat{z} \tag{13.14}$$
$$= \frac{\mu_0}{4\pi} i\omega p(t) \frac{e^{-ikr}}{r}\hat{z}$$

where

$$p(t) = p_0 e^{i\omega t}, \quad p_0 = q_0 l, \quad \text{and} \quad \omega/c = 2\pi/\lambda = k.$$

We may write $\vec{A}_z$ in terms of $\dot{p}(t)$

$$\vec{A}_z(\vec{r},t) = \frac{\mu_0}{4\pi} \dot{p}(t) \frac{e^{-ikr}}{r}\hat{z} \tag{13.15}$$

It is often more convenient to use the symmetry of the problem and switch to spherical coordinates. Putting $\hat{z} = \hat{e}_r \cos\theta - \hat{e}_\theta \sin\theta$ we have the components of $\vec{A}$ in spherical coordinates

$$A_r = \frac{\mu_0}{4\pi} i\omega p(t) \cos\theta \frac{e^{-ikr}}{r} = \frac{\mu_0}{4\pi} \dot{p}(t) \cos\theta \frac{e^{-ikr}}{r} \tag{13.16a}$$

$$A_\theta = -\frac{\mu_0}{4\pi} i\omega p(t) \sin\theta \frac{e^{-ikr}}{r} = -\frac{\mu_0}{4\pi} \dot{p}(t) \sin\theta \frac{e^{-ikr}}{r} \tag{13.16b}$$

$$\vec{A}_\phi = 0 \tag{13.16c}$$

The scalar potential ϕ may be obtained easily from the Lorentz condition

$$\nabla \cdot \vec{A} + \frac{1}{c^2} \frac{\partial \phi}{\partial t} = 0$$

We first calculate $\nabla \cdot \vec{A}$. In rectangular coordinates, the only nonzero component is A_z, so

$$\nabla \cdot \vec{A} = \frac{\partial A_z}{\partial z} = \frac{\partial A_z}{\partial r} \frac{\partial r}{\partial z} = \frac{\mu_0}{4\pi} \dot{p}(t) \frac{\partial}{\partial r} \left(\frac{e^{-ikr}}{r} \right) \cos \theta$$

Thus,

$$\frac{\partial \phi}{\partial t} = -c^2 \nabla \cdot \vec{A} = -\frac{1}{4\pi\varepsilon_0} \dot{p}(t) \cos \theta \frac{\partial}{\partial r} \left(\frac{e^{-ikr}}{r} \right)$$

and

$$\phi = -\frac{1}{4\pi\varepsilon_0} p(t) \cos \theta \frac{\partial}{\partial r} \left(\frac{e^{-ikr}}{r} \right) \tag{13.17}$$

or, after we carry the differentiation,

$$\phi = \frac{1}{4\pi\varepsilon_0} p(t) \cos \theta \left[+\frac{1}{r^2} + \frac{ik}{r} \right] e^{-ikr} \tag{13.17a}$$

The scalar potential can also be expressed as

$$\phi = \frac{1}{4\pi\varepsilon_0} \left\{ \frac{z}{r^3} q(t - r/c) + \frac{z}{cr^2} I(t - r/c) \right\} \tag{13.18}$$

The $\vec{E}$ and $\vec{H}$ fields can be determined from the following equations:

$$\vec{E} = -\nabla \phi - \frac{\partial \vec{A}}{\partial t}, \quad \text{and} \quad \vec{H} = \frac{1}{\mu_0} \nabla \times \vec{A} \tag{13.19}$$

a. The $\vec{E}$ Field

Let us calculate the $\vec{E}$ field first. In spherical coordinates, with $\partial / \partial \varphi = 0$, the grad$\phi$ is

$$\nabla \phi = \frac{\partial \phi}{\partial r} \hat{r} + \frac{1}{r} \frac{\partial \phi}{\partial \theta} \hat{\theta}$$

Now

$$\frac{\partial \phi}{\partial r} = \frac{q_0 l}{4\pi\varepsilon_0} \left\{ \cos \theta \frac{\partial}{\partial r} \left[\left(+\frac{1}{r^2} + \frac{ik}{r} \right) e^{-i(kr - \omega t)} \right] \right\}$$

$$= \frac{q_0 l}{4\pi\varepsilon_0} \cos \theta \left(\frac{-2}{r^3} - \frac{2ik}{r^2} + \frac{k^2}{r} \right) e^{-i(kr - \omega t)}$$

and

$$\frac{\partial \phi}{\partial \theta} = \frac{q_0 l}{4\pi\varepsilon_0} \left[\frac{1}{r^2} + \frac{ik}{r} \right] e^{-i(kr - \omega t)} (-\sin \theta)$$

Next we calculate

$$\frac{\partial \vec{A}}{\partial t} = i\omega\vec{A} = -\omega^2 \frac{\mu_0 q_0 l}{4\pi} \frac{e^{-i(kr-\omega t)}}{r}(\cos\theta\hat{r} - \sin\theta\hat{\theta})$$

Then from equation (13.19) for $\vec{E}$ we find

$$E_r = -\frac{\partial\phi}{\partial r} - \frac{\partial A_r}{\partial t} = \frac{iq_0 l}{2\pi\varepsilon_0}\cos\theta\left(\frac{k}{r^2} - \frac{i}{r^3}\right)e^{-i(kr-\omega t)} \tag{13.20a}$$

$$E_\theta = -\frac{\partial\phi}{r\partial\theta} - \frac{\partial A_\theta}{\partial t} = -\frac{q_0 l}{4\pi\varepsilon_0}\sin\theta\left(\frac{1}{r} - \frac{i}{kr^2} - \frac{1}{k^2 r^3}\right)e^{-i(kr-\omega t)} \tag{13.20b}$$

$$E_\varphi = 0 \tag{13.20c}$$

b. The $\vec{H}$ Field

The $\vec{H}$ field is easier to calculate. In spherical coordinates, with $A_\varphi = 0$ and $\partial/\partial t = 0$, we have

$$\vec{H} = \frac{1}{\mu_0}\nabla\times\vec{A} = \frac{1}{\mu_0}\frac{1}{r}\left[\frac{\partial(rA_\theta)}{\partial r} - \frac{\partial A_r}{\partial\theta}\right]$$

Now

$$\frac{1}{r}\frac{\partial(rA_\theta)}{\partial r} = -\frac{\mu_0}{4\pi r}i\omega q_0 l\sin\theta\frac{\partial}{\partial r}[e^{-i(kr-\omega t)}] = -\frac{\mu_0 k\omega q_0 l}{4\pi r}\sin\theta e^{-i(kr-\omega t)}$$

and

$$\frac{1}{r}\frac{\partial A_r}{\partial\theta} = \frac{\mu_0}{4\pi r^2}i\omega q_0 l e^{-i(kr-\omega t)}\frac{\partial(\cos\theta)}{\partial\theta} = -\frac{\mu_0}{4\pi r^2}i\omega q_0 l e^{-i(kr-\omega t)}\sin\theta$$

Then equation (13.19) for $\vec{H}$ gives

$$H_r = 0 \tag{13.21a}$$

$$H_\theta = 0 \tag{13.21b}$$

$$H_\varphi = -\frac{k\omega q_0 l}{4\pi}\sin\theta\left(\frac{1}{r} + \frac{i}{kr^2}\right)e^{-i(kr-\omega t)} \tag{13.21c}$$

Equation (13.20) and equation (13.21) contain terms falling off as r^{-1}, r^{-2}, and r^{-3}. We keep the latter terms on purpose. Let us now examine these two equations in near zone ($r \ll \lambda$, or $kr \ll 1$) and the radiation zone ($kr \gg 1$)

1. **Near zone: $r \ll \lambda$, i.e., $kr \ll 1$**

$$E_r \cong \frac{q_0 l \cos\theta}{2\pi\varepsilon_0 r^3} e^{i\omega t}, \quad E_\theta \cong \frac{q_0 l \sin\theta}{2\pi\varepsilon_0 r^3} e^{i\omega t}, \quad E_\varphi = 0 \tag{13.22a}$$

$$H_r = 0, \quad H_\theta = 0, \quad H_\varphi \cong \frac{i\omega q_0 l \sin\theta}{4\pi r^2} e^{i\omega t} \tag{13.22b}$$

The electric field is simply the dipole field in Chapter 2, and the magnetic field is transverse. One can show that the ratio of magnetic to electric field dominates in the zone is given by (see homework) kr, i.e., electric fields dominate in the near zone and, in the limit $k \to 0$, the field is purely electric. We will see that E_r term vanishes in the radiation zone.

2. **Radiation zone: $kr \gg 1$**
   In the far-field approximation the dominant terms are now

$$E_r = 0, \quad E_\theta = -\frac{q_0 l k^2}{4\pi\varepsilon_0} \sin\theta \frac{e^{-i(kr-\omega t)}}{r}, \quad E_\varphi = 0 \tag{13.23a}$$

$$H_r = 0, \quad H_\theta = 0, \quad H_\varphi = -\frac{c q_0 l k^2}{4\pi} \sin\theta \frac{e^{-i(kr-\omega t)}}{r} \tag{13.23b}$$

We arrived at these results by neglecting terms involving r^{-2} and its higher order. The fields are outgoing spherical waves, and at large distances from the source the electric and magnetic fields are each transverse and behave as r^{-1}. Moreover, we see, from equation (13.23), that the quotient of the amplitudes at any point is

$$E_\theta/H_\varphi = \left(q_0 l k^2/4\pi\varepsilon_0\right)(4\pi/c q_0 l k^2) = 1/c\varepsilon_0 = (\mu_0/\varepsilon_0)^{1/2}$$

which is the same as that for harmonic plane waves. At large distances the fields look locally to be plane. The only difference is that the amplitudes do not remain constant in the direction of propagation but fall off slowly as $1/r$, because the wavefronts are spherical rather than truly planar. And the electric and magnetic fields are in phase. The fields are mutually perpendicular and their directions are shown in **Figure 13.3** and the energy flow at all points is in the direction of the radius vector. **Figure 13.4** is the plot of the electric field lines in the axial plane of the dipole

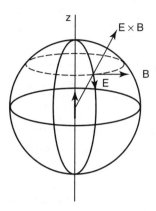

I FIGURE 13.3

for $\omega t = 0, 2\pi. 4\pi \ldots$; they have cylindrical symmetry about the dipole axis. The pattern is the same for $\omega t = \pi, 3\pi, 5\pi$, but with all the directions reversed. The magnetic field lines are azimuthal circles about the vertical axis of the dipole.

The instantaneous total power W crossing the surface S of any sphere of radius r in the radiation zone is given by integrating the Poynting vector $\vec{S}$ over the surface. Using the values obtained for radiation zone in equations (13.23) we obtain

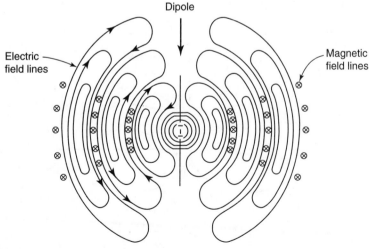

I FIGURE 13.4

$$\left\langle \vec{S} \right\rangle = \frac{1}{2} \operatorname{Re} \left(\vec{E} \times \vec{H}^* \right) = \frac{c p_0^2 k^4}{32\pi^2 \varepsilon_0} \frac{\sin^2 \theta}{r^2} \hat{r} \tag{13.24}$$

and the total radiated power is

$$P = \frac{c p_0^2 k^4}{32\pi^2 \varepsilon_0} \int_0^{2\pi} d\varphi \int_0^\pi \frac{\sin^2 \theta}{r^2} r^2 \sin\theta d\theta = \frac{c p_0^2 k^4}{12\pi \varepsilon_0} = \frac{c p_0^2}{12\pi \varepsilon_0} \left(\frac{2\pi}{\lambda} \right)^4 \tag{13.25}$$

The total average radiated power thus varies directly as the square of the amplitude of the electric dipole and inversely as the fourth power of the wavelength λ.

Equation (13.24) shows that the energy flow varies as $\sin^2\theta$; thus, it vanishes along the dipole axis. The radiant energy in the direction θ is proportional to the length OP and independent of the azimuthal angle φ, so that in three dimensions the polar plot has a doughnut shape with no radiation along the axis and a maximum in the equatorial plane (**Figure 13.5**).

Equation (13.25), the total power radiated, is expressed in terms of dipole moment. Now let us cast it in terms of the current amplitude; to electrical engineers the current flowing in an aerial is a more natural parameter than the dipole moment p. Because $I_0 = i\omega \, p_0/l$, equation (13.25) can be rewritten as

$$P = \frac{c p_0^2}{12\pi \varepsilon_0} \left(\frac{2\pi}{\lambda} \right)^4 = \frac{p_0^2 \omega^4}{12\pi \varepsilon_0 c^3} = \frac{1}{12\pi \varepsilon_0} \frac{I_0^2 l^2 \omega^2}{c^3}$$

$$= \frac{1}{12\pi \varepsilon_0 c} \frac{4\pi^2 l^2}{\lambda^2} I_0^2 = \frac{1}{2} \frac{2\pi}{3} \sqrt{\frac{\mu_0}{\varepsilon_0}} \left(\frac{l}{\lambda} \right)^2 I_0^2 \tag{13.25a}$$

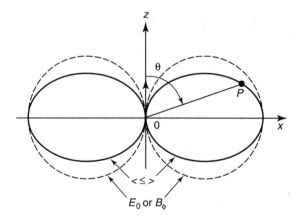

I FIGURE 13.5

This result has an interesting physical interpretation. Dimensionally, l/λ is dimensionless and $(\mu_0/\varepsilon_0)^{1/2}$ has the dimensions of a resistance. And we also know that a resistance R carrying a current $I = I_0 e^{i\omega t}$ dissipates energy at an average rate RI_0^2. Comparison of this with equation (13.25a) leads us to define the radiation resistance of a dipole by

$$R_{rad} = \frac{2\pi}{3}\sqrt{\frac{\mu_0}{\varepsilon_0}}\left(\frac{l}{\lambda}\right)^2 = 789\left(\frac{l}{\lambda}\right)^2 ohms \tag{13.26}$$

Equation (13.26) is valid only for $l \ll \lambda$. It is the resistance presented to the terminals of a transmission line from a signal generator driving the oscillating current. For an antenna, its ratio l/λ is approximately 10^{-2}, and then the radiation resistance is approximately 0.08 ohm. The ohmic resistance of the antenna could be appreciably larger, and this would result in most of the input power being dissipated as heat rather than being radiated as electromagnetic energy. Hence, a short dipole antenna is, in general, a rather inefficient radiator. To get appreciable radiation would require $l \sim \lambda$, but then the dipole approximation is no longer valid.

13.3 Electric Dipole Radiation: Generalization

In obtaining the dipole fields from equation (13.12), we assumed a harmonic time dependence by writing $q = q_0 e^{i\omega t}$ and then the dipole moment $\vec{p}(t) = \vec{p}_0 e^{i\omega t}$, where $\vec{p}_0 = q_0 l$. In fact, the integral in equation (13.12) may be written in terms of the electric dipole moment $\vec{p}(t)$ of the charge distribution, and equation (13.12) reads as

$$\vec{A}(\vec{r}, t) = \frac{\mu_0}{4\pi r}\dot{\vec{p}}(t - r/c) \tag{13.27}$$

It will take a few steps to reach equation (13.27) from equation (13.12). We first consider the x-component of the dipole moment:

$$p_x(t) = \int x\rho(r, t)\,d\tau$$

from this we get

$$\frac{dp_x}{dt} = \int x\frac{\partial\rho}{\partial t}\,d\pi = -\int x\nabla\cdot\vec{j}\,d\tau$$

where we have made use of the equation of continuity. Now

$$\nabla\cdot(x\vec{j}) = (\nabla x)\cdot\vec{j} + x(\nabla\cdot\vec{j}) = j_x + x\nabla\cdot\vec{j}$$

Hence,

$$\frac{dp_x}{dt} = -\int x\nabla \cdot \vec{j}\, d\tau = \int j_x\, d\tau - \int \nabla \cdot (x\vec{j})\, d\tau$$

We now convert the volume integral to surface integral by using the divergence theorem,

$$\int_V \nabla \cdot (x\vec{j})\, d\tau = \int_S (x\vec{j}) \cdot dS$$

the surface integral vanishes because the current distribution is zero at large distances. And finally we obtain the simple result

$$\frac{dp_x}{dt} = \int j_x\, d\tau \tag{13.28}$$

Thus, the vector potential $\vec{A}$ may be written in terms of the electric dipole moment $\vec{p}(t)$ of the charge distribution as given by equation (13.27). In components we have now, in place of equation (13.16),

$$A_r = \frac{\mu_0}{4\pi r}\dot{p}(t - r/c)\cos\theta, \quad A_\theta = \frac{\mu_0}{4\pi r}\dot{p}(t - r/c)\sin\theta, \, A_\varphi = 0 \tag{13.29}$$

The magnetic field is

$$\vec{H} = \mu_0^{-1}\nabla \times \vec{A} = \frac{\mu_0^{-1}}{(1)(r)(r\sin\theta)}\begin{vmatrix} \hat{r} & r\hat{\theta} & r\sin\theta\hat{\varphi} \\ \frac{\partial}{\partial r} & \frac{\partial}{\partial\theta} & \frac{\partial}{\partial\varphi} \\ A_r & rA_\theta & r\sin\theta\, A_\varphi \end{vmatrix}$$

From this we find $H_r = 0 = H_\theta$, and

$$H_\varphi = \frac{1}{\mu_0 r}\left[\frac{\partial}{\partial r}(rA_\theta) - \frac{\partial A_r}{\partial\theta}\right] = \frac{\sin\theta}{4\pi r}\left[-\frac{\partial}{\partial r}[\dot{p}] + \frac{[\dot{p}]}{r}\right]$$

where the brackets [ ] imply that the quantity contained is to be evaluated at the retarded time, $t - r/c$. Now

$$\frac{\partial}{\partial x}\dot{p}_y(t - r/c) = \ddot{p}_y\frac{\partial}{\partial x}\left(-\frac{r}{c}\right) = -\frac{x}{rc}\ddot{p}_y \text{ etc.}$$

and

$$\frac{\partial}{\partial r}\dot{p}(t - r/c) = -\frac{1}{c}\ddot{p}$$

Thus,

$$H_\varphi = \frac{\sin\theta}{4\pi}\left[\frac{\ddot{p}}{cr} + \frac{\dot{p}}{r^2}\right] \tag{13.30}$$

Here again, the brackets [ ] imply that the quantity contained is to be evaluated at the retarded time.

We cannot calculate $\vec{E}$ from the Lorentz condition, because we do not know ϕ. But we can use Maxwell equation

$$\nabla \times \vec{B} - (1/c^2)\, \partial \vec{E}/\partial t = 0$$

From this we find

$$E_\theta = \frac{\sin\theta}{4\pi\varepsilon_0 c}\left[\frac{\dot{p}}{cr} + \frac{p}{r^2} + \frac{cp}{r^3}\right], \quad E_r = \frac{\cos\theta}{4\pi\varepsilon_0 c}\left[\frac{p}{r^2} + \frac{cp}{r^3}\right] \tag{13.31}$$

Again, the brackets [ ] imply that the quantity contained is to be evaluated at the retarded time.

The radiation fields are then given by

$$H_\varphi = \frac{\sin\theta}{4\pi}\left[\frac{\ddot{p}}{cr}\right], \quad E_\theta = \frac{\sin\theta}{4\pi\varepsilon_0 c}\left[\frac{\ddot{p}}{cr}\right] \tag{13.32}$$

or in vector notation

$$\vec{H}_{rad} = \frac{1}{4\pi cr}\left[\ddot{\vec{p}}\right] \times \hat{r} \tag{13.33a}$$

$$\vec{E}_{rad} = \frac{1}{4\pi\varepsilon_0 c^2 r}\left(\left[\ddot{\vec{p}} \times \hat{r}\right]\right) \times \hat{r} = -\frac{\mu_0}{4\pi r}\left[\ddot{\vec{p}} \times \hat{r}(\hat{r} \cdot \ddot{\vec{p}})\right] \tag{13.33b}$$

Again, the brackets [ ] imply that the quantity contained is to be evaluated at the retarded time. In equation (13.33b) we have used the vector identity for three arbitrary vectors $\vec{A}, \vec{B}, \vec{C}$:

$$\left(\vec{A} \times \vec{B}\right) \times \vec{C} = \vec{B}\left(\vec{A} \cdot \vec{C}\right) - \vec{A}\left(\vec{B} \cdot \vec{C}\right)$$

It is evident that

$$\vec{E} = c\vec{B} \times \hat{r} \tag{13.34}$$

Thus, at large distances a spherical wavefront appears locally to be plane, and the fields perpendicular to each other and both are perpendicular to the direction of propagation $\hat{r}$.

The Poynting vector gives the power radiated per unit area in the direction $\hat{r}$ at time t:

$$\vec{S}_{rad} = \vec{E}_{rad} \times \vec{H}_{rad} = (c/\mu_0)\,|\vec{B}|^2\hat{r} = \frac{\mu_0}{16\pi^2 cr^2}\,|\ddot{p}(t - r/c)|^2\sin^2\theta\hat{r}$$

where the $\sin^2\theta$ comes from the cross-product in equation (13.33a).

The energy per unit time radiated in a solid angle $d\Omega$ is $(\vec{S}_{rad} \cdot \hat{r})\, r^2\, d\Omega$, or

$$\frac{dP}{d\Omega} = \frac{\mu_0 \left(\ddot{p}(t - r/c)\right)^2 \sin^2 \theta}{16\pi^2 c}$$

Integrating this over all angles, we find the energy is lost from the source at a rate

$$P = \frac{dE}{dt} = \frac{\mu_0}{16\pi^2 c} \left(\ddot{p}(t - r/c)\right)^2 \iint \sin^2 \theta \cos \theta \, d\varphi$$

or

$$P = \frac{2}{3} \frac{1}{4\pi\varepsilon_0 c^3} \left(\ddot{p}(t - r/c)\right)^2 \tag{13.35}$$

J. J. Larmor first derived this result in 1897, and so it is known as Larmor's formula.

In the dipole radiation, the radiation is governed by the second derivative of the dipole moment, $\vec{p} = \Sigma\, e\vec{l}$, $\ddot{\vec{p}} = \Sigma\, e\ddot{\vec{l}} = \Sigma\, e\dot{\vec{v}}$ so uniformly moving charges do not radiate.

All the results obtained thus far for electric dipole radiation are nonrelativistic approximations to the exact expression. This is easy to see. The dimensions of the dipole were assumed to be smaller than the wavelength of the radiation, $l \ll \lambda$. If we take v as a velocity characteristic of the charges and if T denotes the order of magnitude of the time over which the charge distribution changes significantly, then $T \sim l/v$ and the frequency ν of the radiation is of order T^{-1}. Now $\lambda = c/\nu \sim cT \sim cl/v$, which implies that $v \ll c$ for l is much smaller than λ.

13.4 Magnetic Dipole Radiation

We now consider a current loop of radius a with alternating current $I(t) = I_0 e^{-i\omega t}$. For simplicity, we choose coordinate axes so that the loop is in the xy-plane with center at the origin and consider a field point directly about the x-axis so that it has coordinates $(x, 0, z)$ (**Figure 13.6**). The loop has no electric dipole and is uncharged. But it radiates because of the varying current. If the loop is small (i.e., $a \ll \lambda = 2\pi c/\omega$), the dominant radiation is magnetic dipole radiation. The magnetic moment is

$$\vec{m}(t) = \pi a^2 I(t)\, \hat{z} = \pi a^2 I_0 e^{-i\omega t}\, \hat{z} = \vec{m}_0 e^{-i\omega t}\, \hat{z}$$

As the loop is uncharged, the scalar potential is zero. The retarded vector potential is, according to equation (13.7),

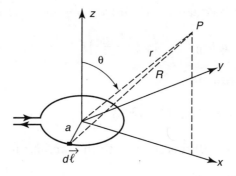

I FIGURE 13.6

$$\vec{A}(\vec{r}, t) = \frac{\mu_0}{4\pi} \int \frac{I_0 e^{-i\omega(t - R/c)}}{R} \, d\vec{l'}$$

Now the field point P is directly above the x-axis, the x-component of $\vec{A}$ is zero because the contributions from both sides of the x-axis cancel each other. Thus, $\vec{A}$ is in the direction of the y-axis, and the above expression for $\vec{A}$ becomes

$$\vec{A}(x, 0, z, t) = \frac{\mu_0 I_0}{4\pi} \int_0^{2\pi} \frac{e^{-i\omega(t - R/c)}}{R} a d\phi'' \cos\phi' \, \hat{j}$$

where ϕ' is the angle between a and the x-axis, and

$$R = (r^2 + a^2 - 2ax\cos\phi')^{1/2} \cong r(1 - ax\cos\phi'/r^2)$$

and therefore

$$\vec{A}(x, 0, z, t) = \frac{\mu_0 I_0 a}{4\pi r} e^{-i\omega(t - r/c)} \hat{j} \int_0^{2\pi} \frac{e^{-i(2\pi/\lambda)(ax\cos\phi'/r)}}{1 - (ax\cos\phi'/r^2)} \cos\phi' \, d\phi' \tag{13.36}$$

To evaluate the integral, we expand both numerator and denominator in Taylor series, and we have

$$\int_0^{2\pi} \left[1 - i(2\pi a/\lambda)(x/r)\cos\phi'\right](1 + ax\cos\phi'/r^2)\cos\phi' \, d\phi'$$

$$= (ax/r)(1/r - 2\pi i/\lambda) \int_0^{2\pi} \cos^2\phi' \, d\phi'$$

Now $\int_0^{2\pi c} \cos^2\phi' \, d\phi' = \pi$. We finally find

$$\int_0^{2\pi} \frac{e^{-i(2\pi/\lambda)(ax\cos\phi'/r)}}{1 - (ax\cos\phi'/r^2)} \cos\phi' \, d\phi' = \frac{\pi ax}{r}\left(\frac{1}{r} - \frac{2\pi i}{\lambda}\right)$$

Substituting this into equation (13.36) we obtain

$$\vec{A}(x,0,z,t) = \frac{\mu_0}{4\pi r}(\pi a^2 I_0)\, e^{-i\omega(t-r/c)}\left(\frac{x}{r}\right)\left(\frac{1}{r} - \frac{2\pi i}{\lambda}\right)\hat{j}$$

Thus, in polar coordinates

$$\vec{A}(r,\theta,t) = \frac{\mu_0 m_0}{4\pi r}\, e^{-i\omega(t-r/c)}\sin\theta\left(\frac{1}{r} - \frac{2\pi i}{\lambda}\right)\hat{\varphi} \tag{13.37}$$

In the static limit ($\omega = 0$) this reduces to the familiar formula

$$\vec{A}(r,\theta t) = \frac{\mu_0}{4\pi r}\, \frac{+2m'_0\sin\theta}{r^2}\,\hat{\varphi}.$$

From $\vec{A}$ we obtain the fields at large distances ($r \gg \lambda$)

$$\vec{E} = -\partial\vec{A}/\partial t = i\omega\vec{A} = (i\omega)\frac{\mu_0 m_0}{4\pi r}\, e^{-i\omega(t-r/c)}\sin\theta\left(\frac{-2\pi i}{\lambda}\right)\hat{\varphi}$$

or, after simplification,

$$\vec{E} = \frac{\mu_0 m_0}{4\pi r}\, e^{-i\omega(t-r/c)}\sin\theta\, \frac{\omega^2}{c}\,\hat{\varphi} \tag{13.38}$$

and $\vec{H} = \mu_0^{-1}\nabla\times\vec{A}$. In spherical coordinates curl $\vec{A}$ is

$$\nabla\times\vec{A} = \frac{1}{(1)(r)(r\sin\theta)}\begin{vmatrix} \hat{r} & r\hat{\theta} & r\sin\theta\hat{\varphi} \\ \frac{\partial}{\partial r} & \frac{\partial}{\partial\theta} & \frac{\partial}{\partial\varphi} \\ A_r & rA_\theta & r\sin\theta\, A_\varphi \end{vmatrix}$$

and for our case, $A_r = A_\theta = 0$ and A_φ are given by equation (13.28). Thus,

$$\begin{aligned} \vec{H} &= \frac{1}{\mu_0 r^2\sin\theta}\left(\hat{r}\,\frac{\partial}{\partial r} - r\hat{\theta}\,\frac{\partial}{\partial r}\right)(r\sin\theta A_\varphi) \\ &= \frac{m_0}{4\pi}\left[\hat{r}\left(\frac{1}{r^2} - \frac{2\pi i}{\lambda r^2}\right)2\cos\theta - \hat{\theta}\left(\frac{-1}{r^3} + \frac{2\pi i}{\lambda r^2} + \frac{2\pi^2 i}{\lambda^2 r}\right)\sin\theta\right]e^{-i\omega(t-r/c)} \end{aligned} \tag{13.39}$$

For $r \gg \lambda$, this reduces to

$$\vec{H} = -\frac{m_0}{4\pi r}\,\frac{\omega^2}{c^2}\sin\theta e^{-i\omega(t-r/c)}\hat{\theta} \tag{13.40}$$

Note that $\vec{E}$ and $\vec{H}$ are in phase, perpendicular to each other, and normal to the direction of propagation ($\hat{r}$) and the ratio of the amplitudes of these two fields is equal to c: $E_0/H_0 = c$. In fact, the fields of a magnetic dipole are very similar in structure to the fields of an oscillating electric dipole, with one exception: $\vec{E}$ now points in the $\hat{\varphi}$ direction and $\vec{H}$ is in the $\hat{\theta}$ direction. From the fields $\vec{E}$ and $\vec{H}$ we can calculate the energy flux as

$$\langle \vec{S} \rangle = \frac{1}{2} \mathrm{Re} (\vec{E} \times \vec{H}^*) = \frac{\mu_0 m_0^2 \omega^4}{32\pi^2 c^3} \frac{\sin^2\theta}{r^2} \hat{r} \tag{13.41}$$

Integrating equation (13.41) over a sphere and simplifying, we obtain the total radiated power

$$W = \frac{\mu_0 m_0^2 \omega^4}{12\pi c^3} = \frac{1}{2} \frac{\pi}{6} \sqrt{\frac{\mu_0}{\varepsilon_0}} \left(\frac{2\pi a}{\lambda} \right)^4 I_0^2 \tag{13.42}$$

The radiation resistance of a magnetic dipole is

$$R_{rad} = \frac{\pi}{6} \sqrt{\frac{\mu_0}{\varepsilon_0}} \left(\frac{2\pi a}{\lambda} \right)^4 \tag{13.43}$$

and it goes the fourth power of the ratio of dimension of the structure to the wavelength; for the electric dipole, it goes the second power. Thus, for radiation of a fairly long wavelength compared with the dimensions of the source, such as radiation of visible light by atoms, the electric dipole radiation is much more effective.

Both magnetic and electric dipoles give the same radiation pattern, having a $\sin^2\theta$ dependence.

13.5 Radiation from a Linear Antenna

In the preceding sections we dealt with radiating systems whose dimensions are much smaller than the wavelength. So it was permissible to treat the currents as constant over the systems. The antennae are generally not short compared with wavelength of the radiation they transmit. For this reason the current in the antenna is not constant and the variation of its amplitude must be taken into account. The problem is then a very difficult one because the radiation fields depend on the currents and the current, in turn, depends in part on the fields. We follow the usual approach to tackle this problem: (1) find the approximately correct current distribution by neglecting the radiation and then (2) calculate the radiation produced by the approximate current distribution.

For simplicity we consider a simple center-driven linear antenna, straight wires fed by a signal generator connected across a gap at the middle of the wire, as shown in **Figure 13.7**. The antenna is assumed to be oriented along the z-axis and has a length L. The current density is assumed to vary harmonically in time and space along the antenna. Standing waves are set up with the conditions that the current at each end is zero. As the current is fed in from a generator at the center, we must allow for a discontinuity of form there. Hence,

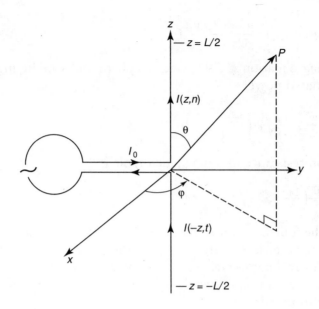

I FIGURE 13.7

$$I(z,t) = \left(A_1 e^{ikz} + B_1 e^{-ikz}\right) e^{i\omega t} \quad \text{for } z > 0$$

$$I(z,t) = \left(A_2 e^{ikz} + B_2 e^{-ikz}\right) e^{i\omega t} \quad \text{for } z < 0$$

where $k = \omega/c = 2\pi/\lambda$. For the current to vanish at $z = L/2$, we have

$$A_1 e^{ikL/2} + B_1 e^{-ikL/2} = 0, \quad \text{or} \quad B_1 = -A_1 e^{ikL}$$

Similarly, at $z = -L/2$ we have

$$B_2 = -A_2 e^{-ikL}$$

At the fed-in point the current is

$$I_0 e^{i\omega t} = I(0,t) = (A_1 + B_1) e^{i\omega t} = (A_2 + B_2) e^{i\omega t}$$

Thus

$$I_0 = A_1 + B_1 = A_1(1 - e^{ikL}) = A_1 e^{ikL/2}\left(-2i \sin \frac{kL}{2}\right)$$

and also

$$I_0 = A_2 e^{-ikL/2}\left(2i \sin \frac{kL}{2}\right)$$

This can be all put together concisely in the form

$$I(z,t) = \frac{I_0}{\sin(kL/2)} \sin k\,(L/2 - |z|)\,e^{i\omega t} \tag{13.44}$$

For example, for $L = \lambda/2$, the form of the current distribution is as shown in **Figure 13.8a,** and for $\lambda/2 < L < \lambda$ it is like the one in Figure 13.8b.

We may now determine the properties of the radiation field from equation (13.7). We first write equation (13.44) as $I(z, t) = I(z)e^{i\omega t}$, so equation (13.7) becomes

$$\vec{A}(\vec{r},t) = \frac{\mu_0}{4\pi}\int \frac{I(z')\,e^{i\omega\,(t-R/c)}}{R}\,dz'\hat{z} = \frac{\mu_0}{4\pi}\,e^{i\omega t}\,\hat{z}\int \frac{I(z')\,e^{ikR}}{R}\,dz' \tag{13.45}$$

In the far-field approximation we may write

$$R = |\vec{r} - \vec{z}'| \cong r - \hat{r}\cdot\vec{z}' = r - z'\cos\theta$$

in the exponent and replace R in the denominator by r, so that

$$\vec{A}(\vec{r},t) = \frac{\mu_0}{4\pi}\,e^{i\omega t}\frac{e^{-ikr}}{r}\,\hat{z}\int I(z')\,e^{ikz'\cos\theta}dz'$$

Now, from equation (13.44), $I(z')$ is

$$I(z') = \frac{I_0}{\sin(kL/2)}\sin k\,(L/2 - |z'|)$$

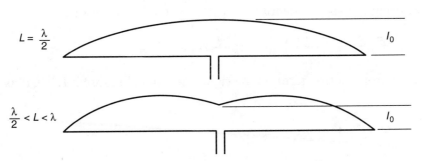

I FIGURE 13.8

and the vector potential becomes

$$\vec{A}(\vec{r},t) = \frac{\mu_0 I_0}{\sin(kL/2)} e^{i\omega t} \frac{e^{-ikr}}{4\pi r} \hat{z} \int_{-L/2}^{L/2} \sin k(L/2 - |z'|) e^{ikz'\cos\theta} dz' \tag{13.46}$$

The integral can split into two parts:

$$\int_{-L/2}^{0} \sin k(L/2 + z') e^{ikz'\cos\theta} dz' + \int_{0}^{L/2} \sin k(L/2 - z') e^{ikz'\cos\theta} dz'$$

In the first integral changing the dummy variable of integration from z' to $-z'$ and then combining with the second integral, we obtain

$$\int_{0}^{L/2} \sin k\left(\frac{L}{2} - z'\right)(2\cos[kCz']) dz' \quad (C = \cos\theta)$$

$$= \int_{0}^{L/2} \left[\sin k\left(\frac{L}{2} - z' + Cz'\right) + \sin k\left(\frac{L}{2} - z' - Cz'\right)\right] dz'$$

$$= \frac{2\left[\cos((kL/2)\cos\theta) - \cos(kL/2)\right]}{k\sin^2\theta}$$

Substituting this into equation (13.46) we obtain the vector potential.

$$\vec{A}(\vec{r},t) = \frac{\mu_0 I_0}{\sin(kL/2)} e^{i\omega t} \frac{e^{-ikr}}{2\pi r} \hat{z} \frac{\cos[(kL/2)\cos\theta] - \cos(kL/2)}{k\sin^2\theta} \tag{13.47}$$

From this we may calculate the fields in the radiation zone: $\vec{B} = \nabla \times \vec{A}$ and $\vec{E} = c\vec{B} \times \hat{r}$. The calculation of $\nabla \times \vec{A}$ is a very formidable task. However, for the radiation field, we may only retain terms that fall off as $1/r$. Then

$$\nabla \times \vec{A} = \frac{\mu_0 I_0 e^{i\omega t}}{2\pi k \sin(kL/2)} \nabla \times \left(\frac{f(\theta) e^{-ikr}}{r} \hat{z}\right) \tag{13.48}$$

with

$$f(\theta) = \frac{\cos[(kL/2)\cos\theta] - \cos(kL/2)}{\sin^2\theta}$$

Now, using the vector identity $\nabla \times \left(\alpha\vec{\beta}\right) = (\nabla\alpha) \times \vec{\beta} + \alpha\left(\nabla \times \vec{\beta}\right)$

$$\nabla \times \left(\frac{f(\theta) e^{-ikr}}{r} \hat{z}\right) = \left(\nabla \frac{f(\theta)}{r}\right) \times \left(e^{-ikr}\hat{z}\right) + \frac{f(\theta)}{r}\left(\nabla \times (e^{-ikr}\hat{z})\right)$$

We may drop the first term on the right-hand side, because it falls off as $1/r^2$. Then

$$\nabla \times \left(\frac{f(\theta) e^{-ikr}}{r} \hat{z}\right) \cong \frac{f(\theta)}{r}\left(\nabla \times e^{-ikr}\hat{z}\right)$$

$$= \frac{f(\theta)}{r}(-ik) e^{-ikr}\left(\hat{x} \frac{\partial r}{\partial y} - \hat{y} \frac{\partial r}{\partial x}\right) = -\frac{f(\theta)}{r} ike^{-ikr}\sin\theta\hat{\phi}$$

Setting this back into equation (13.48), we get the explicit expression for $\vec{B}$ and $\vec{H} = \vec{B}/\mu_0$; then $\vec{E} = c\vec{B} \times \hat{r}$. It is obvious that $\vec{E}$ is proportional to

$$-\frac{f(\theta)}{r} ike^{-ikr}\sin\theta\hat{\theta}$$

The pointing vector is given by

$$\left\langle \vec{S} \right\rangle = \frac{1}{2}\left(\vec{E} \times \vec{H}^* \right)$$

Because $\vec{B}$ and $\vec{E}$ are each proportional to $f(\theta)\sin\theta$, the Poynting vector is proportional to the square of this. For a half-wave antenna, $L = \lambda/2$; then $kL = (2\pi/\lambda)(\lambda/2) = \pi$, and $f(\theta)$ reduces to

$$f(\theta) = \frac{\cos\left[(\pi/2)\cos\theta\right]}{\sin^2\theta}$$

and

$$\left\langle \vec{S} \right\rangle \propto [f(\theta)\sin\theta]^2 = \left(\frac{\cos\left[(\pi/2)\cos\theta\right]}{\sin\theta} \right)^2$$

It is seen that $\left\langle \vec{S} \right\rangle$ has its maximum at $\theta = \pi/2$ and reduces to zero at $\theta = 0$ and π. In general, the shape is not too different from that described by the angular distribution from a dipole, but more of the power goes out in the equatorial plane as indicated in polar diagram **Figure 13.9**, where the dashed line is from a simple dipole.

As the antenna length is increased, the radiation pattern becomes multilobed and quite complicated. Concentrated single-lobe emission can nevertheless be obtained by dividing a long antenna into half-wavelength segments with suitable inductors in between to keep all the currents in phase. A parallel array of such antennas can concentrate the energy into a single directional beam, which is obviously useful for point-to-point communication. The detailed calculations are quite messy, but the basic principles are the same as those we have been discussing.

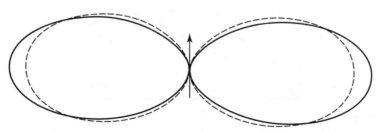

I FIGURE 13.9

13.6 The Lienard-Weichert Potential: Fast-Moving Point Charges

The retarded potentials describe the potentials of arbitrary charge and current distribution. In the last two sections we applied them to cases where the current distributions are bounded and dipole approximations could be made. We now remove these restrictions and consider a single charged particle in arbitrary motion, without invoking the dipole approximation. Because the calculation of the potentials depends on the position and velocity of the charged particle at the retarded time, we must know the path of the particle. **Figure 13.10** shows a path of the particle described by the radius vector $\vec{r}'(t')$. We can use equation (13.6) to calculate the scalar potential. Because we deal with a point charge, there is no direct variation of ρ with r', and hence the charge density can be replaced by Dirac's δ-function. However, because of the appreciable motion of the charge, it essentially looks like an extended charge distribution as a function of time. Thus, we need to consider two factors. The distance between the charge and the point of observation $\vec{R}(t') = \vec{r} - \vec{r}'(t')$ varies with time, and the argument of the δ-function describing the position of the charge depends on time because of the retardation effect. So we write

$$\rho(\vec{r}', t) = q\delta\left(t' - \left[t - \frac{\|\vec{r} - \vec{r}'(t')\|}{c}\right]\right) = q\delta(t'') \qquad (13.49)$$

where

$$t'' = t' - \left[t - \frac{\|\vec{r} - \vec{r}'\|}{c}\right] = t' - \left[t - \frac{\|\vec{R}\|}{c}\right] \qquad (13.50)$$

I FIGURE 13.10

The evaluation of the potential $\phi(\vec{r}, t)$ requires integration with respect to time t' over the entire volume that contains the charge

$$\phi(\vec{r}, t) = \frac{q}{4\pi\varepsilon_0} \int_{-\infty}^{\infty} \frac{\delta(t' - t + \|\vec{r} - \vec{r}'(t')\|/c)}{\|\vec{r} - \vec{r}'(t')\|} \, dt' \tag{13.51}$$

Integration over t' is not quite as straightforward as the $\vec{r}'$ integration because $\vec{R}(t')(=\vec{r} - \vec{r}'(t'))$ appears in the argument of the δ-function. But if we can put the integral in the form $\int_{-\infty}^{\infty} g(x) \delta(x - x') \, dx$, then its integral is readily found and is equal to $g(x')$. To this end, we first introduce a new variable t'':

$$t'' = t' - t + \frac{|\vec{r} - \vec{r}'(t')|}{c}$$

then

$$dt'' = dt' + \frac{1}{c} \frac{d}{dt'} |\vec{r} - \vec{r}'(t)| \, dt' \tag{13.52}$$

Note that the observation is made at a fixed time t, so we have set $dt = 0$. Now let the coordinates of the charged particles be $x_1'(t'), x_2'(t'), x_3'(t')$ and those of the point x_1, x_2, x_3.

Then

$$|\vec{r} - \vec{r}'(t')| = \sqrt{\sum_i [x_i - x_i'(t')]}$$

and

$$\frac{1}{c} \frac{d}{dt'} |\vec{r} - r'(t')| = \frac{1}{c} \sum_i \frac{\partial}{\partial x_i'} |\vec{r} - r'(t')| \frac{dx_i'}{dt'} \tag{13.53}$$

Now $\partial |\vec{r} - \vec{r}'(t')| / \partial t$ are the components of the gradient of $|\vec{r} - \vec{r}'(t')|$ and dx'/dt' are the components of $d\vec{r}'/dt'$. Thus we can write

$$\frac{1}{c} \frac{d}{dt'} |\vec{r} - r'(t')| = \frac{1}{c} \nabla' |\vec{r} - \vec{r}'(t')| \cdot \frac{d\vec{r}'}{dt'}$$

where (∇' operates on the coordinates of the charged particle only. Carrying out the gradient we obtain

$$\nabla' |\vec{r} - \vec{r}'(t')| = -\frac{\vec{r} - \vec{r}'(t')}{|\vec{r} - \vec{r}'(t')|} = -\frac{\vec{R}}{|R|}$$

And $d\vec{r}'/dt' = \vec{v}$ is the velocity of the charged particle. Putting these back into equation (13.53), we get

$$\frac{1}{c}\frac{d}{dt'}|\vec{r}-r'(t')| = -\frac{1}{c}\frac{\vec{R}}{R}\cdot\vec{v}$$

Equation (13.52) now becomes $dt'' = dt'\left[1-\vec{\beta}\cdot\vec{R}/R\right]$
or

$$dt' = \frac{R}{R-\vec{\beta}\cdot\vec{R}}$$

where $\vec{\beta}=\vec{v}/c$. Finally, we can rewrite the potential ϕ in the desired form

$$\phi(\vec{r},t) = \frac{q}{4\pi\varepsilon_0}\int_{-\infty}^{\infty}\frac{\delta(t'')}{R(t')}\frac{R(t')}{R(t')-\vec{\beta}(t')\cdot\vec{R}(t')}\,dt''$$

$$= \frac{q}{4\pi\varepsilon_0}\int_{-\infty}^{\infty}\frac{\delta(t'')}{R(t')-\vec{\beta}(t')\cdot\vec{R}(t')}\,dt''$$

Applying the important property of δ-function $\int_{-\infty}^{\infty}g(x)\delta(x-x')\,dx = g(x')$, we obtain

$$\phi(\vec{r},t) = \frac{q}{4\pi\varepsilon_0}\left[\frac{1}{R(t')-\vec{\beta}\cdot R(t')}\right]_{t''=0}$$

$$= \frac{q}{4\pi\varepsilon_0}\left[\frac{1}{R(t')-\vec{\beta}\cdot\vec{R}(t')}\right]_{t'=t-R(t')/c} \tag{13.54}$$

Our nest task is to find the vector potential $\vec{A}(\vec{r},t)$. Because the current density is equal to the product of the charge density and the velocity of the particle, then a similar derivation to that of the scalar potential can be used to determine $\vec{A}(\vec{r},t)$. The result is

$$\vec{A}(\vec{r},t) = \frac{\mu_0 q}{4\pi}\left[\frac{\vec{v}}{R(t')-\vec{\beta}\cdot\vec{R}(t')}\right]_{t'=t-R(t')/c} \tag{13.55}$$

The potentials in equations (13.54) and (13.55) are known as the Lienard-Weichert potentials. By comparing these equations, we find that

$$\vec{A}(\vec{r},t) = \vec{v}(t')\,\phi(\vec{r},t)/c^2 \tag{13.56}$$

We have not made any assumption about the magnitude of $\vec{v}$ relative to the speed of light c. Equations (13.54) and (13.55) are in fact relativistically correct. In the nonrelativistic limit we find the familiar results

$$\phi \to \frac{1}{4\pi\varepsilon_0}\left[\frac{q}{R(t')}\right]_{t'=t-R(t')/c}, \quad \vec{A} \to \frac{\mu_0}{4\pi}\left[\frac{\vec{j}(t')}{R(t')}\right]_{t'=t-R(t')/c}$$

It is possible now to calculate the corresponding electric and magnetic fields of an arbitrarily moving point charged particle from $\vec{E} = -\nabla\phi - \partial\vec{A}/\partial t, \vec{B} = \nabla\times\vec{A}$, with ϕ and $\vec{A}$ given by equations (13.54) and (13.55). The actual calculation is quite complicated, and we perform these in the next section.

13.7 Fields of an Accelerated Point Charge

Again, $\vec{r} = (x_1, x_2, x_3)$ is the radius vector of the point of observation P at time t, $\vec{r}'(t')$ (t') is the radius vector of the charge at t' and $\vec{R} = \vec{r} - \vec{r}'$. First let us rewrite equations (13.54) and (13.55) as

$$\phi(\vec{r}, t) = \frac{q}{4\pi\varepsilon_0}\left[\frac{1}{R(t') - \vec{\beta}\cdot\vec{R}(t')}\right]_{t'=t-R(t')/c}$$ (13.54)

$$= \frac{q}{4\pi\varepsilon_0}\left[\frac{1}{S}\right]_{t'=t-R(t')/c}$$

$$\vec{A}(\vec{r}, t) = \frac{\mu_0 q}{4\pi}\left[\frac{\vec{v}}{R(t') - \vec{\beta}\cdot\vec{R}(t')}\right]_{t'=t-R(t')/c}$$ (13.55)

$$= \frac{q}{4\pi\varepsilon_0 c^2}\left[\frac{\vec{v}}{S}\right]_{t'=t-R(t')/c}$$

where, for convenience, we have set $S = R - \vec{R}\cdot\vec{v}/c$.

The fields are calculated from the following two relations

$$\vec{E} = -\nabla\phi - \partial\vec{A}/\partial t, \quad \vec{B} = \nabla\times\vec{A}$$

Caution must be exercised in applying the operator ∇. The components of ∇ are partial derivatives at time t, which is "fixed" at observation; they are not at time t'. At time t' a signal propagated with c is emitted at $\vec{r}'$ and arrives at the observation point P at time t. Because the time variation with respect to t' is given, to compute the fields we have to transform $\partial/\partial t|_{x_i}$ and ∇ to expressions in terms of $\partial/\partial t'|_{x_i}$. This can be done and we need to be a little patient. We start with the relation

$$R = c(t - t') = |\vec{r} - \vec{r}'| = \left[\sum(x_i - x_i')^2\right]^{1/2}$$ (13.57)

We see immediately

$$\frac{\partial R}{\partial t} = c\left(1 - \frac{\partial t'}{\partial t}\right).$$

and

$$\frac{\partial R}{\partial t} = \frac{\partial R}{\partial t'}\frac{\partial t'}{\partial t}$$

But, as shown in Section 13.6,

$$\frac{\partial R}{\partial t'} = \sum_i \frac{\partial R}{\partial x_i'}\frac{dx_i'}{dt'} = -\frac{\vec{R}\cdot\vec{v}}{R}$$

Thus,

$$c\left(1 - \frac{\partial t'}{\partial t}\right) = -\frac{\vec{R}\cdot\vec{v}}{R}\frac{\partial t'}{\partial t}$$

or

$$\frac{\partial t'}{\partial t} = \frac{R}{R - \vec{v}\cdot\vec{R}/c} = \frac{R}{S} \tag{13.58}$$

and so

$$\frac{\partial}{\partial t} = \frac{R}{S}\frac{\partial}{\partial t'} \tag{13.59}$$

We now transform the del operator, ∇. Because $R = R(x_i, x_i'(t'))$, we can write

$$\nabla R = \nabla_1 R + \frac{\partial R}{\partial t'}\nabla t' = \frac{\vec{R}}{R} - \frac{\vec{R}\cdot\vec{v}}{R}\nabla t' \tag{13.60}$$

where ∇_1 implies differentiation with respect to x_i at time t'. But from equation (13.57) we have

$$\nabla R = -c\nabla t'$$

Combining this with equation (13.60), we obtain

$$\nabla t' = -\vec{R}/Sc \tag{13.61}$$

Substituting this into equation (13.60) we obtain

$$\nabla R = \nabla_1 R + \frac{\partial R}{\partial t'}\nabla t' = \nabla_1 - \frac{\vec{R}}{Sc}\frac{\partial R}{\partial t'}$$

and, in general, for ∇

$$\nabla = \nabla_1 - \frac{\vec{R}}{Sc}\frac{\partial}{\partial t'} \tag{13.62}$$

We are ready to compute $\vec{E}$ and $\vec{B}$, with ϕ and $\vec{A}$ given by equations (13.54) and (13.55):

$$\vec{E} = -\frac{q}{4\pi\varepsilon_0}\nabla\left(\frac{1}{S}\right) - \frac{q}{4\pi\varepsilon_0}\frac{\partial}{\partial t}\left(\frac{\vec{v}}{c^2 S}\right)$$

Now

$$\nabla\left(\frac{1}{S}\right) = \nabla_1\left(\frac{1}{S}\right) - \frac{\vec{R}}{Sc}\frac{\partial}{\partial t'}\left(\frac{1}{S}\right) = -\frac{1}{S^2}\nabla_1 S + \frac{\vec{R}}{S^3 c}\frac{\partial S}{\partial t'}$$

and

$$\frac{\partial}{\partial t}\left(\frac{\vec{v}}{Sc^2}\right) = \frac{R}{S}\frac{\partial}{\partial t'}\left(\frac{\vec{v}}{Sc^2}\right) = \frac{R}{S}\left[\frac{1}{c^2 S}\dot{\vec{v}} + \frac{\vec{v}}{c^2}\frac{\partial}{\partial t'}\frac{1}{S}\right]$$

$$= \frac{R}{S}\left[\frac{1}{c^2 S}\dot{\vec{v}} - \frac{\vec{v}}{c^2 S^2}\frac{\partial S}{\partial t'}\right]$$

$$\nabla_1 S = \nabla_1\left(R - \vec{v}\cdot\vec{R}/c\right) = \frac{\vec{R}}{R} - \frac{\vec{v}}{c}$$

Putting these back into the expression for $\vec{E}$ we finally obtain, after rearrangement and combining the terms,

$$\vec{E} = \frac{q}{4\pi\varepsilon_0}\left[\frac{1}{S^3}\left(\vec{R} - \frac{R\vec{v}}{c}\right)\left(1 - \frac{v^2}{c^2}\right) + \frac{1}{c^2 S^3}\left\{\vec{R}\times\left[\left(\vec{R} - \frac{R\vec{v}}{c}\right)\times\dot{\vec{v}}\right]\right\}\right] \tag{13.63}$$

The first term on the right side falls off as R^{-2} and is known as the velocity field. The second term falls off as R^{-1} and is called the acceleration field or radiation field.

Similarly, the magnetic field is

$$\vec{B} = \frac{q}{4\pi\varepsilon_0 c^2}\left[\frac{\vec{v}\times\vec{R}}{S^3}\left(1 - \frac{v^2}{c^2}\right) + \frac{1}{cS^3}\frac{\vec{R}}{R}\times\left\{\vec{R}\times\left[\left(\vec{R} - \frac{R\vec{v}}{c}\right)\times\dot{\vec{v}}\right]\right\}\right] \tag{13.64}$$

which can be expressed as

$$c\vec{B} = \hat{R}\times\vec{E} \tag{13.65}$$

That is, the magnetic field $\vec{B}$ is always perpendicular to $\vec{E}$ and $\vec{R}$.

We just pointed out above that

$$\vec{E}_v \propto 1/R^2, \quad \vec{E}_a \propto 1/R$$

Now let us compute the Poynting vector for the fields; we find that the contribution to this vector due to the two components is

$$\vec{S}_v \propto 1/R^4, \quad \vec{S}_a \propto 1/R^2$$

To find the energy radiated by the charge, we have to integrate the normal component of the Poynting vector $\vec{S}$ over the surface of a sphere of radius R. Because the element of the surface area involves R^2, the integral containing $\vec{S}_v$ varies as $1/R^2$ whereas that involving $\vec{S}_a$ remains finite. Thus, for large R, the contribution due to $\vec{S}_v$ tends to zero whereas that due to $\vec{S}_a$ is finite. This implies that a charged particle moving with a uniform velocity cannot radiate energy. Energy can be radiated only by accelerated charged particles.

Acceleration could be positive and negative; a negative acceleration is also known as deceleration. A charged particle in deceleration also emits radiation, and it is known as bremsstrahlung radiation. This happens when, for example, an electron moves passing by a nucleus and is decelerated by the nucleus.

13.8 The Fields of a Point Charge in Uniform Motion

As a simple application of equations (13.63) and (13.64), we calculate the electric and magnetic fields of a point charge in uniform motion, i.e., moving with constant velocity along a straight line. As $\vec{a} = 0$, we have

$$\vec{E} = \frac{q}{4\pi\varepsilon_0} \frac{R}{[\vec{R} \cdot (c\hat{R} - \vec{v})]^3} [(c^2 - v^2)(c\hat{R} - \vec{v})] \tag{13.66}$$

Our task is to find $R(c\hat{R} - \vec{v})$ and $\vec{R} \cdot (c\hat{R} - \vec{v})$. Now the fields of the moving charged particle are propagated with the velocity of light; this means that $R = c(t - t')$. With this basic relation, let us first calculate $R(c\hat{R} - \vec{v})$:

$$R(c\hat{R} - \vec{v}) = c\vec{R} - R\vec{v} = c(\vec{r} - \vec{r}') - c(t - t')\vec{v} = c(\vec{r} - \vec{v}t) \tag{13.67}$$

It is little tedious to calculate the quantity $\vec{R} \cdot (c\hat{R} - \vec{v})$:

$$\vec{R} \cdot (c\hat{R} - \vec{v}) = cR - \vec{R} \cdot \vec{v} = Rc(1 - \hat{R} \cdot \vec{v}/c) \tag{13.68}$$

Now

$$\hat{R} = \frac{\vec{R}}{R} = \frac{\vec{r} - \vec{r}'}{c(t - t')} = \frac{\vec{r} - \vec{v}t'}{c(t - t')}$$

then equation (13.68) becomes

$$\vec{R} \cdot (c\hat{R} - \vec{v}) = c^2(t - t') \left[1 - \frac{(\vec{r} - \vec{v}t') \cdot \vec{v}}{c^2(t - t')} \right] = c^2(t - t') - \vec{r} \cdot \vec{v} + v^2 t'$$

(13.69)

$$= (c^2 t - \vec{r} \cdot \vec{v}) - (c^2 - v^2) t'$$

The particle is moving, so it is a little tricky to find the retarded time t', which can be found from the relation

$$R = |\vec{r} - \vec{r}'| = |\vec{r} - \vec{v}t'| = c(t - t')$$

or $|\vec{r} - \vec{v}t'| = c(t - t')$. Squaring both sides, we obtain

$$r^2 - 2\vec{r} \cdot \vec{v}t' + v^2 t'^2 = c^2(t^2 - 2tt' + t'^2)$$

which is a quadratic equation for the retarded time t'. Solving for t' we find that

$$t' = \frac{(c^2 t - \vec{r} \cdot \vec{v}) \pm \sqrt{(c^2 t - \vec{r} \cdot \vec{v})^2 + (c^2 - v^2)(r^2 - c^2 t^2)}}{c^2 - v^2}$$

What sign should we take? We should take the minus sign, because in the limit $v = 0$ the above equation reduces to $t' = t \pm r/c$ and the minus sign gives the retarded time. Thus,

$$t' = \frac{(c^2 t - \vec{r} \cdot \vec{v}) - \sqrt{(c^2 t - \vec{r} \cdot \vec{v})^2 + (c^2 - v^2)(r^2 - c^2 t^2)}}{c^2 - v^2}$$

Substituting this into equation (13.68) we obtain

$$\vec{R} \cdot (c\hat{R} - \vec{v}) = \sqrt{(c^2 t - \vec{r} \cdot \vec{v})^2 + (c^2 - v^2)(r^2 - c^2 t^2)}$$

(13.70)

The term under the square root is

$$I = c^4 t^2 - 2c^2 t(\vec{r} \cdot \vec{v}) + (\vec{r} \cdot \vec{v})^2 + c^2 r^2 - c^4 t^2 - v^2 r^2 + v^2 c^2 t^2$$

$$= (\vec{r} \cdot \vec{v})^2 + (c^2 - v^2) r^2 + c^2 (vt)^2 - 2c^2 (\vec{r} \cdot \vec{v}t)$$

Now

$$\vec{\Re} = \vec{r} - \vec{v}t \quad \text{or} \quad \vec{v}t = \vec{r} - \vec{\Re}$$

where $\vec{\Re}$ is the vector from the present location of the particle to the field point P whose position vector is $\vec{r}$. We can now eliminate t from I:

$$I = (\vec{r} \cdot \vec{v})^2 + (c^2 - v^2)\,r^2 + c^2(r^2 + \Re^2 - 2\vec{r} \cdot \vec{\Re}) - 2c^2(r^2 - \vec{r} \cdot \vec{\Re})$$

$$= (\vec{r} \cdot \vec{v})^2 - r^2 v^2 + c^2 \Re^2$$

We can express the first two terms on the right-hand side in terms of $\Re$ and angle θ between $\vec{\Re}$ and $\vec{v}$:

$$(\vec{r} \cdot \vec{v})^2 - r^2 v^2 = \left((\vec{\Re} + \vec{v}t) \cdot \vec{v} \right)^2 - (\vec{\Re} + \vec{v}t) \cdot (\vec{\Re} + \vec{v}t)\,v^2$$

$$= -\Re^2 v^2 (1 - \cos^2\theta) = -\Re^2 v^2 \sin^2\theta$$

Then

$$I = -\Re^2 v^2 \sin^2\theta + c^2 \Re^2 = \Re^2 c^2 (1 - v^2 \sin^2\theta/c^2)$$

and

$$\vec{R} \cdot (c\hat{R} - \vec{v}) = \Re c \sqrt{1 - v^2 \sin^2\theta/c^2} \tag{13.71}$$

In terms of $\Re$, equation (13.67) becomes

$$R(c\hat{R} - \vec{v}) = c(\vec{r} - \vec{v}t) = c\vec{\Re}$$

Substituting this and equation (13.71) into equation (13.67) for $\vec{E}$, we obtain

$$\vec{E}(\vec{r}, t) = \frac{q}{4\pi\varepsilon_0} \frac{1 - v^2/c^2}{(1 - v^2 \sin^2\theta/c^2)^{3/2}} \frac{\vec{\Re}}{\Re^2} \tag{13.72}$$

Because of the factor $\vec{\Re}$, the electric field $\vec{E}$ points along the line from the present position of the particle. Furthermore, because of the $\sin^2\theta$ in the denominator, the $\vec{E}$ field is strongly concentrated in the plane perpendicular to its motion. This is in agreement with earlier results found in Chapter 10.

We can calculate the electric and magnetic fields of a point charge in uniform motion directly from the Lienard-Weichert potentials. We leave this as homework.

13.9 Radiation from an Accelerated Charge

The calculation of radiation from an accelerated charge at high velocity is quite complicated, and so for simplicity we consider here only the case of a slowly moving charge, so that v/c can be neglected. Then $S \cong R$ and the fields as given by equations (13.63) and (13.64) reduce to

$$\vec{E}_a = \frac{q}{4\pi\varepsilon_0 c^2 R^3} \left[\vec{R} \times \left(\vec{R} \times \dot{\vec{v}} \right) \right] \tag{13.73}$$

$$\vec{B}_a = \frac{q}{4\pi\varepsilon_0 c^3 R^2} \left[\dot{\vec{v}} \times \vec{R} \right] = \frac{\vec{R} \times \vec{E}}{Rc} \tag{13.74}$$

From these field vectors the Poynting vector is found to be

$$\vec{S}_a = \vec{E}_a \times \vec{H}_a = \vec{E}_a \times \frac{\vec{B}_a}{\mu_0} = \vec{E}_a \times \frac{1}{\mu_0 c} \left(\frac{\vec{R} \times \vec{E}_a}{R} \right)$$

Because $\vec{E}_a \perp \vec{R}$, this reduces to

$$\vec{S}_a = \frac{1}{\mu_0 c} E_a^2 \hat{R} \tag{13.75}$$

Now

$$\vec{E}_a = \frac{q}{4\pi\varepsilon_0 c^2 R^3} \left[\vec{R} \times \left(\vec{R} \times \dot{\vec{v}} \right) \right]$$

$$= \frac{q}{4\pi\varepsilon_0 c^2 R^3} \left[\left(\vec{R} \times \dot{\vec{v}} \right) \vec{R} - (\vec{R} \cdot \vec{R}) \dot{\vec{v}} \right] = \frac{q}{4\pi\varepsilon_0 c^2 R^3} \left(R\dot{v} \cos\theta \, \vec{R} - R^2 \dot{\vec{v}} \right)$$

where θ is the angle between $\vec{R}$ and $\dot{\vec{v}}$. Substituting this into equation (13.75) we obtain

$$\vec{S}_a = \frac{1}{\mu_0 c} \frac{q^2}{16\pi^2 \varepsilon_0 c^4 R^6} \left[R\dot{v} \cos\theta \, \vec{R} - R^2 \dot{\vec{v}} \right]^2 \hat{R}$$

$$= \frac{q^2 \dot{v}^2}{16\pi^2 \varepsilon_0 c^3} \frac{1}{R^2} \sin^2\theta \hat{R} \tag{13.76}$$

The power radiated per unit solid angle can be found by multiplying equation (13.76) by R^2 that is the area per unit solid angle:

$$\frac{dP}{d\Omega} = \frac{q^2 \dot{v}^2}{16\pi^2 \varepsilon_0 c^3} \sin^2\theta \tag{13.77}$$

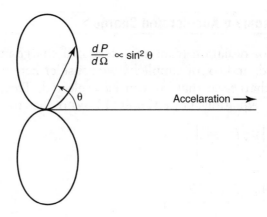

I FIGURE 13.11

We see that the angular distribution of the radiated energy is just the $\sin^2\theta$ distribution, as shown in **Figure 13.11**.

Integrating over the surface of a sphere surrounding the charge, we obtain the total radiated power

$$P_r = \frac{q^2 \dot{v}^2}{16\pi^2 \varepsilon_0 c^3 R^2} \int_0^{2\pi} \int_0^{\pi} \sin^2\theta R^2 \sin\theta d\theta d\varphi$$

$$= \frac{q^2 \dot{v}^2}{16\pi^2 \varepsilon_0 c^3 R^2} \int_0^{2\pi} \int_0^{\pi} (1 - \cos^2\theta) R^2 \sin\theta d\theta d\varphi$$

from which we get

$$P_r = \frac{q^2}{4\pi\varepsilon_0} \frac{2}{3} \frac{\dot{v}^2}{c^3} \tag{13.78}$$

for the power radiated from a slowly moving accelerated charge, in agreement with our earlier result of equation (13.35), which is Larmor's formula.

13.10 Radiation Damping

The energy lost into radiation by an accelerating charged particle must come at the expense of the particle's kinetic energy, as energy is conserved. So the accelerating particle must be subject to an additional damping force due to its own radiation. This reaction of the radiation on the motion of the charged particle is called radiation reaction or radiation damping, which has not been included in our previous discussion of particle dynamics.

We consider a nonrelativistic particle ($v \ll c$). The power radiated by it is given by Larmor's formula, equation (13.78). Because energy is conserved, the energy radiated in a time interval, say, t_1 to t_2, is equal to the rate at which the charge loses its kinetic energy under the influence of the radiation reaction force in this time interval:

$$\int_{t_1}^{t_2} \vec{F}_r \cdot \vec{v} \, dt = -\frac{q^2}{6\pi\varepsilon_0 c^3} \int_{t_1}^{t_2} \dot{\vec{v}} \cdot \dot{\vec{v}} \, dt$$

Integrating the right-hand side by parts gives

$$\int_{t_1}^{t_2} \vec{F}_r \cdot \vec{v} \, dt = -\frac{q^2}{6\pi\varepsilon_0 c^3} \left\{ \left[\dot{\vec{v}} \cdot \dot{\vec{v}} \right] \Big|_{t_1}^{t_2} - \int_{t_1}^{t_2} \dot{\vec{v}} \cdot \ddot{\vec{v}} \, dt \right\}$$

If the motion is periodic or if the time interval is short so that the state of the system is approximately the same at t_1 and t_2, then we may neglect the integrated term, and the above equation becomes

$$\int_{t_1}^{t_2} \vec{F}_r \cdot \vec{v} \, dt = \frac{q^2}{6\pi\varepsilon_0 c^3} \int_{t_1}^{t_2} \dddot{\vec{r}} \cdot \dot{\vec{v}} \, dt$$

or

$$\int_{t_1}^{t_2} \left(\vec{F}_r - \frac{q^2}{6\pi\varepsilon_0 c^3} \dddot{\vec{v}} \right) \cdot \vec{v} \, dt = 0 \qquad (13.79)$$

Thus, over a cycle or over a short time interval we can take, for the radiation reaction force,

$$\vec{F}_r = \frac{q^2}{6\pi\,\varepsilon_0 c^3} \dddot{\vec{v}} \qquad (13.80)$$

This is known as the Abraham-Lorentz formula for the radiation reaction force. It is the time average of the parallel component over very special time interval. The use of equation (13.80) is valid only when the reaction force is small compared with the external force. Then, the motion of an accelerating particle of charge q and mass m in a conservative force field $\vec{F}_e$ is governed by the following Newtonian equation of motion:

$$m \frac{d^2\vec{r}}{dt^2} = \vec{F}_e + \frac{q^2}{6\pi\varepsilon_0} \frac{d^3\vec{r}}{dt^3} \qquad (13.81)$$

which is known as the Abraham-Lorentz equation. Remember that equation (13.81) is valid provided that $|\vec{F}_r| \ll |\vec{F}_e| \cong |m\dot{\vec{v}}|$.

The Abraham-Lorentz equation has some peculiarities: It is a differential equation of third order in t and possesses unphysical "runaway" solutions as well as solutions that are physically reasonable.

We have learned that the radiation reaction force is a recoil effect of the particle's own fields acting back on the charge, but no attempt was made to identify the mechanism responsible for this radiation reaction force. It is due to the time delay effect. A stationary charge distribution has no net electric force on itself, and all the internal repulsions cancel out. When this charge distribution is accelerating, one part Δq_A of it feels the presence of another part Δq_B of the same charge distribution at its retarded position. As a result, the overall internal repulsions no longer cancel out. Lorentz calculated the electromagnetic self-force using a spherical charge distribution, and the mathematics is very cumbersome. Griffiths used a less realistic "dumbbell" model to elucidate the mechanism involved. Even Griffiths' calculation is tedious, and we do not intend to repeat here. We refer interested reader to his book (David J. Griffths, Sec.11.2.3, *Introduction to Electrodynamics*, third edition. Prentice Hall, 1999).

13.11 Scattering of Radiation

When an electromagnetic wave impinges on a small body consisting of charged particles, the charges are set into oscillatory motions. They radiate. The result is that energy is extracted from the incident wave and is then re-emitted into space. We describe this process as scattering of the original incident radiation by the charged particles. The effectiveness of a given body as a scatter is described by its scattering cross-section that is defined as

$$\sigma = \frac{\text{total scattered power}}{\text{incident energy flux}} \tag{13.82}$$

the cross-section σ is seen to have the dimensionality of an area. Similarly, the differential cross-section is defined as

$$\frac{d\sigma}{d\Omega} = \frac{\text{scattered energy/unit time/unit solid angle}}{\text{incident energy/unit area/unit time}} \tag{13.83}$$

Consider a plane monochromatic wave incident on a particle of charge q. The electric field vector of the incident plane wave can be written as

$$\vec{E} = \alpha \vec{E_0} e^{i(\vec{k} \cdot \vec{r} - \omega t)}$$

where α is the unit polarization vector of the incident wave. The subsequent force on q is given by Newton's second law

$$\vec{F} = m\dot{\vec{v}} = q(\vec{E} + \vec{v} + \vec{B})$$

For the nonrelativistic case in which $v \ll c$, the term $\vec{v} \times \vec{B}$ in the Lorentz force may be ignored. Then we have

$$\vec{F} = m\dot{\vec{v}} = q\vec{E} = \alpha q \vec{E}_0 e^{i(\vec{k}\cdot\vec{r} - \omega t)}$$

from which we obtain

$$\dot{\vec{v}}(t) = \alpha\,(q/m)\,\vec{E}_0\,e^{i(\vec{k}\cdot\vec{r} - \omega t)} \tag{13.84}$$

The time-average power radiated per unit solid angles is given by equation (13.77), which gives, after substituting equation (13.84) into it,

$$\frac{dP}{d\Omega} = \frac{q^4 E_0^2}{16\pi^2 m^2 \varepsilon_0 c^3}\sin^2\theta$$

and from equation (13.75) we have the incident energy flux: $S = 1/\mu_0 c E_a^2$. Hence, the differential scattering cross-section is

$$\frac{d\sigma}{d\Omega} = \frac{dP/d\Omega}{S} = \left(\frac{q^2}{4\pi\varepsilon_0 mc^2}\right)^2 \sin^2\theta \tag{13.85}$$

where the angle θ is the angle between $\dot{\vec{v}}$ (i.e., $\vec{E}$) and the direction of the outgoing radiation. The quantity in the bracket is the classical radius of the charge "particle" r_0:

$$r_0 = \frac{q^2}{4\pi\varepsilon_0 mc^2} \tag{13.86}$$

The total cross-section is obtained by integrating over all angles, and the reader should easily find

$$\sigma_T = \int \frac{d\sigma}{d\Omega}\,d\Omega = \frac{8\pi}{3}r_0^2 \tag{13.87}$$

where σ_T is known as the Thomson scattering formula. For scattering by a free electron, $\sigma_T = 0.665 \times 10^{-28}$ m^2, $r_0 = 2.8 \times 10^{-15}$ m. Only electrons in a dense plasma are likely to produce significant Thomson scattering, one example being the corona seen around the sun as a bright ring during a total eclipse of the sun.

We now consider the scattering of the radiation from an electron bound in atom and estimate the total power radiated from it. For bound electrons we have electric dipole polarizability $\alpha(\omega)$, as given by equation (7.71), and a displacement x given by equation (7.70):

$$x = \frac{-eE_0 e^{-i\omega t}}{m(\omega_0^2 - \omega^2 - i\beta\omega)}$$

where $\omega_0\,(=\omega_r)$ is the natural frequency and $\beta = b/m$ is the damping constant. The induced dipole moment is

$$\vec{p}(t) = -e\vec{x} = \frac{e^2/m}{\omega_0^2 - \omega^2 - i\beta\omega}\vec{E}_0 e^{-i\omega t}$$

for incident wavelength larger than the size of the atom. The real part of this gives

$$\vec{p}(t) = \frac{e^2/m}{[(\omega_0^2 - \omega^2)^2 + \beta^2\omega^2]^{1/2}}\vec{E}_0 \cos(\omega t - \delta) \tag{13.88}$$

where $\delta = \tan^{-1}[\beta\omega/(\omega_0^2 - \omega^2)]$. And so, from Larmor's formula, the atom radiates power at a mean rate

$$\frac{2}{3}\frac{1}{4\pi\varepsilon_0 c^3}\left\langle\left|\dddot{p}\right|^2\right\rangle = \frac{e^4 E_0^2}{12\pi\varepsilon_0 c^3 m^2}\frac{\omega^4}{[(\omega_0^2 - \omega^2)^2 + \beta^2\omega^2]}$$

And the mean incident power per unit area is

$$\left\langle\vec{S}\right\rangle = \frac{1}{2}\varepsilon_0 c E_0^2.$$

Hence, the scattering cross-section σ is

$$\sigma = \frac{8\pi}{3}\left(\frac{e^2}{4\pi\varepsilon_0 mc^2}\right)^2\frac{\omega^4}{[(\omega_0^2 - \omega^2)^2 + \beta^2\omega^2]}$$

$$= \sigma_T\frac{\omega^4}{[(\omega_0^2 - \omega^2)^2 + \beta^2\omega^2]} \tag{13.89}$$

where σ_T is the Thomson scattering cross-section.

Because $\beta \leq \omega_0$, at low frequencies ($\omega \ll \omega_0$), and equation (13.89) reduces to

$$\sigma_R = \sigma_T\left(\frac{\omega}{\omega_0}\right)^4 \tag{13.90}$$

It is known as the Rayleigh scattering formula, first investigated by Lord Rayleigh in 1891, which shows scattering varies as λ^{-4}: short wavelengths are scattered much more than long wavelengths. Thus, the sky is blue in the day, because the blue light is scattered more than red light, and the sky is red at sunrise and sunset when seen in the reflection of sun's rays from high clouds. At high frequencies, $\omega \gg \omega_0$, $\beta \ll \omega_0$, the scattering cross-section σ reduces to σ_T, the Thomson scattering cross-section. This is expected, because in this case the electron is effectively free. At still higher frequencies classical theory no longer is valid and we get photon–electron interaction or Compton scattering. From quantum theory, the classical limit for Thomson scattering is $\leq\omega \ll mc^2$.

Figure **13.12** shows the frequency dependence of the scattering cross-section σ, where σ_0 is known as resonance scattering (or resonance fluorescence in quantum theory) that occurs when $\omega = \omega_0$.

Problems

1. Find the retarded potentials of an infinite, straight, filamentary current.
2. An electric dipole of constant strength p_0 rotates about a perpendicular axis through its center at constant angular velocity ω_0. Find a formula for its rate of radiation of energy.
3. Find the expressions for the fields produced by a charge moving with uniform velocity.
4. Consider an accelerated charged particle at low velocity. Using equations (13.73) and (13.74), find the angular distribution of radiated energy when the velocity of the particle is collinear.
5. Consider an electron in a circular orbit of radius r about a proton. Show that the total energy of the electron is

$$E = \frac{1}{2}\frac{e^2}{4\pi\varepsilon_0 r}$$

The electron radiates according to Larmor's formula and spirals toward the proton. Assume that at any time the orbit is approximately circular and

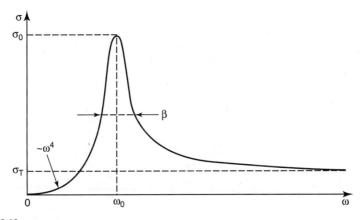

I **FIGURE 13.12**

(a) Estimate the time to fall from $r = 10\ A$ to $r = 1\ A$.

(b) Show that the energy spectrum of the radiation is approximately

$$\frac{dE}{dt} = \frac{1}{3}\left(\frac{e^2}{4\pi\,\varepsilon_0}\right)^{2/3}\left(\frac{m}{\omega}\right)^{1/3}$$

6. Show that in the absence of external forces the Abraham-Lorentz equation has the "runaway" solution $\ddot{\vec{r}}\,(t) = \ddot{\vec{r}}\,(0)\,e^{t/\tau}$ and the physical solution $\ddot{\vec{r}} = 0$, where

$$\tau = \frac{2}{3}\left(\frac{q^2}{4\pi\varepsilon_0}\right)\frac{1}{mc^3}$$

14 | Motion of Charged Particles in Electric and Magnetic Fields

In previous chapters the discussion has centered mainly on the properties of electromagnetic fields and their interaction with matter. We now devote this chapter to a discussion on the motion of charged particles in electric and magnetic fields. We first examine the trajectories of individual charged particles in such fields and then consider the behavior of a system of charged particles, the plasma, and hydromagnetics.

14.1 Motion of a Charged Particle in Electromagnetic Fields

a. Equations of Motion

When a charged particle q moves in an electromagnetic field at a velocity $\vec{v}$, the Lorentz force

$$\vec{F} = q\,(\vec{E} + \vec{v} \times \vec{B}) \tag{14.1}$$

acts upon it, and Newton's equation of motion takes the form

$$\frac{d}{dt}\,(m\vec{v}) = q\,(\vec{E} + \vec{v} \times \vec{B}) \tag{14.2}$$

where m is mass of the charged particle. We first consider the nonrelativistic case, i.e., $v \ll c$. Then the mass m may be taken to be constant.

b. Motion in Magnetic Field

In this case, the equation of motion has the form

$$m\frac{d\vec{v}}{dt} = q\vec{v} \times \vec{B} \tag{14.3}$$

Taking the scalar product of both sides by $\vec{v}$, we obtain

$$m\vec{v} \cdot \frac{d\vec{v}}{dt} = q\vec{v} \cdot (\vec{v} \times \vec{B}) \tag{14.4}$$

Using the vector identity $\vec{A} \cdot (\vec{B} \times \vec{C}) = \vec{C} \cdot (\vec{A} \times \vec{B}) = \vec{B} \cdot (\vec{C} \times \vec{A})$, equation (14.4) becomes

$$\frac{1}{2} m \frac{d(\vec{v} \cdot \vec{v})}{dt} = q\vec{B} \cdot (\vec{v} \times \vec{v}) = 0$$

and therefore

$$\frac{1}{2} mv^2 = const., \quad v^2 = const. \tag{14.5}$$

Equations (14.5) state that the magnetic field does no work and produces no change in the kinetic energy of the particle nor in its speed. Why is this so? This is because the magnetic force is always perpendicular to the direction of motion. The force is, therefore, perpendicular to the displacement and hence does no work. The magnetic force only changes the direction of motion of the particle.

If the magnetic field is homogeneous and constant in time, the velocity of the particle $\vec{v}$ can then be written as the sum of two components:

$$\vec{v} = \vec{v}_\perp + \vec{v}_\parallel \tag{14.6}$$

where $\vec{v}_\perp$ is the component of velocity perpendicular to the magnetic field and $\vec{v}_\parallel$ in the direction of the magnetic field. In terms of $\vec{v}_\parallel$ and $\vec{v}_\perp$, equation (14.3) becomes

$$md\vec{v}_\perp/dt + md\vec{v}_\parallel/dt = q\vec{v}_\perp \times \vec{B} + q\vec{v}_\parallel \times \vec{B}$$

or

$$md\vec{v}_\perp/dt = q\vec{v}_\perp \times \vec{B} \tag{14.7}$$

$$md\vec{v}_\parallel/dt = q\vec{v}_\parallel \times \vec{B} \tag{14.8}$$

Equation (14.8) gives

$$\vec{v}_\parallel = const. \tag{14.9}$$

i.e., the velocity of the particle is constant in the direction of the magnetic field. It is easy to show that $v_\perp^2$ is also a constant. From equation (14.5) we have

$$v^2 = v_\perp^2 + \vec{v}_\parallel = \text{const.}$$

and hence

$$v_\perp^2 = v^2 - v_\parallel^2 = \text{const.} \tag{14.10}$$

Equation (14.7) is a very useful equation. The angle between $\vec{v}_\perp$ and $\vec{B}$ remains constant throughout the motion and equal to $\pi/2$; $\vec{v}_\perp$ and $\vec{B}$ do not change in magnitude. Hence, the force in equation (14.7) is constant in magnitude and is perpendicular to the velocity. Consequently, equation (14.7) describes the motion with constant acceleration (in magnitude) that is always perpendicular to the velocity. This is motion on a circle, and the Lorentz force provides a centripetal force. Thus,

$$\frac{mv_\perp^2}{r} = qv_\perp B$$

where r is the radius of the orbit and is given by

$$r = \frac{mv_\perp}{qB} \tag{14.11}$$

The radius r is often called the Larmor radius of the particle. The angular frequency of revolution ω is given by

$$\omega = \frac{2\pi}{T} = \frac{2\pi}{2\pi r/v_\perp} = \frac{v_\perp}{r} = \frac{qB}{m} \tag{14.12}$$

which is independent of the linear velocity, and it is opposite to B when q is positive, as shown in **Figure 14.1**. Considered as a vector, the angular velocity is directed along the axis of rotation in the sense of advance of a right-handed screw:

$$\vec{\omega} = -(q/m)\vec{B} \tag{14.12a}$$

The complete motion of the charged particle is described as a gyration of the particle in an orbit (the Larmor orbit) superimposed on the uniform motion of the orbit center, or the guiding center, along a magnetic field line. The resulting helical motion is shown in **Figure 14.2**.

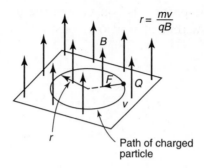

$$r = \frac{mv}{qB}$$

Path of charged
particle

I FIGURE 14.1

An interesting quantity, which we use later, is the magnetic moment of the gyrating particle. By definition, the magnetic moment μ is given by "current × area":

$$\mu = \frac{qv_\perp}{2\pi r}\,\pi r^2 = \frac{K_\perp}{B}, \quad K_\perp = \frac{1}{2}\,mv_\perp^2 \tag{14.13}$$

It is directed opposite to the magnetic field and is thus a diamagnetic moment.

As an example of application, allow a stream of charged particles passing through a uniform magnetic field, where the velocities of the individual particles can be calculated by equation (14.11) from the observed deflections. This method is used in obtaining the *magnetic spectrum* of β-rays (electrons) from radioactive sources. The rays, proceeding from the source S (**Figure 14.3**), pass through the slit L into a magnetic field at right angles to the plane of the paper, and, after completing a semicircle, impinge on the photographic plate PQ. On account of the finite angular width of the beam, different electrons describe different circular paths, but all those of the same velocity cross at approximately the same point after completing a half revolution. This focusing action results in a single line at P corresponding to a

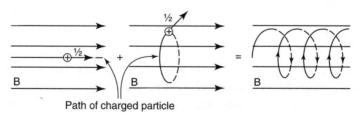

Path of charged particle

I FIGURE 14.2

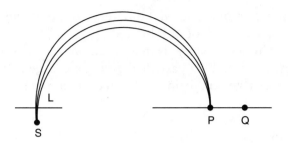

I FIGURE 14.3

group of electrons of a definite initial velocity, another group of greater velocity coming to a focus at Q and so on.

c. Motion in a Constant Electric Field

Another simple case is the motion of a charged particle in an electric field $\vec{E} = -\nabla\phi$ that is constant in time. Then

$$m d\vec{v}/dt = -q\nabla\phi \qquad (14.14)$$

Taking the scalar product of both sides with $\vec{v} = d\vec{r}/dt$, we obtain

$$m\frac{d\vec{v}}{dt} \cdot \vec{v} = \frac{d}{dt}\left(\frac{mv^2}{2}\right) = -q\nabla\phi \cdot \frac{d\vec{r}}{dt} = -q\frac{d\phi}{dt}$$

$$\frac{d}{dt}\left(\frac{mv^2}{2}\right) = -q\frac{d\phi}{dt}$$

Integrating

$$mv^2/2 + q\phi = \text{const.} \qquad (14.15)$$

i.e., energy is conserved for motion in an electrostatic field. Thus, if an electron is initially at rest and is then subjected to a potential difference ϕ, then

$$mv^2/2 = |e|\phi$$

or

$$\phi = \sqrt{\frac{2|e|}{m}}\,\phi \cong 6 \times 10^5 \sqrt{\phi} \; m/s = 600 \sqrt{\phi} \; km/s$$

Hence, the application of a potential difference of only 1 volt to an electron gives the electron a velocity of 600 km/sec.

In modern physics, energy is measured in eV (electron volt). One eV is the energy acquired by a particle carrying a charge equal in magnitude to the electron charge, on being subjected to a potential difference of 1 volt:

$$1 \, eV = 1.6 \times 10^{-19} \text{ coulomb} \cdot 1 \text{ volt} = 1.6 \times 10^{-19} \text{ joule}.$$

d. Drift of Charged Particle in Crossed $\vec{E}$ and $\vec{B}$ Fields

We now consider the motion of a charged particle q in a region where a uniform electric field and a uniform magnetic field are mutually perpendicular. Because the magnetic field exerts no force in the direction of the field lines, the motion parallel to $\vec{B}$ is uniformly accelerated motion produced by the component of $\vec{E}$ parallel to the magnetic field lines. The motion in the plane perpendicular to the magnetic field requires special attention. For convenience, let the particle velocity $\vec{v}$ be written as

$$\vec{v} = \vec{u} + \vec{v}' \tag{14.16}$$

Then equation (14.1), the Lorentz force, may be written as

$$\vec{F} = q\,(\vec{E} + \vec{u} \times \vec{B} + \vec{v}' \times \vec{B}) \tag{14.17}$$

Now let

$$\vec{u} = \frac{\vec{E} \times \vec{B}}{B^2} = \vec{u}_d \tag{14.18}$$

then equation (14.17) becomes

$$\vec{F} = q\left(\vec{E} + \frac{1}{B^2}(\vec{E} \times \vec{B}) \times \vec{B} + \vec{v}' \times \vec{B}\right) \tag{14.19}$$

Expanding the triple vector product, with the help of the vector identity

$$(\vec{A} \times \vec{B}) \times \vec{C} = (\vec{A} \cdot \vec{C})\vec{B} - (\vec{B} \cdot \vec{C})\vec{A}$$

and taking into account that $\vec{B} \cdot \vec{E} = 0$, we have

$$\frac{1}{B^2}(\vec{E} \times \vec{B}) \times \vec{B} = -\vec{E}$$

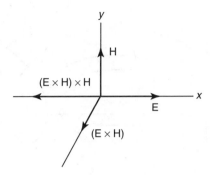

I FIGURE 14.4

as illustrated in **Figure 14.4**. Thus, this particular choice of $\vec{u}$ causes the first two terms on the right of equation (14.19) to cancel each other, and the remaining force $q\vec{v}' \times \vec{B}$ is just what was studied under case A:

$$m \frac{d\vec{v}'}{dt} = q\vec{v} \times \vec{B} \tag{14.20}$$

The motion defined by this equation is independent of the electric field and consists of a gyration around the magnetic field lines. The total velocity $\vec{v}$ is the sum of $\vec{v}'$ and a translation velocity $\vec{u}_d$, perpendicular to both $\vec{E}$ and $\vec{B}$. Thus, the total motion of the particle is made up of three terms: (1) constant velocity $\vec{v}_\parallel$ parallel to $\vec{B}$, (2) gyration about the magnetic field lines at the angular frequency $\omega = v'_\perp / r' = qB/m$, and (3) a constant translation velocity $u_d = E/B$ at right angles to both $\vec{E}$ and $\vec{B}$. Therefore, the path of the particle is a cycloid described by a point at a distance r' from the center of a circle rolling on a line parallel to $\vec{u}_d$. As the velocity of the center of the rolling circle is u_d, its radius a is

$$a = u_d / \omega = mE/qB^2$$

independent of the velocity of the particle. The ratio of r' to a is

$$r'/a = v'/ u_d$$

According as v' is greater than, equal to, or less than u_d, the generating point lies outside, on the circumference of, or inside the rolling circle. Typical paths are shown in **Figure 14.5** for the case of a positive charged particle, $\vec{B}$ being directed out of the page (i.e., toward the reader). The velocity u_d is

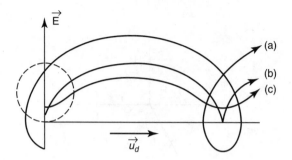

I FIGURE 14.5

called the drift velocity of the particle. Because u_d depends only on $\vec{E}$ and $\vec{B}$, positive and negative particles drift in the same direction.

The deflection of a stream of charged particles in crossed electric and magnetic fields at right angles to the initial velocity of the stream has many applications. Here we give two examples.

Example 1: J. J. Thomson's e/m Experiment

In 1897, J. J. Thomson used it to determine the ratio of *e/m* of electrons coming from the cathode of an evacuated tube through which a discharge is passing. The electric field is produced by the parallel plate capacitor *AB* (**Figure 14.6**) between the plates of which the stream of electrons *CD* passes, the magnetic field (indicated by dots) at right angles to the plane of the figure being produced by a solenoid.

First, the potential difference of the capacitor is adjusted until the stream suffers no deflection. Under this condition, the downward force due to the magnetic field just balances the upward force due to the electric field, and the velocity of the electrons is equal to the drift velocity *u*:

$$v = u = \frac{1}{\mu_0} \frac{E}{H}$$

Then, the electric field is suppressed; the electrons describe the circular arc *CF* of radius

$$r = -\frac{v}{(e/m)\mu_0 H}$$

The radius *r* is obtained from the observed deflection, and as *v* is known, *e/m* can be calculated by the following equation:

$$\frac{e}{m} = -\frac{1}{r}\left(\frac{v}{\mu_0 H}\right)$$

obtained from the preceding equation.

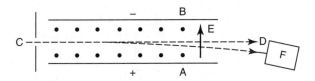

I FIGURE 14.6

The deflection may be observed by placing at F a metal cylinder with a narrow opening, which is connected to an electrometer. The strength of the magnetic field is varies until the electrometer shows a maximum rate of deflection, indicating that all the electrons in the stream are entering the chamber. We can also measure the deflection by means of a photographic plate at D at right angles to the stream.

In Thomson's experiments the velocity of the electrons was of the order of one-tenth the velocity of light.

The most recent value of e/m of the electron is

$$e/m = 1.759 \times 10^{11} \text{ coulomb/kg.}$$

The charge of the electron can be determined from Millikan's oil drop experiment. From this measurement and that of the ratio of charge to mass, we can calculate the mass of the electron as well as certain other important physical constants. For Millikan's oil drop experiment we refer interested readers to textbooks on Modern Physics.

Example 2: The Mass Spectrometer

The mass of an atom is one of its most characteristic properties, and an accurate knowledge of atomic masses provides considerable insight into nuclear phenomena. A variety of instruments with the generic name of mass spectrometers has been devised to measure atomic masses, and we now describe the operating principles of the particularly simple one shown in Figure 14.7.

We first produce ions of the substance under study. If the substance is a gas, ions can be formed readily by electron bombardment; if it is a solid, it is often convenient to incorporate it into an electrode that is used as one terminal of an electric arc discharge. The ions emerge from their source through a slit with the charge $+e$ and are then accelerated by an electric field.

When the ions enter the spectrometer, they are traveling in slightly different directions with slightly different speeds. A pair of slits serves to collimate the beam, that is, to eliminate those ions not moving in the desired direction. Then the beam passes through a velocity selector. The velocity selector consists of uniform electric and magnetic fields that are perpendicular to each

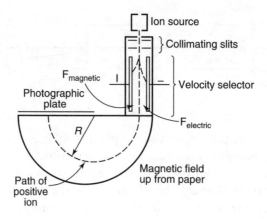

I FIGURE 14.7

other and to the beam of ions. The electric field E exerts the force eE on the ions to the right, whereas the magnetic field B exerts the force evB on them to the left. For an ion to reach the slit at the far end of the velocity selector, it must suffer no deflection inside the selector, which means that the condition for escape is

$$eE = evB$$

Hence, the ions that escape all have the velocity

$$v = E/B$$

Once past the velocity selector, the ions enter a uniform magnetic field and follow circular path whose radius is given by equation (14.11). As e, v, and B are known, a measurement of the radius yields a value of the ion mass m.

e. Particle Drift in a Converging Magnetic Field

A very interesting phenomenon occurs when a charged particle is moving in a region in which the magnetic field is space dependent, one in which the field lines are slowly converging in space. The force exerted on the charged particle now has a backward component as well as the inward component that leads to its helical path, as shown in **Figure 14.8**. If the backward force becomes strong enough and extends over long enough distance, it could reverse the particle's direction of motion. A converging magnetic field can thus act as a magnetic mirror.

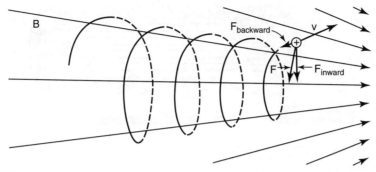

The particle motion may be treated as a perturbation of the helical orbit in Figure 14.2. To specify the problem more precisely, we assume that the flux line through the guiding center coincides with the z-axis and the magnetic field has azimuthal symmetry about the z-axis $\vec{B} = B_r \hat{r} + B_z \hat{z}$. Because the magnetic force is perpendicular to both $\vec{B}$ and $\vec{v}$, we see from Figure 14.5 that there is a backward force (i.e., the force component in the negative z direction). Taking the z-component of the equation of motion (14.3), we obtain

$$F_z = m \frac{dv_z}{dt} = -qv_\theta B_r \tag{14.21}$$

We can relate B_r to B_θ by evaluating $\nabla \cdot \vec{B} = 0$ for the case in point:

$$\frac{1}{r}\frac{\partial}{\partial r}(rB_r) + \frac{\partial B_z}{\partial z} = 0$$

Because the magnetic field lines converge slowly, any variation of B_z with z is small, and so $\partial B_z / \partial z$ may be taken constant over the orbit cross-section. Then we obtain, from the last equation,

$$B_r = -\frac{1}{2} r \frac{\partial B_z}{\partial z}$$

Substituting this into equation (14.20) we have

$$F_z = m \frac{dv_z}{dt} = qv_\theta B_r = \frac{1}{2} qrv_\theta \frac{\partial B_z}{\partial z}$$

By using equation (14.12a) we write r as

$$r = \frac{v_\theta}{\omega} = \frac{v_\theta}{-qB_z/m} = -\frac{mv_\theta}{qB_z}$$

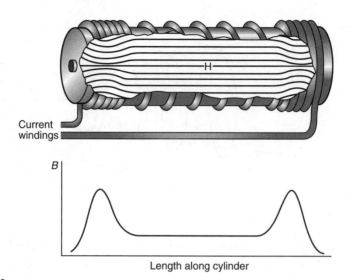

I FIGURE 14.9

then eliminating r from last expression for Fz, we obtain

$$m\frac{dv_z}{dt} = -\frac{1}{2}\frac{mv_z^2}{B_z}\frac{\partial B_z}{\partial z} = -\mu\frac{\partial B_z}{\partial z}$$ (14.22)

where $\mu = -1/2\, mv_z^2/B_z$ is the magnetic moment, given by equation (14.13). We show later that μ is a constant of the motion.

Magnetic Mirror and Magnetic Bottle

The parallel component (z-component) of the force, equation (14.22), is always in such a direction that accelerates particles toward the weaker part of the field. Thus, gyrating particles that are moving into regions of stronger magnetic field are slowed down, i.e., v_z is decreased. As v_z is decreased, conservation of energy requires that simultaneously the vertical component of the velocity be speed up. If the magnetic field converges sufficiently, the gyrating particle approaches an evertighter helical spiral path until it finally reverse its direction of motion, i.e., it is reflected back into the region of weaker field. A converging magnetic field can thus act as a *magnetic mirror*.

Magnetic mirrors are found both in the laboratory and in nature. In the laboratory a pair of them is used as a *magnetic bottle* (**Figure 14.9**) to contain hot plasma (highly ionized gas) in research on thermonuclear fusion. If a solid container were used, contact with its walls would cool the plasma and the charged particles would not have enough energy to interact. Magnetic bottles of this kind encounter leaky problem, because particles moving along

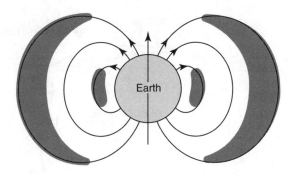

I FIGURE 14.10

the axis of a magnetic mirror experience no backward force and hence are able to escape.

The earth's magnetic field traps high energetic electrons and protons from space in two Van Allen belts (**Figure 14.10**), discovered by Prof. James Van Allen in 1957. The inner torus contains mostly protons and is approximately 9000 km above the earth surface, and the outer one contains mostly electrons and is about 20,000 km above the earth's surface.

The trapping of the charged particles is easily understood on the basis of the mirror effect. **Figure 14.11** is a sketch of the way in which the converging lines of magnetic field produce reflections of the helical motion of charged particles and hence effectively trap them.

f. Magnetic Moment, a Constant of the Motion

For slow variations of the magnetic field in space and time, the magnetic moment, μ, of a charged particle is nearly constant and provides an approximate integral of the motion. To see this, let us consider the total kinetic energy, K, of the particle, which is constant in the magnetic field, because the Lorentz force does not do work on the particle, but neither $K_\perp$ nor $K_\parallel$ is constant, where

$$K_\perp = \frac{1}{2}\,mv_\perp^2 \quad \text{and} \quad K_\parallel = \frac{1}{2}\,mv_\parallel^2 = \frac{1}{2}\,mv_z^2$$

Now let us consider

$$\frac{dK_\parallel}{dt} = \frac{d}{dt}(K - K_\perp) = -\frac{dK_\perp}{dt} \tag{14.23}$$

From equation (14.13), we find

$$\mu B_z = K_\perp$$

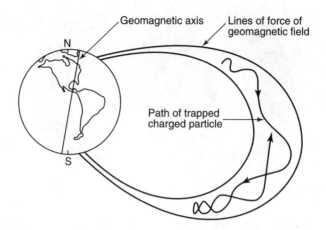

I FIGURE 14.11

We can now rewrite equation (14.23) as

$$\frac{d}{dt}\left(\frac{1}{2}\,mv_z^2\right) = -\frac{d}{dt}\left(\mu B_z\right) \tag{14.24}$$

Let us revisit equation (14.22), and multiply it by $v_z = \partial z/\partial t$:

$$\frac{d}{dt}\left(\frac{1}{2}\,mv_z^2\right) = -\mu\,\frac{\partial B_z}{\partial z}\frac{\partial z}{\partial t} = -\mu\,\frac{dB_z}{dt}$$

where d/dt represents the time derivative taken along the dynamical path. Comparing this equation with equation (14.24), we see that the magnetic moment μ is a constant of the motion. This is an approximate result that holds so long as B_z varies slowly. The constancy of μ for motion in a slowly varying magnetic field is an example of *adiabatic invariance*.

It is interesting to note that the particle is constrained to move on the surface of a flux tube. This follows from the fact that the magnetic flux through the orbit is given by

$$\Phi = B_z \pi R^2 = \pi B_z \frac{m^2 v_\perp^2}{q^2 B_z^2} = \frac{2\pi m}{q^2}\frac{K_\perp}{B_z} = \frac{2\pi m}{q^2}\,\mu$$

and μ is constant.

g. Magnetic Lens With Axial Symmetry

In this section we see that a slowly varying axially symmetric magnetic field can act as a lens for parallel beams of charged particles. We limit our discussion to the case where all charged particles move close to the axis of symmetry. This eliminates the aberration problems of the lens due to its finite "thickness." We further assume that the radial component of the velocity is negligible.

Newtonian equations of motion in cylindrical coordinates (ρ, φ, z) are

$$F_\rho = m \left[\frac{d^2\rho}{dt^2} - \rho \left(\frac{d\varphi}{dt} \right)^2 \right] \qquad (14.25a)$$

$$F_\varphi = \frac{m}{\rho} \frac{d}{dt} \left(\rho^2 \frac{d\varphi}{dt} \right) \qquad (14.25b)$$

$$F_z = m \frac{d^2z}{dt^2} \qquad (14.25c)$$

where the force is the Lorentz force.

The magnetic field is axial symmetric, i.e.,

$$\vec{B}(\vec{r}) = B_\rho \hat{\rho} + B_z \hat{z}$$

so the Lorentz force is

$$q\vec{v} \times \vec{B} = q \left\{ \rho \frac{d\varphi}{dt} B_z \hat{\rho} + \left[\frac{dz}{dt} B_\rho - \frac{d\rho}{dt} B_z \right] \hat{\varphi} + \rho \frac{d\varphi}{dt} B_\rho \hat{z} \right\}$$

$$\cong q \left\{ \rho \frac{d\varphi}{dt} B_z \hat{\rho} + \left[\frac{dz}{dt} B_\rho - \frac{d\rho}{dt} B_z \right] \hat{\varphi} \right\}$$

The term in $\rho(d\varphi/dt) = v_\varphi B_\rho$ is disregarded, because the major components of $\vec{B}$ and $\vec{v}$ are the axial ones.

Equations (14.25) now become

$$m \left[\frac{d^2\rho}{dt^2} - \rho \left(\frac{d\varphi}{dt} \right)^2 \right] = qB_z \rho \frac{d\varphi}{dt} \qquad (14.26a)$$

$$\frac{m}{\rho} \frac{d}{dt} \left(\rho^2 \frac{d\varphi}{dt} \right) = q\rho \left[B_\rho \frac{dz}{dt} - B_z \frac{d\rho}{dt} \right] \qquad (14.26b)$$

$$m \frac{d^2z}{dt^2} = 0 \qquad (14.26c)$$

We now solve equation (14.26b) for $d\varphi/dt$. To this end, we need a relation between B_ρ and B_z. Let us construct a small cylinder of length Δz with its axis as the axis of symmetry of a nonuniform axially symmetric $\vec{B}$ field (**Figure 14.12**). Gauss' law applied to it yields

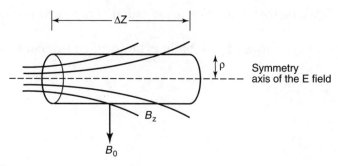

I FIGURE 14.12

$$(2\pi\rho)\Delta z B_\rho + \pi\rho^2[B_z(z + \Delta z) - B_z(z)] = 0$$

where the first term is the radial flux through cylinder sides and the second term is the axial flux through ends. This gives a relation between B_ρ and B_z:

$$B_\rho = -\frac{1}{2}\rho\frac{\partial B_z}{\partial z}$$

Substituting this into equation (14.26b) we get

$$\frac{d}{dt}\left(\rho^2\frac{d\varphi}{dt}\right) = -\frac{q}{m}\rho\left[\frac{1}{2}\rho\left(\frac{dz}{dt}\right)\left(\frac{\partial B_z}{\partial z}\right) + \left(\frac{d\rho}{dt}\right)B_z\right]$$

Now

$$\frac{d}{dt} = \frac{dz}{dt}\frac{d}{dz} = v\frac{d}{dz}$$

Thus, the above equation of motion becomes

$$\frac{d}{dz}\left(\rho^2 v\frac{d\varphi}{dt}\right) = -\frac{q}{m}\left[\frac{\rho^2}{2}\left(\frac{dz}{dt}\right)\left(\frac{\partial B_z}{\partial z}\right) + \rho\left(\frac{d\rho}{dt}\right)B_z\right]$$

where it has been assumed that radial derivatives are negligible compared with axial derivatives. Integrating this equation yields

$$\rho^2\frac{d\varphi}{dz}\bigg|_{z_0}^{z} = -\frac{q}{2mv}(\rho^2 B_z)_{z_0}^{z}$$

Here we limit the $\vec{B}$ field to a finite region, which means the lens is "short." At z_0 where the particle enters the lens, $B_z = 0$ and $d\varphi/dt = 0$, i.e., the particle is traveling in a straight line. Therefore at z, a typical point within the lens,

$d\varphi/dz = -qB_z/2mv$

or

$$d\varphi/dt = -qB_z/2m \tag{14.27}$$

That is, the angular frequency of a charged particle within the lens is just the Larmor frequency.

Substituting equation (14.27) into equation (14.26a) we obtain

$$\frac{d^2\rho}{dt^2} = -\left(\frac{q}{2m}\right)^2 \rho B_z^2$$

Using the relation $d(\ )/dt = v\, d(\ )/dz$, this transforms into

$$\frac{d^2\rho}{dz^2} = -\left(\frac{q}{2mv}\right)^2 \rho B_z^2$$

Integrating this and assuming that ρ changes little within the lens:

$$\frac{d\rho}{dz} = -\left(\frac{q}{2mv}\right)^2 \rho \int_{-\infty}^{\infty} B_z^2\, dz \tag{14.28}$$

where the integral extends over all z because B_z is zero outside the lens. Because $d\rho/dz = 0$ at the near side of the lens where the particle enters the lens, the left-hand side of equation (14.28) is evaluated at a value of z on the far side of the lens. We see that $d\rho/dz$ is the angle of deflection α of the trajectory, and because α is proportional to ρ, the lens brings all axial rays to a point focus. The lens is a converging one, because the right side of equation (14.28) is negative. The focal length of the lens is defined by the following equation

$$\frac{|\alpha|}{\rho} = -\left(\frac{q}{2mv}\right)^2 \int_{-\infty}^{\infty} B_z^2\, dz \tag{14.29}$$

which depends on the velocity of the charged particles.

h. The Cyclotron

The cyclotron is based on the principle that a charged particle performs circular motion in a static uniform magnetic field with an angular frequency ω_c independent of the velocity of the particle [see equation (14.12), $\omega_c = qB/m$]. This device, constructed first by E. O. Lawrence in 1932, consists of a metal pillbox cut into two semicircular "Ds," D_1 and D_2 (**Figure 14.13**), which are placed between the poles of a strong electromagnet designed to produce a very uniform magnetic field $\vec{B}$ throughout the region occupied by the Ds.

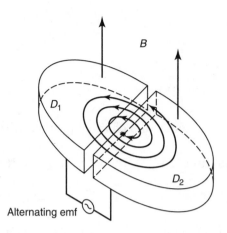

I FIGURE 14.13

The latter are connected to a high-frequency electric oscillator that gives rise to an alternating electric field of frequency ω_c between the Ds. Charged particles are produced in the gap between the Ds near the center, where they are accelerated by the electric field so as to enter one of the Ds. Here, protected from the electric field by the shielding effect of the metal D, their paths are curved by the magnetic field so as to bring them back to the gap between the Ds when the electric field is reversed. There they get an added increment in velocity before entering the other D. Continuing in this fashion in ever widening semicircles, they emerge finally under the influence of an auxiliary deflecting electric field.

The cyclotron frequency ω_c of a charged particle is independent of its velocity only in the nonrelativistic limit. Because ω_c is proportional to the charge-to-mass ratio of the particle being accelerated, as the particle speed approaches the speed of light, its mass and, therefore its cyclotron frequency, becomes dependent on its velocity. Unless steps are taken to alter the frequency of the oscillating $\vec{E}$ field in the D-gap as the particle velocity approaches c, the circulating particles will rapidly get out of phase with the accelerating field and the particles will become decelerated. The critical velocities occur at particle energies in the neighborhood of 100–200 MeV for protons. To operate proton cyclotrons at these energies, the accelerating field must be modulated to compensate for the change in ω_c with velocity. Such frequency-modulated cyclotrons are operative and produce protons of energy in the hundreds of MeV region.

The D cavities of the cyclotron must be evacuated to prevent loss of energy of the particles in the accelerator to the residual gas in the D cavities.

Any momentum imparted to the accelerated particles in the radial direction or in the direction perpendicular to the plane of the Ds causes the particle beam to blow up. To examine the stability of the particle beam, two effects must be considered: (1) motion in the radial direction (in-and-out stability) and (2) motion perpendicular to the D-plane (up-and-down stability).

The radial equation of motion for the charged particle is

$$m\left(\frac{d^2\rho}{dt^2} - \frac{v^2}{\rho}\right) = -qvB \tag{14.30}$$

the confining magnetic field in the neighborhood of ρ_0 is represented by the expression

$$B = B_0\left(\rho_0/\rho\right)^n \tag{14.31}$$

where B_0 is the value at $\rho = \rho_0$. Using Taylor's expansion and keeping the first terms, we obtain

$$B \cong B_0\left(1 - n\frac{\rho - \rho_0}{\rho_0}\right) = \frac{m}{q}\,\omega_0\left(1 - n\frac{\rho - \rho_0}{\rho_0}\right) \tag{14.32}$$

where $\omega_0 = qB_0/m$.

Equation (14.30), the radial equation of motion, now becomes

$$\frac{d^2\rho}{dt^2} - \frac{v^2}{\rho} = -v\omega_0\left(1 - n\frac{\rho - \rho_0}{\rho}\right) \tag{14.33}$$

In the neighborhood of ρ_0, we may write

$$v \approx \omega_0\rho_0, \quad \text{and} \quad \rho = \rho_0 + \Delta\rho$$

from which it follows that

$$\rho^{-1} = \frac{1}{\rho_0 + \Delta\rho} = \frac{1}{\rho_0}\left(\frac{1}{1 + \Delta\rho/\rho_0}\right) = \frac{1}{\rho_0}\left(\frac{1}{1 + (\rho - \rho_0)/\rho_0}\right) \approx \frac{1}{\rho_0}\left(1 - \frac{\rho - \rho_0}{\rho}\right)$$

Substituting these into equation (14.33) we obtain, after rearranging terms,

$$\frac{d^2(\rho - \rho_0)}{dt^2} + \omega_0^2(1 - n)(\rho - \rho_0) = 0 \tag{14.34}$$

which is a differential equation for small oscillations in the radial motion about $\rho = \rho_0$ of angular frequency

$$\omega_\rho = \omega_0\sqrt{1 - n} \tag{14.35}$$

Thus, the cyclotron orbits are stable against small radial disturbances as long as $1 < 1 - n > 0$. If $n = 0$, the period of radial oscillation is in resonance with the cyclotron frequency in the uniform field.

We now examine the up-and-down stability. The effect of a radial dependence of the magnetic field must be considered, because for radial stability the B field is weaker near the edges of the Ds than it is in the center (**Figure 14.14**). The guiding field is chosen to have reflectional symmetry about the plane of the equilibrium orbit of the accelerating charges. This equilibrium or symmetry plane is designated by $z = -0$. For z near the equilibrium plane, B_0 is in general nonvanishing but small; that is,

$$B_\rho(z) \cong B_\rho(0) + z\left(\frac{\partial B_\rho}{\partial z}\right)_0$$

But $z = 0$ in the equilibrium or symmetry plane, $B_\rho(0) = 0$, and therefore

$$B_\rho(z) \cong z\left(\frac{\partial B_\rho}{\partial z}\right)_0$$

Now $\nabla \times \vec{B} = 0$, which gives $\partial B_\rho/\partial z = \partial B_z/\partial \rho$, and thus

$$B_\rho(z) \cong z\left(\frac{\partial B_\rho}{\partial z}\right)_0 = z\left(\frac{\partial B_z}{\partial \rho}\right)_0 \tag{14.36}$$

Near the symmetry plane the components of $\vec{B}$ other than the z-component are small, and to a good approximation

$$B_z = B_0\left(\frac{\rho_0}{\rho}\right)^n \cong B_0\left(1 - n\frac{\rho - \rho_0}{\rho_0}\right)$$

Substituting this into equation (14.36) we obtain

$$B_\rho(z) = -\frac{nz}{\rho_0}B_0$$

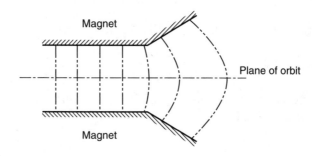

I FIGURE 14.14

The equation of motion perpendicular to the symmetry plane becomes

$$m \frac{d^2 z}{dt^2} = -qvB_p(z) = \frac{n}{\rho_0} B_0$$

But

$$\omega_0 = qB_0/m \quad \text{and} \quad v = \omega_0 \rho_0$$

Using these relations we can write the equation of motion perpendicular to the symmetry plane in a more familiar form:

$$\frac{d^2 z}{dt^2} + n\omega_0^2 z = 0 \tag{14.37}$$

This is a differential equation for small oscillations of the orbit about the equilibrium plane, and the frequency of oscillation is

$$\omega_z = \omega_0 \sqrt{n} \tag{14.38}$$

which is identical to the frequency of radial oscillations when $n = 1/2$.

From the above discussion we see that the cyclotron orbits are stable against small disturbances in the radial and axial directions, if departure from uniformity in the guiding field is small and corresponds to a weakening of the field as ρ increases.

14.2 Hydromagnetics

So far our discussion has been on single-particle motion in electromagnetic fields. We now turn to a consideration of some of the aspects of the motion of system of charged particles in $\vec{E}$ and $\vec{B}$ fields. We consider the situation where the number of particles in the system is very large and the time scale of the phenomena is very long, so that a macroscopic approach is valid. Then the system can be assigned such macroscopic quantities as current density, mass density, conductivity, and so forth. Furthermore, we consider the situation where the electrons and the positive ions in the system move as a single fluid. This system is usually referred to as the magnetohydrodynamic or hydromagnetic system. If the positive ions and the electrons in the system move as separate fluid, then it is referred to as plasma. The basic physical principles are identical for hydromagnetics and plasma.

Within the fluid, the electromagnetic fields satisfy Maxwell equations. For simplicity we assume that the fields vary slowly enough in time so that the displacement current can be ignored with respect to conduction current. In this low frequency limit we have

$$\nabla \times \vec{E} = -\frac{\partial \vec{B}}{\partial t} \tag{14.39a}$$

$$\nabla \times \vec{B} = \mu_0 \vec{j} \tag{14.39b}$$

$$\nabla \cdot \vec{j} = 0 \tag{14.39c}$$

There is another equation that determines $\vec{j}$ in terms of the electromagnetic fields. When the charged fluid moves in a magnetic field, electric fields are induced in the fluid and current flows; this induced electric field is given by $\vec{v} \times \vec{B}$ (see Chapter 5). A generalized Ohm's law determines the current density $\vec{j}$

$$\vec{j} = \sigma \left(\vec{E} + \vec{v} \times \vec{B} \right) \tag{14.39d}$$

where $\vec{v}$ is the fluid velocity. This is the other equation that we are looking for.

The magnetic field is capable of acting on the currents through the $\vec{j} \times \vec{B}$ force, and the resultant forces change the state of motion. This interaction between fields and flow is the concern of hydromagnetics.

a. Magnetic Pressure

We now explore the concept of magnetic pressure by considering the force acting on unit volume of the fluid. Writing the fluid pressure as P, the total force acting on unit volume of the fluid is

$$\vec{F} = \vec{j} \times \vec{B} - \nabla P \tag{14.40}$$

Using Ampere's law we can recast the magnetic force term $\vec{j} \times \vec{B}$ as

$$\vec{j} \times \vec{B} = -\frac{1}{\mu_0} \vec{B} \times (\nabla \times \vec{B}) \tag{14.41}$$

Now

$$\vec{B} \times (\nabla \times \vec{B}) = -(\vec{B} \cdot \nabla) \vec{B} + \frac{1}{2} \nabla B^2$$

where

$$\left(\vec{B} \cdot \nabla \right) \vec{B} = \left[B_x \frac{\partial}{\partial x} + B_y \frac{\partial}{\partial y} + B_z \frac{\partial}{\partial z} \right] \vec{B}$$

Substituting these into equation (14.41) we obtain

$$\vec{j} \times \vec{B} = -\nabla \left(\frac{1}{2\mu_0} B^2 \right) + \frac{1}{\mu_0} (\vec{B} \cdot \nabla) \vec{B} \tag{14.42}$$

Substituting these into equation (14.40) we obtain

$$\vec{F} = -\nabla \left(P + \frac{B^2}{2\mu_0} \right) + \frac{1}{\mu_0} (\vec{B} \cdot \nabla) \vec{B} \tag{14.43}$$

We see at once that the term in the bracket may be regarded as magnetic pressure P_B:

$$P_B = \frac{1}{2\mu_0} B^2 \tag{14.44}$$

This is only part of the effect of the $\vec{j} \times \vec{B}$ force, as indicated clearly by equation (14.42). For many simple geometric configurations, for example, a unidirectional field, the second term on the right-hand side of equations (14.42) and (14.43) vanishes. In the static limit we then have the relation

$$P + P_B = \text{constant} \tag{14.45}$$

For the fluid to remain in static equilibrium, any variation in the mechanical fluid pressure P must be compensated by an opposite variation in the magnetic pressure P_B.

The concept of magnetic pressure was first introduced by Faraday, who pictured the tubes of force as elastic filaments, these filaments being under tension in the direction of the field and compressed in the transverse directions. We see now the tension along the magnetic field lines amounts to be B^2/μ_0, as shown in **Figure 14.15**. Stretching the tube of force increases the tension, which means that the field is increased.

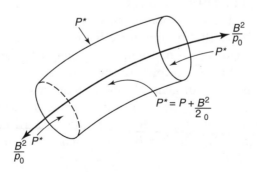

I FIGURE 14.15

b. Alfven Waves

In 1942 Alfven extended this analogy a little further. Transverse waves can be set up using an elastic string, Alfven asked the following question: by "plucking" a tube of force might it not be possible to create transverse waves propagating along the magnetic field lines? The existence of the hydromagnetic waves was subsequently confirmed and they are now more commonly known as Alfven waves.

The velocity of a transverse wave on a string is given by $(T/\rho)^{1/2}$ where T is the tension in the string and ρ is the mass density. By analogy, the phase velocity of Alfven waves v_A is just

$$v_A = B/\sqrt{\mu_0 \rho} \tag{14.46}$$

where ρ is now the mass density of the plasma, the conducting fluid. This formula for v_A can be derived from a theoretical consideration.

c. Magnetic Confinement

The existence of a magnetic pressure makes it possible for a plasma to confine itself. Returning to equation (14.40), the condition for equilibrium of a plasma in a magnetic field is

$$\vec{F} = 0 \quad \text{or} \quad \nabla P = \vec{j} \times \vec{B} \tag{14.47}$$

Taking the scalar product of this equation with $\vec{B}$ and $\vec{j}$, respectively, we obtain

$$\vec{B} \cdot \nabla P = \vec{B} \cdot (\vec{j} \times \vec{B}) = \vec{j} \cdot (\vec{B} \times \vec{B}) = 0 \tag{14.48}$$

$$\vec{j} \cdot \nabla P = \vec{j} \cdot (\vec{j} \times \vec{B}) = \vec{B} \cdot (\vec{j} \times \vec{j}) = 0 \tag{14.49}$$

Equations (14.48) and (14.49) indicate that both $\vec{B}$ and $\vec{j}$ lie on surfaces of constant pressure. If these isobaric surfaces happen to be closed, then equation (14.48) states that no magnetic field line can cross them, and they may be viewed as made up from a winding of magnetic field lines. Similarly, from equation (14.49) we may picture the same isobaric surface as a winding lines of $\vec{j}$; these lines in general intersect the magnetic field lines arbitrarily. **Figure 14.16** shows a set of such nested surfaces on which the pressure increases toward the axis and $\vec{j} \times \vec{B}$ also points toward the axis. Thus, the plasma is contained by the $\vec{j} \times \vec{B}$ force. This is known as magnetic confinement.

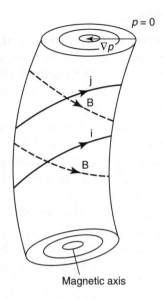

I FIGURE 14.16

d. Pinch Effect

An example of the magnetic confinement is provided by the "linear pinch." As shown in **Figure 14.17,** an electric current flows through a cylindrical plasma axially. Associated with this is an azimuthal magnetic field lines and a $\vec{j} \times \vec{B}$ force directed toward the axis that tends to squeeze or "pinch" the plasma. The plasma contracts until the compressing electromagnetic force is balanced by the increased kinetic pressure $p = NkT$. This is known as the "pinch effect."

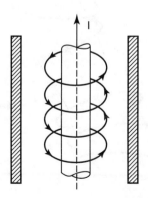

I FIGURE 14.17

e. Kink and Sausage Instabilities

The equilibrium of the linear pinch is inherently unstable. A slight deviation from the equilibrium state tends to increase and results into the disintegration of the form of plasma. **Figure 14.18** illustrates what is called a kink instability that occurs when a linear pinch is bent due to some perturbation. The magnetic field lines crowd together on the concave side of the kink and so the magnetic pressure is increased, but on the convex side the magnetic pressure is decreased because the magnetic field lines spread apart. Thus, the plasma column is unstable with respect to bending. **Figure 14.19** shows the so-called sausage instability, which might be induced by random fluctuations. Because for a given current the magnetic field at the surface of the plasma column is inversely proportional to r, the magnetic pressure $B^2/2\mu_0$ is greater in the region where the perturbation squeezes the plasma into a neck and decreases where the plasma column fattens into a bulge. Consequently, there is no longer equilibrium between the magnetic and plasma pressures and the perturbation grows rapidly.

A great deal of work is now being done on magnetic confinement in connection with attempts to produce controlled nuclear fusion energy. For nuclear fusion to take place, we have to raise the temperature of plasma to a value high enough ($\approx 10^8$ K). The hope is that by means of a magnetic field, the plasma can be confined to a volume away from the surface of the container, so as to prevent the heat loss that occurs if the plasma is in contact with the walls. Practical schemes for containing plasma have to be much more sophisticated than the simple linear pinch discussed above, which has

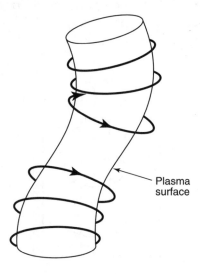

Plasma
surface

I FIGURE 14.18

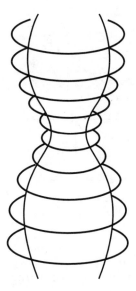

I FIGURE 14.19

various stability problems. One such system is the tokamak, an axially sym-
metrical toroidal system. The plasma is contained by the magnetic field of
the current that flows along its axis. A powerful magnetic field is applied
parallel to the current that suppresses the instability. It is beyond the scope
of the book to discuss the operational principle of the tokamak.

Problems

1. A charged particle of charge q and mass m moves in constant
   uniform electric and magnetic fields, $\vec{E} = (0, E, 0)$, $\vec{B} = (0, 0, B)$.
   Show that, in rectangular coordinates, the drift velocity is given by
   $\vec{v}_D = (E/B)\,\hat{x}$ and the particle coordinates can be expressed as

   $$x = x_0 + v_D t + \frac{v_{0_x} - v_D}{\omega_c}\sin\omega_c t, \quad y = y_0 + \frac{v_{0_x} - v_D}{\omega_c}(\cos\omega_c t - 1)$$

   where $\omega_c = qB/m$.

2. A region contains a magnetic field $\vec{B} = 5.0 \times 10^{-4}\,T\,\hat{z}$ and an electric
   field $\vec{E} = 5.0\hat{z}$ V/m. A proton ($q = 1.602 \times 10^{-19}$ C, $m = 1.673 \times 10^{-27}$
   kg) enters the fields at the origin with an initial velocity
   $\vec{u}_0 = 2.5 \times 10^5\,m/s\,\hat{x}$. Describe the proton's motion and give its position
   after three complete revolutions.

3. A current loop placed in a magnetic field $\vec{B}$ experiences a torque $\vec{\tau}$
   (**Figure 14.20**). Show that $\vec{\tau} = \vec{\mu} \times \vec{B}$, where $\vec{\mu}$ is the magnetic moment of

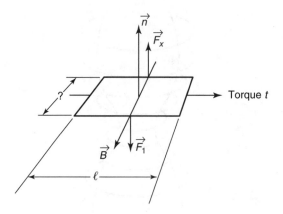

I FIGURE 14.20

the loop. You may consider the following special case, and then generalize your result.

4. The rectangular coil in **Figure 14.21** is in a magnetic field

$$\vec{B} = 0.05 \frac{\hat{x} + \hat{y}}{\sqrt{2}} \, T$$

Find the torque about the z-axis when the coil is in the position shown and carriers a current of 5.0 A.

5. A uniform magnetic field $\vec{B} = 85.3 \, \mu T \hat{z}$ exists in the region $x \geq 0$. An electron enters this field at the origin with a velocity $\vec{v}_0 = 450 km/s \, \hat{x}$; find the position where it exits.

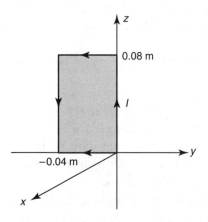

I FIGURE 14.21

6. A proton is fixed in position and an electron revolves about it in a circular path of radius 0.35×10^{-10} cm. What is the magnetic field at the proton?

7. A current flows in a cylindrical tube of plasma. The plasma pressure vanishes at $r = r_0$ and $P = N(r)kT$. Show that

$$2NkT = \frac{\pi}{2\mu_0} (rB)^2_{r = r_0}$$

where

$$N = \int_0^{r_0} n(r) \, 2\pi r \, dr$$

8. The solar corona consists of ionized hydrogen with approximately 10^{12} particles/m³. Assuming the magnetic field in the corona to be 1000 gauss, calculate the phase velocity of Alfven wave.

Solutions of Laplace's Equation in Spherical Polar Coordinates

In spherical polar coordinates Laplace's equation has the form

$$\frac{1}{r^2}\frac{\partial}{\partial r}\left(r^2\frac{\partial V}{\partial r}\right) + \frac{1}{r^2\sin\theta}\frac{\partial}{\partial\theta}\left(\sin\theta\frac{\partial V}{\partial\theta}\right) + \frac{1}{r^2\sin^2\theta}\frac{\partial^2 V}{\partial\phi^2} = 0 \tag{A-1}$$

Its general solutions, known as spherical harmonic functions, are given in Chow, T.L. *Mathematical Methods for Physicists* Cambridge, UK: Cambridge University Press, 2000, or other textbooks on mathematical physics. Here we are interested in the case in which the potential V is a function of r and θ only; then the last term in equation (A-1) drops out, and if $\mu = \cos\theta$, (A-1) becomes

$$\frac{\partial}{\partial r}\left(r^2\frac{\partial V}{\partial r}\right) + \frac{\partial}{\partial\mu}\left\{(1-\mu^2)\frac{\partial V}{\partial\mu}\right\} = 0 \tag{A-2}$$

This differential equation has a solution of the form

$$V = r^n P_n$$

where P_n is a function of μ alone. Substituting in (A-2) we find that P_n satisfies the differential equation

$$\frac{d}{dt}\left\{(1-\mu^2)\frac{dP_n}{d\mu}\right\} + n(n+1)P_n = 0 \tag{A-3}$$

which is known as Legendre's equation. If we replace n in (A-3) by $-(n+1)$, the coefficient of the last term becomes

$$[-(n+1)][-(n+1)+1] = n(n+1)$$

Hence Legendre's equation for $P_{-(n+1)}$ is the same as that for P_n. Consequently, P_n and $P_{-(n+1)}$ are identical. Thus, every P_n satisfying Legendre's equation provides us with two solutions of Laplace's equation, namely,

$$r^n P_n \quad \text{and} \quad P_n/r^{n+1}$$

Evidently the coefficients of the P's as well as of the V's are arbitrary. Hence, we may introduce any numerical factors. Those included in the P's and V's below are the ones conventionally employed. Now

$$V_0 = 1/r, \quad P_0 = 1 \tag{A-4}$$

satisfies Laplace's equation, as is seen at once by substitution in (A-2). Let us take the z axis in the direction of the polar axis of the spherical coordinates, so that $z = r\cos\theta$. We can get a second solution from (A-4) by differentiating partially with respect to z. Remember that $\partial r/\partial z = z/r = \cos\theta$,

$$V_1 = \frac{z}{r^3} = \frac{\cos\theta}{r^2}, \quad P_1 = \cos\theta \tag{A-5}$$

Differentiating again with respect to z to get a third solution

$$V_2 = \frac{1}{2}\left(\frac{3z^2}{r^5} - \frac{1}{r^3}\right) = \frac{3\cos^2\theta - 1}{2r^3}, \quad P_2 = \frac{1}{2}(3\cos^2\theta - 1) \tag{A-6}$$

and repeating the process for a fourth solution,

$$\left. \begin{aligned} V_3 &= \frac{1}{2}\left(\frac{5z^3}{r^7} - \frac{3z}{r^5}\right) = \frac{5\cos^3\theta - 3\cos\theta}{2r^4}, \\ P_3 &= \frac{1}{2}(5\cos^3\theta - 3\cos\theta), \end{aligned} \right\} \tag{A-7}$$

and so on. When n is an integer, the solutions of Legendre's equation are polynomials in $\cos\theta$, known as Legendre polynomials, and n is called the degree of the polynomials. The first six terms of $P_n(\cos\theta)$ are given in **Table A-1**.

The solutions of Laplace's equation we have obtained are of the form P_n/r^{n+1}, but, because we have found that, the P_n's we can write down at once are solutions of the form $r^n P_n$. **Table A-2** contains the first few solutions of Laplace's equation in spherical polar coordinates with axial symmetry.

These functions are known as zonal harmonics.

Table A-1   First Six Legendre Polynomials

n	$P_n(\cos \theta)$
0	1
1	$\cos \theta$
2	$\frac{3}{2} \cos^2 \theta - \frac{1}{2}$
3	$\frac{5}{2} \cos^3 \theta - \frac{3}{2} \cos \theta$
4	$\frac{35}{8} \cos^4 \theta - \frac{15}{4} \cos^2 \theta + \frac{3}{8}$
5	$\frac{63}{8} \cos^5 \theta - \frac{35}{4} \cos^3 \theta + \frac{15}{8} \cos \theta$

Table A-2   First Four Solutions of Laplace's Equation

n	$r^n P_n (\cos \theta)$	$r^{-(n+1)} P_n \cos \theta$
0	1	r^{-1}
1	$r \cos \theta$	$r^{-2} \cos \theta$
2	$\frac{1}{2} r^2 (3 \cos^2 \theta - 1)$	$\frac{1}{2} r^{-3} (3 \cos^2 \theta - 1)$
3	$\frac{1}{2} r^3 (5 \cos^3 \theta - 3 \cos \theta)$	$\frac{1}{2} r^{-4} (5 \cos^3 \theta - 3 \cos \theta)$
4	$\frac{1}{8} r^4 (35 \cos^4 \theta - 30 \cos^2 \theta + 3)$	$\frac{1}{8} r^{-5} (35 \cos^4 \theta - 30 \cos^2 \theta + 3)$

Index